Paradigm Shift in Technologies and Innovation Systems

John Cantwell · Takabumi Hayashi
Editors

Paradigm Shift in Technologies and Innovation Systems

 Springer

Editors
John Cantwell
Rutgers University
Newark, NJ, USA

Takabumi Hayashi
Rikkyo University
Tokyo, Japan

ISBN 978-981-32-9352-6   ISBN 978-981-32-9350-2 (eBook)
https://doi.org/10.1007/978-981-32-9350-2

This Springer imprint is published by the registered company Springer Nature Singapore Pte Ltd.
The registered company address is: 152 Beach Road, #21-01/04 Gateway East, Singapore 189721, Singapore

Preface

Innovation is essential to maintain market competitiveness and sustainable growth, as well as to drive socioeconomic development. Accordingly, at all times, larger companies have been committed to research and development (R&D) investment, securing outstanding R&D personnel, and building an effective R&D management system.

However, there is no universally desirable R&D strategy that a firm should pursue. For example, the nature of the qualities required for R&D talent appropriate to drive innovation will change over time and will vary across industries. The importance of cooperative relationships with other companies in R&D also changes, and the purpose of cooperation also varies. A critical reason is that technology continues to evolve and the technological structure required by firms and industries changes. As the set of technologies at the core of the technological system changes, not only the size of the R&D project required and the composition of researchers will change, but so too the combination of proprietary development and technical cooperation with other institutions. In other words, the paradigm of innovation in a systemic sense continues to evolve. The sustained growth of a firm's market competitiveness depends upon its capacity to reflect, respond and adapt to this evolving paradigm. The most common reason for the decline of an excellent company in the past has been the failure to adapt to a shift in the prevailing paradigm. Understanding the changing character of the technological paradigm is, therefore, an important research theme in management and economics. The purpose of this book is to elucidate various unexplored aspects of a technological paradigm, and of paradigm change over time.

As is well known, information technology (IT) has been at the core of technological architecture since the late-twentieth century. IT has streamlined factories, office labor, and social infrastructure not only in the electrical and electronics industries, but in all industries, and it has spurred numerous product innovations to create a variety of innovative new products, thereby creating a new kind of consumer society. It appears that Japanese companies were initially successful in adapting successfully to the IT-centered innovation paradigm and strengthening their market competitiveness.

However, the technological architecture has evolved further, now becoming centered on digital technology, and the paradigm of innovation has changed accordingly. As symbolized by the concept of a knowledge-based economy at the end of the twentieth century, innovations in the massive and highly sophisticated use and processing of data and knowledge have continuously emerged. Within IT, software technologies such as computer programming, computer system design, and artificial intelligence have become the core of the technology system rather than hardware. The foundations for the primacy of these software technologies include the explosive evolution of the internet, the digitization of data, and the dramatic improvement in computer information processing capabilities.

This new technological architecture is in the process of further transforming the paradigm of innovation. For example, we have witnessed the emergence of an open modular industrial system in place of a closed integral system as in the past, which had taken the form of a vertically integrated business model in which all technologies were developed and produced internally. The current shift toward an open modular business model is based on the premise of a global division of labor, and therefore, a platform-type business model that acquires global standards through an increasingly open international division of labor, and this has been gradually undermining strategies of concealing and monopolizing technologies in specific firms or locations. Software development and the explosive spread of the internet, which do not require large-scale equipment and funds, have enabled some areas of knowledge to be disseminated throughout the world including in developing countries, as well as facilitating the sharing and integration of excellent knowledge from around the world. As a result, the globalization of R&D systems, which can take advantage of geographically diversified talented human resources, has become an indispensable R&D strategy. Furthermore, since basic research has become more important for the development of these advanced technologies, a strengthening of the collaboration between private companies, universities, and other research institutions has become an important strategic issue.

Firms that failed to adapt successfully to this new paradigm have lost their competitive advantage and have stagnated, whereas those that have succeeded in adapting have gained a competitive advantage.

What kinds of strategic adaptation will enterprises and industries need to adapt to the new paradigm in the future? From this viewpoint, the book will analyze the state of internationalization of corporate R&D, the new development of global standardization strategies, and the role played by large cities, where research institutes are concentrated, in the exchange of technical knowledge, the development of human resources for knowledge, and methods for formulating research and development strategies.

Accordingly, the book consists of two parts. The first part, which consists of the following five chapters, mainly discusses the concept of a techno-paradigm from the perspective of its historical and macroeconomic sides. Then, in the second part, paradigm shifts are examined from the perspective of specific industrial sectors and companies.

Here we would like to highlight some of the key themes to emerge from each chapter.

Chapter 1 of the first part, entitled "The Philosophy of Paradigm Change in the History of Social Evolution" by John Cantwell, begins by defining and explaining the meaning of a technological paradigm, and of paradigm shift. The chapter places the idea of paradigm change in a philosophical context, arguing that a paradigm shift can be seen as representing a Hegelian evolutionary process in a social or cultural setting. The development of a techno-socio-economic paradigm at a society-wide level is a process of continual co-evolution between three sets of factors: knowledge, institutions, and technology in production.

In Chap. 2, "Paradigm Changes in Technological Knowledge Connections in Urban Innovation Systems" by Salma Zaman, it is argued that innovation today cannot be realized by only one company or even by one location in isolation, and often there is a need for cross-city and cross-firm knowledge exchange. Therefore, taking the world's largest cities for innovation leading to patenting, this study examines between which cities knowledge flows occur more intensively. For this purpose, by analyzing the pattern of connections between cities from a mapping of the locations of the inventors of pairs of citing and cited patents, based on the data of the United States Patent and Trademark Office, this chapter elucidates the city-based knowledge network that shows the geography of knowledge exchange relationships. The chapter discusses the effect of paradigm shift on this subnational geography of knowledge sourcing in innovation.

Chapter 3, "World-Wide Dispersion of Research and Development (R&D) Capabilities" by Takabumi Hayashi and Atsuho Nakayama, examines to what extent R&D capabilities have been geographically disseminated and dispersed worldwide over the past 40–50 years, analyzing scientific papers and US patents as R&D outputs. The analysis shows that the number of different nationalities of author affiliations on papers and the range of locations of the first-named inventors of U.S. patents have increased and diversified.

Chapter 4, "International Standardization of the New Technology Paradigm: A Strategy for Royalty-Free Intellectual Property" by Yasuro Uchida, starts from the point that in the past, the international standardization of intellectual property (IP) was an important strategy. One of the roles of IP has been to generate a source of revenues from royalties, but what is now increasing in IT-related business fields is actually royalty-free exchange. This means that we have witnessed a paradigm shift in IP strategy in the IT-related fields. This chapter examines in detail why such royalty-free cases are increasing, and the background to this emergent phenomenon.

Chapter 5, "New Roles for Japanese Companies at the Knowledge-Based Economy" by Fumio Komoda, examines how the technology paradigm has shifted from being hardware-oriented to becoming software-oriented in accordance with trends in the progress and adaptation of IT. The author argues here that one of the reasons for the downturn of Japanese companies is that they had not been successful in adapting themselves to the transition from hardware to software technologies, remaining overly committed to their hardware heritage.

In Chap. 6, moving into the second part of the book, "Paradigm Shifts in the TFT-LCD Industry and Japan's Competitive Position in East Asia", by Kazuhiro Asakawa, identifies paradigm shifts in the TFT-LCD industry that have directly or indirectly led to a change in Japan's competitive position. The chapter shows that the declining competitiveness of a country and a firm cannot be fully understood without closely relating it to the shifts in the paradigm of the industry. In addition, this chapter shows the mutual alignment between paradigm shifts in the industry, the changing competitive positions of a country and a firm, as well as the changing locus of innovation, in terms of the distribution of efforts between domestic versus international or in-house versus collaborative activities.

In Chap. 7, "Business-University Collaboration in a Developing Country in the Industry 4.0 Era—The Case of Hungary" by Annamaria Inzelt, discusses how in recent years the majority of Hungarian business R&D expenditure has come from companies wholly, or majority-owned by foreign interests. This high proportion indicates the substantial role of foreign companies in the Hungarian research agenda and in business-university collaboration. This chapter focuses on how foreign companies are shaping business-university collaboration in research and experimental development and touches upon the role of government as facilitator.

Chapter 8, "Text Mining Method for Building New Business Strategies" by Fumio Komoda, Yoshihiro Muragaki and Ken Masamune, examines how under a new technology paradigm, identifying market needs and finding appropriate technical ideas become much more unpredictable. In order to solve this problem, using pinpoint focus type text mining may successfully provide useful solution methods for companies. To demonstrate this, the authors use the illustration of a neurosurgical robot and discuss how well-founded ideas for the development of a surgical robot can successfully be discovered.

Chapter 9, "Paradigm Change in the History of the Pharmaceutical Industry" by Sarah Edris, argues that with the advancement of molecular biology as a science, we see a paradigm shift in the nature of R&D in the development of new drugs from around the 1990s. As a result, many new forms of cooperation in R&D among private companies, universities, and public research institutions have now become unavoidable.

Chapter 10, "Knowledge Transfer and Creation Systems: Perspectives on Corporate Socialization Mechanisms and Human Resource Management" by Tamiko Kasahara, explores the role played by corporate socialization mechanisms (CSMs) and human resource management (HRM) practices in knowledge sharing and creation in multinational corporations. Drawing on a longitudinal case study of Cambridge Technology Partners (CTP), the findings of this study show that CSMs were incorporated into HRM practices. Specifically, training and development practices played a role in transferring knowledge from the headquarters to a focal subsidiary at the corporate level, and from senior to junior consultants at the individual level. At present, CSMs have laid the foundation for further knowledge creation for CTP.

Finally, Chap. 11, "Redefining the Internationalization of R&D Activities: How Far Have the Firms' R&D Members of US and Japanese Companies Been Diversified?", by Takabumi Hayashi, examines how the role played by foreign researchers and engineers engaged in R&D activities in the US and in the overseas R&D activities of US multinational corporations are no longer negligible. The chapter also examines the extent to which the internationalization of R&D by US companies would be affected if the outcomes of their activities in the US were included in the internationalization of R&D. Finally, analyzing the case of Canon Inc., the most active Japanese-based company in patenting, it is shown how much the internationalization of R&D activities differs between IBM and Canon.

The further evolution of IT, and of international knowledge creation and sharing will doubtless continue to lead to further gradual changes in the innovation paradigm. The various themes presented in this book help us to better understand at least some aspects of this evolution. We encourage further research on these topics in the future, and hope to contribute to such efforts ourselves.

Finally, in editing the book, we would like to express our gratitude to the following editing staff members of Springer publishing: Ms. Swetha Divakar, Mr. Yutaka Hirachi, and Ms. Shinko Mimura, for their valuable comments and kind cooperation.

Newark, USA   John Cantwell
Tokyo, Japan   Takabumi Hayashi

Contents

Editors and Contributors

About the Editors

John Cantwell (Co-editor, Chap. 1) (Ph.D., Reading) is Distinguished Professor of International Business at Rutgers University (New Jersey, USA) since 2002. He was previously Professor of International Economics at the University of Reading in the UK. His early work helped to launch a new literature on multinational companies and international networks for technology creation, beyond merely international technology transfer. Professor Cantwell's total citation count on Google Scholar is well over 15,000. His published research spans the fields of International Business and Management, Economics, Economic History and Philosophy, Economic Geography, and Innovation Studies. He served as the Editor-in-Chief of the *Journal of International Business Studies (JIBS)* from 2011–16. He was the elected Dean of the European International Business Academy (EIBA) Fellows from 2015–18. He is also an elected Fellow of the Academy of International Business (AIB) since 2005.

Takabumi Hayashi (Co-editor, Chap. 9 and Chap. 10) (Ph.D. in Economics, Rikkyo University) is Professor Emeritus of Rikkyo University, Tokyo. He successively filled the position of Senior Lecturer at Fukuoka University, Associate Professor, and Professor of International Business at Rikkyo University, and Professor at Kokushikan University, Tokyo. His recent research areas are innovation systems and R&D management, focusing on knowledge creation and diversity management. His works have been widely published in books and journals. His book "Multinational Enterprises and Intellectual Property Rights" (in Japanese; Moriyama Shoten, Tokyo, 1989.)" is widely cited, and "Characteristics of Markets in Emerging Countries and New BOP Strategies" (in Japanese; Bunshindo, Tokyo,

2016) received the award from Japan Scholarly Association of Asian Management (JSAAM) in 2018. He has been sitting on the editorial board of several academic journals.

Contributors

Kazuhiro Asakawa Graduate School of Business Administration, Keio University, Yokohama, Japan

John Cantwell Rutgers University, Newark, NJ, USA

Sarah Edris Rutgers Business School, Rutgers University, Newark, NJ, USA

Takabumi Hayashi Rikkyo University, Tokyo, Japan; Tokyo Fuji University, Tokyo, Japan

Annamaria Inzelt IKU Innovation Research Centre, Financial Research Co., Budapest, Hungary

Tamiko Kasahara School of Information and Management, University of Shizuoka, Shizuoka, Japan

Fumio Komoda Honorary Professor, Saitama University, Saitama, Japan

Ken Masamune Institute of Advanced Bio-Medical Engineering and Science, Tokyo Women's Medical University, Tokyo, Japan

Yoshihiro Muragaki Institute of Advanced Bio-Medical Engineering and Science, Tokyo Women's Medical University, Tokyo, Japan

Atsuho Nakayama Marketing Sience, Tokyo Metropolitan University, Tokyo, Japan

Yasuro Uchida Graduate School of Business, University of Hyogo, Kobe, Hyogo, Japan

Salma Zaman Lahore University of Management Sciences (LUMS), Lahore, Pakistan

Chapter 1
The Philosophy of Paradigm Change in the History of Social Evolution

John Cantwell

Abstract A technological paradigm identifies the coherent features consistently present in the evolution of an innovation system over time. These shared characteristics refer to a widespread cluster of innovations during a given era that rely on a common set of scientific principles and on similar organizational methods. The idea of an overarching paradigm that depicts commonalities in innovation efforts may be applied at the level of an industry, a technical field, or in society as a whole, as in the case of a techno-socio-economic paradigm. Occasional paradigm shifts entail some change in the framework for innovation, while preserving certain features of the old ways in a new synthesis. Thus, paradigm shift takes the form of an Hegelian evolutionary or dialectical process. The evolution of an innovation system as a whole derives from the interaction or co-evolution of its central elements: knowledge, institutions, and technology in production. While conventional science isolates causal associations between specific parts of a system, Hegelian conceptual reasoning addresses the combined and interconnected movement of a complex relational system with multiple interdependencies. In the light of this contention, I argue that we should move away from age-old debates over whether social evolution or development is driven primarily by knowledge, by institutions or by the forces of production. Our attention should now turn instead to how these parts move together in an evolving system, and how their mutual goodness of fit adjusts during phases of paradigm shift and realignment.

Keywords Technological paradigm · Techno-socio-economic paradigm · Paradigm shift · Hegelian approach · Social evolution · Technological evolution

J. Cantwell (✉)
Rutgers University, Newark, NJ, USA
e-mail: cantwell@business.rutgers.edu

© Springer Nature Singapore Pte Ltd. 2019
J. Cantwell and T. Hayashi (eds.), *Paradigm Shift in Technologies and Innovation Systems*, https://doi.org/10.1007/978-981-32-9350-2_1

1.1   The Meaning of a Technological Paradigm, and of Paradigm Shift

Based on the original contribution of Perez (1983), Christopher Freeman and various of his co-authors and associates at the Science Policy Research Unit (SPRU) in Sussex developed the concept of a techno-socio-economic paradigm (Freeman 1987; Freeman and Perez 1988; von Tunzelmann 1995, 2003; Freeman and Louçã 2001; Perez 2002, 2009). While one critical element of the emergence of a new techno-socio-economic paradigm is some radical breakthrough 'macroin-vention' that forms the basis for a newly pervasive strand of general purpose technology (Bresnahan and Trajtenberg 1995; Lipsey Carlaw and Bekar 2006)—such as machinery, electrical equipment or information technology in each of the three industrial revolutions respectively—in the SPRU tradition a paradigm refers to considerations that range beyond the immediate characteristics of technological innovation itself to the wider social and economic context in which a given technological trajectory is embedded. In this tradition a technological paradigm can be defined as a system of scientific and productive activity based on a widespread cluster of innovations that rely on a common set of scientific principles and on similar organizational methods (as suggested by Dosi 1984). Thus, a prevailing paradigm regulates the cognitive frames shared by technological practitioners in knowledge-based communities, establishes the heuristics of search for solutions to technological problems, and encompasses normative considerations such as the criteria for assessing potential solutions, and goals for the improvement of practice (Dosi and Nelson 2010). In the sense of a paradigm as a society-wide techno-socio-economic paradigm, each paradigmatic system characterizes an epoch, of which there have been three since the first industrial revolution: the mechanical age, the science-based mass production age, and the information age.

The concept of a technological paradigm is bound up with a certain set of social institutions, and with selected organizational forms or methods in the economic domain of production, distribution and exchange. It is a concept that can be applied at various levels of analysis. In particular, the notion of a technological paradigm may also be applied in a microtechnological sense such as the semiconductor paradigm (Dosi 1982), or in the sense of a technological regime that characterizes the nature of innovation among the firms of a given industry (Nelson and Winter 1977) such as the pharmaceutical industry (which is the sense of a paradigm used by Edris 2019). The broader sense of a techno-socio-economic paradigm as used by Freeman, Perez or von Tunzelmann constitutes a constellation of technological paradigms in these narrower senses of the term (Dosi and Nelson 2010), and in this chapter when speaking of a paradigm I will refer primarily to the more macro or integrative concept of a techno-socio-economic paradigm. This conceptualization helps us to explain how a new paradigm emerges once an earlier one has run out of steam, and so innovation becomes increasingly constrained, difficult and costly under the old paradigm. However, it also helps to account for an often lengthy gestation period of slower growth and impaired economic development in the transition between paradigms, since social institutions must be adapted to the new requirements, and this creates problems and resistance from those who believe their interests to be threatened by the change. Another key implication of the gradual adaptation of social institutions during phases of paradigm change, allied to the cumulative nature of technological knowledge which always builds upon and recombines some elements of prior knowledge (Pavitt 1987; Cantwell 1989; Arthur 2007), is that each new paradigm partially incorporates and synthesizes earlier paradigms in the process of reconstituting a new social formation. Thus, for instance, in the realm of technology we move from the centrality of the mechanical to the electro-mechanical, and then to the electronic, as we move from one age to another. In other words, shifting from one paradigm to another is an evolutionary process, not a revolutionary overthrow of the paradigms that preceded it.

Another frequently used and somewhat related construct in the innovation studies literature is the concept of a system of innovation. This has been variously used to refer to a national system of innovation (Nelson 1993; Freeman 1995), a regional system of innovation (Asheim and Gertler 2005), or a sectoral system of innovation (Malerba 2004). This conceptualization might be extended to include the emergence of a global system of innovation in the information age of 'intellectual capitalism' (Granstrand 2018). Just as with the idea of a paradigm, an innovation system generally supposes a framework of interconnectedness across multiple actors, a given institutional context, and a prevalence of certain kinds of business organization and relationships between organizations engaged in productive activities. However, the difference with paradigm shift is that an innovation system is a comparative construct intended to analyze variation across some characteristic forms of innovation arranged in a typology, whether across countries, regions or industries, or for the more (versus less) globally connected forms of activities. It is designed to answer the question 'what is distinctive about the structure of innovation in this country, region or industry, or kind of activity, compared to others?'. In this respect e.g. the national systems of innovation literature bears some relationship to the varieties of capitalism literature, or that on national business systems, and so on.

In contrast, the notion of paradigm shift is intended to focus attention on an evolutionary process over time. In particular, a paradigm shift refers to the occasionally punctuated character of evolutionary change. The counterpart of this social process in evolutionary biology is the normal stability of a species once it has been established (corresponding to the conduct of normal science within some established settled framework in Kuhn's (1962) terminology of the workings of scientific communities), but with occasional processes of speciation (akin to the occasional scientific revolutions that episodically transform the nature of science according to Kuhn 1962). Once a new species emerges it shares many of the characteristics of the species from which it has descended, and yet it also has some new distinctive features that ensure that there is a structural shift in the way in which it interacts with its environment, thereby implying a shift in its path of evolutionary development. Likewise, occasional scientific revolutions lay the foundation for a new theoretical and methodological framework for science, setting it on some new direction that leads to new kinds of discoveries. In the context of technological change, as one technological paradigm comes to succeed another it is not just the leading edge of technological knowledge and the focus of innovative problem-solving that shifts (as would also occur in a scientific revolution), but the associated governance structures, organizational forms and business relationships shift as well (von Tunzelmann 2003).

The distinction between change within a paradigm, and change due to a shift in paradigm, delineates the two major components of an evolutionary process. First is the within-paradigm element of continuity or persistence that is a consequence of path dependence, and which is sometimes associated with lock-in and positive feedback effects (e.g. Arthur 1989). Within a technological paradigm, the cumulative and incremental character of technological change is the basis for the conceptualization of technological accumulation in which new knowledge discovery and recombination is interconnected with knowledge use and application (Cantwell 1989). The biological equivalent of a sustained path is the role of inheritance in reproduction, and in the progress of science it is the consistency of the received theoretical framework and the kinds of experiments to which this regularly gives rise in normal scientific endeavor. Second is the occasional paradigm-shifting emergence of some radical new departure that supercedes and yet also absorbs many aspects of the previously established path, thereby altering the character and behavior of a system. The biological representation of such a path-shifting synthesis of older and newer elements generating a related but novel form begins with genetic mutation (which may be reinforced by environmental diversity or divergence) that then leads to variety in characteristics and ultimately to speciation. In science such novelty is brought about by occasional scientific revolutions that alter some of the fundamental assumptions or premises on which the science in question builds, and so set a new and usually more complex agenda for the subject.

In the case of a techno-socio-economic paradigm we are addressing evolutionary processes at the level of a society or community, rather than in a particular stream of technology or in a specific industry. So when the paradigm shifts the society and its system of technology, production methods and social institutions must be expected to interact differently in some respects. Thus, as we move from the mechanical age to the science-based era of mass production, and then to the information age, the nature of capitalism changes, and society and the economy function somewhat differently than they did in the past. For this reason attempts to derive some fundamental 'laws of capitalist development' or some 'stylized facts of economic development' that are universally applicable from the time of Ricardo or Marx through to the present day are likely to be flawed. One such widely discussed recent attempt has been by Piketty (2013), in his contention that a rising inequality of income and wealth is an inherent tendency in a capitalist economy, which of course echoes some of the claims of Marx in the nineteeth century and calls to mind the so-called gilded age of the late nineteenth century. To maintain a kind of continuity between the gilded age and the trends of the early twenty-first century Piketty argues that the period between 1930 and 1975 in which inequality fell was a kind of temporary aberration.

Yet we should recall that during that same period Kuznets (1955) had convinced many with his 'stylized facts' that income inequality tends to fall with economic development, while the capital-output ratio remains roughly constant—which trends again fitted well the context of the period in which he was writing. So rather than referring to a lengthy period of exceptional circumstances, it seems more plausible to simply say that the techno-socio-economic paradigms of the mechanical age, the science-based mass production age, and the information age are distinct, and aggregate tendencies change as we move from one to the other. One feature shared by the mechanical age and the information age is an increase in rents from land and natural resources, associated with a rising price of land and hence substantial gains by landowners in their holdings of wealth. Yet the institutional context for this commonality varies considerably between the 19th and 21st centuries: in the 19th century there remained an aristocracy, or those who aspired to join an aristocracy, whereas in the 21st century holdings of land and natural resources are more linked to financial institutions.

1.2 Paradigm Shift as an Hegelian Evolutionary Process

I will argue that in philosophical terms the course of paradigm change can be depicted as an Hegelian evolutionary or dialectical process. In an Hegelian perspective as humans or human societies or groups interact with our environment we may consolidate or change our way of life, our knowledge about ourselves and about our environment. As we reflect on the relationship between ourselves and our environment we may also come to reconfigure and to challenge not just our prior knowledge, but the very way in which we think about ourselves, our values or beliefs, our actions, and our way of life. Thus, a contradiction may arise between our old way of looking at the world and some new way of thinking about it. Since the new way of thinking and acting is likely to also imply challenges to our existing norms or institutions, to the prevailing social or organizational structures, and to the interests that may be embedded within them, the contradiction is liable to give rise to a period of conflict or tension between the old and the new, as well as giving rise to increased uncertainty. Yet ultimately some form of reconciliation is generally found between the old and the new approaches, some way of combining the two that preserves aspects of both in some form of synthesis.

Any synthesis between the old ways and the new provides a generally more profound and distinctive way of thinking about the world, which tends to facilitate the scope for sustaining greater diversity or pluralism both within and between societies or social structures such as organizations, by increasing the extent of mutual recognition of relational interdependencies between the various elements of a complex social system. In the Hegelian perspective an evolutionary process of social transformation is inherently systemic or interconnected, the different parts or particulars of a system must move together or co-evolve. Thus, it is not just our way of thinking about the world that comes to be reshaped, but the nature of the reasoning we give for our actions (how we interpret rationality), and our system of practice. Much of this has been reflected in the re-thinking of Hegel that has been under way in recent years among philosophers in the English language literature. These scholars have shown how, in an Hegelian view, through mutual recognition and acceptance of common norms or institutions we can exercise free and rational agency in collaborative endeavors, which may lead to social transitions if crises arise in the authority of norms (Pippin 2008); how human knowledge and ideas are rooted in practice (Westphal 2003); and how there is always a social or relational content in the exercise of rational intentionality (Pinkard 1994), requiring us to move beyond methodological individualism.[1]

An Hegelian approach to paradigms and paradigm change is therefore grounded in the historical evolution of technological or social practice, and is concerned with how the parts of an innovation system interact with one another, and thereby move the system as a whole. Since, as we have seen, a techno-socio-economic paradigm is an interconnected or complex social system, we can think of a paradigm as an historically specific form of interaction between knowledge, institutions, and technology and production, and paradigm shift as the emergence of a new form of capitalism characterized *inter alia* by a shift in the nature of innovation. The nature of knowledge, beliefs or institutions, and practice or conduct of actors within a paradigm is shaped by the system in which they live and operate (which Hegel referred to as a shape of consciousness, or the spirit of an age as a self-conscious way of life). Thus, a paradigm incorporates an outlook on innovation in terms of what are regarded as being the most relevant problems that need to be addressed, and how they can be solved, which includes some understanding about how prevailing practice works, and its limits (Dosi and Nelson 2013). As an outlook, a paradigm includes both the cognitive frames shared by practitioners that orient their innovative or problem-solving activity (Constant 1980), and the normative considerations involved in judgments about how to search for and to share knowledge, and how to set and assess the goals of innovation. In this kind of Hegelian evolutionary process there is a continual interaction and interdependence between the subjective side of knowledge, values and beliefs, and the objective side of the use of technologies in production and their impacts on organizations, society and the natural world.

---

[1]I note in passing that these re-readings of Hegel have the effect of bringing our interpretation of his work closer to that of the early Marx, whose writings were deeply Hegelian in these senses. This calls into question the exaggerated distinction between Hegel's 'idealism' and Marx's 'materialism' that remains commonplace in the Marxist literature. For a critique of the view that Marx was an ontological materialist, in the sense that our ideas are essentially just a reflection of the character of 'matter in motion' or what Engels termed the 'dialectics of nature', see Kline (1988).

In political philosophy the focus of Hegelian scholarship has been on the historical evolution of justice and freedom (e.g. Pinkard 2017). Part of that process of evolution has been a change over time in the very meaning of these terms as institutions have evolved. My focus here is on another feature, namely the improvement and transformation of our knowledge base in the course of social evolution. We may think of what Hegel called 'absolute knowledge' as that kernel which is held in common among various societies or organizations (or other social units), and which has been preserved or inherited from the past. Of course, this kind of foundational knowledge base will not necessarily be held by every society or organization, and some forms of knowledge may atrophy, recede or be lost or forgotten over time. In the case of technological knowledge the relevant distinction is between private knowledge which is held mainly by firms, and public knowledge that is shared and held in common in the public domain in some social context or setting (Nelson 1989). This latter public knowledge base constitutes the foundation for any paradigm. The public aspect of technology is associated with knowledge diffusion and enables the community as a whole to benefit from technological innovation. As indicated earlier, a paradigm is constituted by some common set of scientific principles and organizational methods.

It is worth touching here on the role of teleology in an Hegelian perspective. It is necessary to distinguish between two senses of teleology, one of which relates to the present day, and the other to an imagined future hypothetical state of the world, which carry very different meanings and should not be confused.[2] With respect to the present day, an Hegelian perspective is exclusively backward looking or historical. This is the contention that we can trace out from the past how we came to have the kinds of knowledge or ways of thinking, the institutions, and the technologies that we have today. In other words, we can find the origins or roots of our current techno-socio-economic paradigm, in the form of an historical evolutionary path that led us to where we are now. It is important to appreciate that a social evolutionary framework is not predictive, but explanatory. At any critical juncture in the past where the path has shifted it did so in ways that were conditional at the time they happened. A complex social system is open and creative, and not closed or pre-determined. However, with respect to an imagined future hypothetical state of the world an Hegelian perspective is forward looking, but considers a desirable condition that can never be actually realized. This is what Hegel referred to as the absolute, in which knowledge would be infinite, and we would enjoy perfect relationships between ourselves and with the world. Hegel associated this state of perfection with God. For an evolutionary approach, the relevance of this aspirational state, rather like the notion of utopia (Hodgson 1999), is that it identifies our ultimate consistent goals and purposes. While we can never attain perfect justice,

---

[2]Confusing these two is responsible for various false attributions, such as the assertion that for Hegel the Prussian state of the 1820s was an ideal state, or Fukuyama's claiming as Hegelian his notion that liberal capitalism represented the 'end of history'.

freedom, or comprehensive knowledge, we can strive towards these goals. Seeking to continuously improve our knowledge becomes a goal of self-conscious rational actors.

Following an Hegelian typology, knowledge tends to develop first as an understanding of some processes as phenomena through some form of abstraction that isolates a regular cause and effect relationship. In the case of technological knowledge this relates to the acquisition of know-how and then usually later know-why, sometimes guided by science. Yet in the case of technological knowledge even when there is some know-why, know-how cannot be easily replicated across different contexts, whether across firms or across locations (Winter and Szulanski 2001). To the extent that there is a richer pool of relevant shared public knowledge, the costs of technology transfer or diffusion may be reduced, but not eliminated. Within a paradigm these costs can be somewhat offset when the relevant public knowledge pool lies in and around the core technologies of that paradigm, in which area of problem-solving search there is a greater common awareness of the properties of a wider range of distinct applications across a variety of contexts. This feature of a paradigm also helps to perpetuate the paradigm itself, since it encourages innovation that connects to these core fields of understanding that mark out the paradigm, since these are likely to have a greater potential spread of applications. In such core domain activities within a paradigm, in Hegelian terms understanding passes over into more fundamental conceptual reason, in the sense of a capacity to conceptually recognize and to relate (and hence to learn from) differences in particular behaviors associated with individual concrete applications across specific contexts.

Turning from the development of technological knowledge within paradigms to the wider social and economic characteristics of a techno-socio-economic paradigm that tend to remain relatively stable within a paradigm, but may be disturbed during a paradigm shift, the relevance of a distinction between Hegelian understanding and reason is equally germane. In understanding some social or economic relationship through the abstraction entailed in a formal theory, or a mathematical representation, or in claiming a fixed law or consistent tendency in observed data, a claim is usually made for the generalization (or universality) of that relationship over time and across contexts. As suggested earlier, this kind of generalization may carry some justification within a paradigm, by representing the stylized facts of an era— e.g. the existence of a rising capital-output ratio in the mechanical age (as described in Marx's version of the law of the tendency of the rate of profit to fall), or the constancy of the capital-output ratio in the science-based mass production era that prevailed for much of the 20th century (as described by Kuznets 1955). Using the terminology of evolutionary economics (Nelson 2018), formal theory is relatively better suited to an analysis of these kinds of regularities within paradigms that provide an Hegelian understanding of some selected prominent and consistently observed phenomena. Instead, appreciative theory relatively better allows for complex interdependencies, and for the kinds of systemic movements or shifts associated with the emergence of contradictions or tensions within a complex social system that moves us into the Hegelian realm of conceptual reason.[3] As explained earlier, when a techno-socio-economic paradigm occasionally shifts, not only is there a shift in the commonly observed regularities of behavior at a systemic level, but allied to this there is a shift in our ways of thinking about the world and of reasoning about it. From a philosophical perspective it is important to try and better grasp the generally subtle shifts in modes of reasoning or rationality that provide a kind of foundational underpinning to the occasional shifts in techno-socio-economic paradigms.

1.3 The Continual Co-evolution of Knowledge, Institutions, and Technology in Production

In any social and economic system there is a continual interaction between the knowledge base of a society or community, its institutions—ranging from values and beliefs through informal norms to formal laws or regulations—and its technology developed and used in production. In a capitalist society of the kind that took root after the first industrial revolution from the techno-socio-economic paradigm of the mechanical age onwards, in which innovation in production became central, all three of these elements (knowledge, institutions and technology) have been changing and interacting with one another continuously. In the heyday of a paradigm the three elements advance together faster and more effectively, as the paradigm ensures a good fit between them. The nature of the interaction between the three elements follows a consistent pattern or structure which reflects the essential outlook of the age, and which ensures some degree of regularity of progress within that structure, although the precise form of that progress is always characterized by some ex ante uncertainty and unpredictability. However, once a paradigm runs out of steam, experimentation with different kinds of practical solutions and new ways of thinking begin to spread more widely, and as they do the three elements tend to fit less well together, creating greater frictions and conflicts at a systemic level. Within a paradigm the structure of interaction between the three elements of the system is largely consistent, but during windows of paradigm shift the very nature of the structure of that interaction changes. The change in the way in which knowledge, institutions and technology interact in a new paradigm is

---

[3]It may be worth noting in passing the sharp difference between the simplistic and impoverished interpretation of reason or rationality found in contemporary Economics as a pure economic self-interest, and the much more sophisticated accounts of rationality found in the philosophical literature of the Enlightenment era, culminating in Kant's critical philosophy. Hegel's critique of Kant moved us yet another step forward in allowing for the social evolution of reason or rationality, in the way that I echo here, and which is precluded if we treat human rationality as being fixed or given in nature.

manifested in some differences in behaviors or tendencies in the workings of a system as we move from one paradigm to another, as discussed earlier.

So, I am arguing, in any social system knowledge, institutions and technology continually co-evolve. Within a paradigm the evolution of each of these three elements is largely consistent, and so as they change they still continue to fit relatively well with one another, hence leading to mutually reinforcing positive feedback effects on each part of the system. Instead, during periods of paradigm shift, the three elements evolve in ways that become potentially less consistent, and so the form of co-evolution of the three parts involves an increasing number of negative feedback effects that may disrupt and stymie progress in the system as a whole. Establishing a new paradigm is essentially a process of finding some new form of re-alignment between the three elements that once more ensures a better fit between these parts in the course of their mutual co-evolution. The overall evolution of a social system depends on the development of a satisfactory structure of mutual relational interdependencies between its various parts. A structure of interconnected social co-evolution that is satisfactory for at least a time can be described as a paradigm, while the necessity of an occasional shift in that structure can be represented as a paradigm shift. Once this is appreciated, then we should focus our attention on the changing goodness of fit between the parts of a social and economic system as that system evolves over time.

Unfortunately, most of the relevant social science and economic history literature has not focused on this co-evolutionary goodness of fit issue. Rather, much of this literature has debated what is the primary direction of causality between the parts of an economic and social system—which debate, from a co-evolutionary perspective, is rather like trying to resolve the 'chicken and egg' problem. In Hegelian terms, this debate remains trapped at the level of attempting to discern a simple understanding of an observed pattern of behavior in a nutshell (in the deterministic form that 'A causes B'), rather than moving on to a more mature form of conceptual reasoning that can better accommodate complex relational interactions. It is perhaps especially sad that Hegel and Marx are today often interpreted purely through this kind of identification of a primary causal chain of directionality. In the usual Marxist account, the economic structure or the forces of production are believed to drive the superstructure of social institutions, and the ideas and beliefs that prevail in a society are said to reflect the interests of the dominant class in the existing mode of production (see e.g. Cohen 1978). Indeed, the conventional perspective of neoclassical economics is somewhat similar in the primacy it attributes to the economic realm, although this is interpreted more in terms of exchange rather than production, and it stresses in particular the role of markets in coordinating the decisions of independent utility maximizing actors. However, within economics there has been a gradual shift in recent years towards the alternative perspective of institutional economics, which argues the other way round that institutions tend to drive and to regulate the way in which an economy functions (see e.g. Acemoglu and Robinson 2012, on the effects of extractive vs. inclusive institutions, or the similar distinction between limited access vs. open access orders by North et al. 2009). While the first two approaches argue for the primacy either of the economy or of institutions, a third makes the case for the primacy of knowledge or ideas. Among economic historians, Mokyr (2002) contends that the development of knowledge and scientific breakthroughs can explain industrialization and economic development, while McCloskey (2016) claims that it is ideas and values that have driven economic growth.

These kinds of debates over the causal sequencing of the elements of social evolution can never really be resolved, since each of the three camps can refer to legitimate historical evidence that is supportive of their claims. From a co-evolutionary perspective the arguments are akin to the story of the blind men and the elephant: each school of thought provides a valid way of thinking about the evolution of a social system historically, so long as one shares their starting point for analysis and their framing of the critical issues. What is more, their accounts of a complex reality are potentially consistent, even though none of these contributors is generally willing to countenance the possibility of a synthesis, since they remain trapped in a common belief that the fundamental question is one of establishing causal primacy. Addressing that question all depends on how one sets up the problem at hand. The antecedents of most knowledge or ideas can generally be traced back a very long way. So if the question posed is 'which came first?' the case for knowledge or values or beliefs as the driver is at its strongest. However, when knowledge or ideas first arise they may be relatively little related to practice, or at least narrowly constrained in their practical applicability. An extreme illustration would be da Vinci's helicopter, which was realized in practice only around 450 years later. So if instead the appropriate question is 'what drives the development of practice?' the answer is the accumulation of technological knowledge and expertise through a continual process of innovative problem-solving in and around production (Rosenberg 1982). So this takes us back to the story of the system of production and technology as the primary driver of social evolution. Yet at this stage we are reminded of how Freeman and Perez (1988) showed how the selection of viable new technologies, organizational methods and products depends on the adaptability of institutions. So if the question becomes 'what drives the take-up, diffusion and survivability of lines of innovation once they have been hatched?' we are more likely to turn to an institutional story, such as whether institutions are more extractive or inclusive in nature.

Once we step back in Hegelian fashion and attempt to conceptualize the movement of a social system as a whole in history, it becomes necessary to recognize the continual relational interdependence between knowledge, institutions and technology in production in the course of development, rather than a neat primary causal direction of change. This is especially true if our focus of attention is on the evolution of a holistic techno-socio-economic paradigm, as opposed to an attempt to explain some certain specific chain of events or phenomena, or some particular phase of history. An encompassing techno-socio-economic paradigm evolves organically and interdependently, thereby exhibiting a kind of inner life or vitality from within itself, rather than through a simple deterministic response to some underlying primary exogenous mechanism or driver. The movement of the whole entails a necessary co-evolution of the parts, such that we must consider the complex relational interaction between each of the parts and the overall system. The behavior of a system, whether within a paradigm or in a period of paradigm shift, is related to the ever-changing and continuously tested degree of fit or compatibility between the elements of that system, and the extent to which the goodness of fit can be retained or improved. In an Hegelian co-evolutionary approach we must consider the movement of both the whole system and the particulars within it.

Hegel once said famously that a philosopher can only present a philosophy that reflects the context of his or her own time, and the spirit of the age in which he or she lives. This follows from his backward-looking evolutionary approach to explain the roots of the present and the selective absorption of ideas taken from the past, and a belief that the future is unpredictable. It is also related to his even better known saying that 'the owl of Minerva spreads its wings only with the falling of the dusk', meaning that the philosophical knowledge that captures the spirit of an age can only reach full fruition once we have passed the zenith of that age. In other words, only once a society or a way of life has reached maturity can we fully conceptualize the course of its evolution over its entire lifetime. However, on this point I believe that Hegel was not entirely right. As commented upon a moment ago in a different context, ideas can sometimes arise well ahead of their greatest applicability to practice. Indeed, some of the ideas that Hegel took from Aristotle are arguably of this kind, and they were more relevant in his era than they had been in ancient Greece. I would contend that despite his own opinion on the matter, Hegel is more relevant to our age than he was to his own. This would help to account for the shunning of Hegel's work among English speaking philosophers following his death for most of the 19th and 20th centuries, and yet the revival of interest and rediscovery of his writings today.

The spirit of the Enlightenment era was perhaps given its ultimate expression in the classical political economy of Adam Smith and the critical philosophy of Kant. The progress in scientific thought and methods also inherited from the Enlightenment laid the foundations for modern science and social science, grounded on the principle of examining deterministic cause-effect relationships by isolating phenomena and separating them (controlling for) any other extraneous influences. This has led to the dramatic growth and an ever faster expansion of increasingly specialized bodies of knowledge. Yet although these methods continue to be highly useful and productive, we have surely now reached the stage at which the relevance of complex interdependencies must also be given greater attention. With regard to the topic of this paper and this book, we need to be able to analyze the entirety of a techno-socio-economic paradigm and its constituent elements of knowledge, institutions and technology in production, rather than treating the parts separately or continuing to debate which of these parts should be accorded primacy (which is just another variant of the traditional scientific focus on causality). In today's techno-socio-economic paradigm in the information and digital age, not only are knowledge, institutions and technology more interconnected than ever before, but in our way of life we have begun to increasingly appreciate the interrelationship between our society, our production, and the natural world around us and the planet on which we live. In our current paradigm in which co-evolution is more pronounced than ever before, we need to be able to better conceptualize the movement of a complex relational system. This calls for an Hegelian approach, as I have argued here.

References

Acemoglu, D., & Robinson, J. A. (2012). *Why nations fail: The origins of power prosperity and poverty*. New York: Crown Business.

Arthur, W. B. (1989). Competing technologies, increasing returns and lock-in by historical events. *Economic Journal, 99*(1), 106–131.

Arthur, W. B. (2007). The structure of invention. *Research Policy, 36*(2), 274–287.

Asheim, B., & Gertler, M. S. (2005). The geography of innovation: Regional innovation systems. In J. Fagerberg, D. C. Mowery, & R. R. Nelson (Eds.), *The Oxford handbook of innovation*. Oxford: Oxford University Press.

Bresnahan, T. F., & Trajtenberg, M. (1995). General purpose technologies: Engines of growth? *Journal of Econometrics, 65*(1), 83–108.

Cantwell, J. A. (1989). *Technological innovation and multinational corporations*. Oxford: Basil Blackwell.

Cohen, G. A. (1978). *Karl Marx's theory of history: A defence*. Oxford: Oxford University Press.

Constant, E. (1980). *The origins of the turbojet revolution*. Baltimore: John Hopkins University Press.

Dosi, G. (1982). Technological paradigms and technological trajectories: A suggested interpretation of the determinants and directions of technical change. *Research Policy, 11*(2), 147–162.

Dosi, G. (1984). *Technical change and industrial transformation*. London: Macmillan.

Dosi, G., & Nelson, R. R. (2010). Technological change and industrial dynamics as evolutionary processes. In B. Hall & N. Rosenberg (Eds.), *Handbook of the economics of innovation* (Vol. 1). Amsterdam: Elsevier.

Dosi, G., & Nelson, R. R. (2013). The evolution of technologies: An assessment of the state-of-the-art. *Eurasian Business Review, 3*(1), 3–46.

Edris, S. (2019). Paradigm change in the history of the pharmaceutical industry. In J. A. Cantwell & T. Hayashi (Eds.), *Paradigm shift in technologies and innovation systems*. Berlin: Springer.

Freeman, C. (1987). *Technology policy and economic performance: Lessons from Japan*. London: Frances Pinter.

Freeman, C. (1995). The national system of innovation in historical perspective. *Cambridge Journal of Economics, 19*(1), 5–24.

Freeman, C., & Louçã, F. (2001). *As time goes by: From the industrial revolutions to the information revolution*. Oxford: Oxford University Press.

Freeman, C., & Perez, C. (1988). Structural crises of adjustment, business cycles and investment behaviour. In G. Dosi, C. Freeman, R. R. Nelson, G. Silverberg, & L. L. G. Soete (Eds.), *Technical change and economic theory*. London: Frances Pinter.

Granstrand, O. (2018). *Evolving properties of intellectual capitalism: Patents and innovations for growth and welfare*. Cheltenham: Edward Elgar.

Hodgson, G. M. (1999). *Economics and utopia: Why the learning economy is not the end of history*. London: Routledge.

Kline, G. L. (1988). The myth of Marx' materialism. In H. Dahm, T. J. Blakeley, & G. L. Kline (Eds.), *Philosophical sovietology*. Dordrecht: D. Reidel.

Kuhn, T. (1962). *The structure of scientific revolutions*. Chicago: University of Chicago Press.

Kuznets, S. (1955). Economic growth and income inequality. *American Economic Review, 45*(1), 1–28.

Lipsey, R. G., Carlaw, K. I., & Bekar, C. T. (2006). *Economic transformations: General purpose technologies and long-term economic growth*. Oxford: Oxford University Press.

Malerba, F. (Ed.). (2004). *Sectoral system of innovation: Concepts, issues, and analyses of six major sectors in Europe*. Cambridge: Cambridge University Press.

McCloskey, D. N. (2016). *Bourgeois equality: How ideas, not capital or institutions*. Enriched the World, Chicago: University of Chicago Press.

Mokyr, J. (2002). *The gifts of Athena: Historical origins of the knowledge economy*. Princeton: Princeton University Press.

Nelson, R. R. (1989). What is private and what is public about technology? *Science, Technology and Human Values, 14*(3), 229–241.

Nelson, R. R. (Ed.). (1993). *National innovation systems: A comparative analysis*. Oxford: Oxford University Press.

Nelson, R. R. (2018). Economics from an evolutionary perspective. In R. R. Nelson, G. Dosi, C. E. Helfat, A. Pyka, P. P. Saviotti, K. Lee, K. Dopfer, F. Malerba, & S. G. Winter (Eds.), *Modern evolutionary economics: An overview*. Cambridge: Cambridge University Press.

Nelson, R. R., & Winter, S. G. (1977). In search of a useful theory of innovation. *Research Policy, 6*(1), 36–76.

North, D. C., Wallis, J. J., & Weingast, B. R. (2009). *Violence and social orders: A conceptual framework for interpreting recorded human history*. Cambridge: Cambridge University Press.

Pavitt, K. L. R. (1987). The objectives of technology policy. *Science and Public Policy, 14*(4), 182–188.

Perez, C. (1983). Structural change and the assimilation of new technologies in the economic and social systems. *Futures, 15*(5), 357–375.

Perez, C. (2002). *Technological revolutions and financial capital: The dynamics of bubbles and golden ages*. Cheltenham: Edward Elgar.

Perez, C. (2009). The double bubble at the turn of the century: Technological roots and structural implications. *Cambridge Journal of Economics, 33*(4), 779–805.

Piketty, T. (2013). *Capital in the twenty-first century*. Cambridge: Harvard University Press.

Pinkard, T. (1994). *Hegel's phenomenology: The sociality of reason*. Cambridge: Cambridge University Press.

Pinkard, T. (2017). *Does history make sense: Hegel on the historical shapes of justice*. Cambridge: Harvard University Press.

Pippin, R. B. (2008). *Hegel's practical philosophy: Rational agency as ethical life*. Cambridge: Cambridge University Press.

Rosenberg, N. (1982). *Inside the black box: Technology and economics*. Cambridge: Cambridge University Press.

von Tunzelmann, G. N. (1995). *Technology and industrial progress: The foundations of economic growth*. Cheltenham: Edward Elgar.

von Tunzelmann, G. N. (2003). Historical coevolution of governance and technology in the industrial revolutions. *Structural Change and Economic Dynamics, 14*(4), 365–384.

Westphal, K. R. (2003). *Hegel's epistemology: A philosophical introduction to the phenomenology of spirit*. Indiannapolis: Hackett.

Winter, S. G., & Szulanski, G. (2001). Replication as strategy. *Organization Science, 12*(6), 730–743.

John Cantwell (Ph.D., Reading) is Distinguished Professor of International Business in Rutgers University (New Jersey, USA) since 2002. He was previously Professor of International Economics at the University of Reading in the UK. His early work helped to launch a new literature on multinational companies and international networks for technology creation, beyond merely international technology transfer. Professor Cantwell's total citation count on Google Scholar is well over 15,000. His published research spans the fields of International Business and Management, Economics, Economic History and Philosophy, Economic Geography, and Innovation Studies. He served as the Editor-in-Chief of the *Journal of International Business Studies* (*JIBS*) from 2011–16. He was the elected Dean of the European International Business Academy (EIBA) Fellows from 2015–18. He is also an elected Fellow of the Academy of International Business (AIB) since 2005.

Chapter 2
Paradigm Changes in Technological Knowledge Connections in Urban Innovation Systems

Salma Zaman

Abstract The late twentieth century marked the advent of a new information based and internationally networked paradigm, much different from its old science-based predecessor. As we enter the diffusion stage of this paradigm, we expect to see changes in the way organizations and locations interact with one another. We envision that cities and clusters will not rely exclusively on local knowledge sources but will need to combine local with complementary geographically distant (trans-local) knowledge sources. This chapter contributes to the literature on the changing geographic composition of knowledge connections in the new paradigm. Using social network analysis techniques, we construct a unidirectional network of 62 selected cities, since backward citations point in just one direction to prior knowledge sources. We observe how the spatial distribution of our technological network changes during our time period, both in the aggregate and at the level of five selected sectors. The nodes in our network represent cities while the edges represent citations from one city to another. We calculate network statistics such as degree strength and eigenvector centrality to determine which cities have gained influence over time and which cities have become relatively less important. Overall, we observe that the technological knowledge network between our cities is getting denser and more dispersed over our time period. We see that many developing cities are gradually increasing in their centrality to our constructed network and that this increase in centrality is more pronounced in certain sectors, characteristic of the new paradigm, such as the ICT sector. We also observe that while developing cities have become important sources of technological knowledge, they still lag in terms of the knowledge they receive from external sources.

S. Zaman (✉)
Lahore University of Management Sciences (LUMS), Lahore, Pakistan
e-mail: salma.zaman@lums.edu.pk

© Springer Nature Singapore Pte Ltd. 2019
J. Cantwell and T. Hayashi (eds.), *Paradigm Shift in Technologies and Innovation Systems*, https://doi.org/10.1007/978-981-32-9350-2_2

2.1 Introduction

Despite expectations that new information and communication technologies will result in 'the death of distance' (Cairncross 1997) and a 'flat world' (Friedman 2005), economic activity around the world remains 'spiky' (Florida 2005) and geography continues to play an important role (Hortz-Hart 2000; Scott 2001). Improvements in transport and communication have actually reinforced clustering of economic activity because they widen the range of accessible markets for any given region (Scott and Storper 2003).

Innovation, the driver of economic growth is even more concentrated than economic activities (Florida 2005). Large innovative city regions or 'super agglomerations' are coming into being and they play a foundational role in the current world system (Scott 1988, 2001; Veltz 1996). For most of the twentieth century, the United States housed most of the world's innovative regions with a few scattered in Europe and Japan (Florida 2005). But now, major new urban areas which resemble each other are starting to emerge. Hong Kong, Tokyo, Sao Paulo, Singapore, Toronto, Sydney and many other cities are growing bigger and richer while former rich, middle sized cities such as Pittsburgh, St. Louis and Cleveland, built around manufacturing are starting to decline (Storper and Manville 2006; Storper 2013).

A reason for this is, innovation potential of regions depends on internationally mobile factors including research institutions, highly skilled labor and niche markets (Hotz-Hart 2000). Locations, thus, specialize around their national knowledge base and comparative advantages. With increasing globalization, factors of production move towards regions which are the most efficient for certain activities thereby reinforcing their specialization (Hoz-Hart 2000). This increases heightened geographic differentiation and local specialization (Scott 2001).

Additionally, despite the lowered cost of communication, importance for face to face contact for the transmission of complex and ambiguous messages still exists (Leamer and Storper 2001). This idea has been demonstrated by several scholars. Jaffe (1989) and Jaffe et al. (1993) used patent citations to show how distance limits the flow of ideas. Audretsch and Feldman (1996) showed that intellectual innovations are strongly concentrated in urban areas. The recent growth of Silicon Valley further shows that spatial agglomeration is conducive to creating cutting edge technology (Saxenian 1994). Often these face to face meetings are necessary to establish mutual confidence and trust and to evaluate potential partners in a constantly changing business relationship (Storper and Venables 2004).

In fact, rather than seeing a flat world, large city regions or 'super agglomerations' are coming into being which play a foundational role in the current world system (Veltz 1996; Scott 1988, 2001). These large cities possess some similar characteristics. They typically consist of several urban areas and extended suburban surroundings (Hall 2001; Scott and Storper 2003). They are also characterized by high degrees of centrality and interconnectedness in the global networks that provide an infrastructure for the global economy (Sassen 1991, 2012; Wall and van der Knaap 2011; Goerzen et al. 2013).

In many developed countries, evidence shows that major metropolitan areas are growing faster than other areas within the same country (Frey and Alden 1988; Forstall and Fitzsimmons 1993; Summers et al. 1993, Scott and Storper 2003). In many developing countries, economic growth is also seen to be fastest in the large metropolitan areas (Scott and Storper 2003).

Our aim in this chapter is to study these cities of today and their changing technological knowledge connections. Using patent data of 62 innovative cites around the world, we observe how patterns of technological knowledge connectivity between them are changing. We can expect that these new innovative cities are more connected with each other in this current information age (Florida 2005). This is partly because labor is able to move freely between the innovative centers (Florida 2005). It is also because, it is important for the city regions to form useful technological knowledge connections with other regions in order to remain innovative (Uzzi 1997; Bramanti and Ratti 1997; Maillat 1998; Scott 1998; Bresnahan et al. 2001; Bathelt 2007). Advancements in information and communication technologies (ICT) have accelerated the process of knowledge creation and diffusion (Foss and Pederson 2004). In addition, ICT has made previously distant technological combinations possible (Cantwell and Santangelo 2002). Therefore, this age is characterized by an increase in inter-organizational collaboration and openness (Chesbrough 2003).

We also expect that new cities will emerge as important technological knowledge sources and will be more central to the overall technological knowledge network around the world. We also expect some previously rich cities to decline in importance and centrality in the current world system.

2.2 Local-Global Linkages

The success and importance of cities is often attributed to agglomeration economies. Marshall (1890) argued that the reason innovative activity is concentrated is because individuals in close proximity learn from each other when they interact frequently. Therefore, dense urban interactions increase the flow of ideas. According to Marshall, in these dense areas, there exists 'something in the air' since the co-location of firms, employees, individuals and researchers facilitate the generation and the transfer of knowledge, especially tacit knowledge. According to Lucas (1988), ideas flow more freely and can be put into practice quicker if large number of diverse actors are in constant contact with one another. Recent studies have confirmed that there is some truth to these claims.

Jacobs (1969) advanced the idea that cities enjoy an advantage because of their economic and social diversity. This diversity, packed into limited space, facilitates haphazard serendipitous contact amongst people. According to her, it is the exchange of complementary knowledge sources across diverse firms and economic agents which yields a greater return on new economic knowledge. She hypothesizes that the variety of industries within a geographic region promote knowledge spillovers and eventually innovative activity and economic growth. Florida (2002) argued that in cosmopolitan cities the openness of their networks also facilitates innovative activity.

More recently, Jaffe (1989) and Jaffe et al. (1993) use patent citations to show how distance limits the flow of ideas. Audretsch and Feldman (1996) show that innovations are strongly concentrated in urban areas. The recent growth of Silicon Valley also shows that spatial agglomeration is conducive to intellectual innovation and to creating cutting edge technology (Saxenian 1994).

Economic geographers including Storper and Venables (2004), Owen-Smith and Powell (2004) and Grabher (2002) have also argued for the importance of local interactions to transmit knowledge. Storper and Venables (2004) use the term 'local buzz' for the technological knowledge communicated by face to face contact and facilitated by co-presence and co-location of people in a region. This buzz consists of intended and accidental technological knowledge transfer through organized or accidental meetings. This sharing of information allows actors within the region to combine and re-combine similar and non-similar resources to produce innovations and new knowledge (Bathelt et al. 2004).

Although local technological knowledge sharing remains important, an increasing number of studies have questioned the seemingly dominant local learning processes (Malecki and Oinas 1999; Bathelt 2001; Gertler 2001; Vatne 2001). Much of this literature points to the importance of combining both close and distant interactions for the creation of new technological knowledge (Uzzi 1997; Bramanti and Ratti 1997; Oinas 1999; Maillat 1998; Bresnahan et al. 2001; Bathelt 2007).

These external linkages, referred to as 'pipelines', a term coined by Owen-Smith and Powell (2004), are considerably different from local buzz which is characterized as largely unstructured, frequent and broad. The knowledge transferred over pipelines is much more planned, with the amount to be disclosed being decided beforehand (Bathelt et al. 2004). Typically, in a pipeline, procedural rules are set out in the beginning and initial small risks are taken which may be followed by larger risks and commitments (Lorenz 1999). The knowledge that is transferred through these pipelines is rather decisive, non-incremental knowledge flows (Owen-Smith and Powell 2004).

The external technological knowledge connections are important because they offer actors within a region knowledge from disperse sources. When knowledge is continuously reused in different contexts, new knowledge generation processes may be triggered. Geographical separation may, thus, be conducive to innovation (Bathelt and Glucker 2011). An increase in external linkages will therefore lead to an increase in innovation and an increase in local knowledge exchange as well (Cantwell and Zaman 2018).

Indeed, local networks that are too closed, exclusive and rigid may affect the competitiveness of the region (Kern 1996; Bathelt et al. 2004). Uzzi (1996, 1997) coined the term 'over-embeddedness' to describe the resulting technological lock in that occurs when groups of suppliers are embedded with the same set of customers for long periods of time. Burt (1992) discusses 'structural holes' that exist within any region that can only be overcome by non-redundant linkages to external sources of technological knowledge. He refers to network relations as 'plumbing' through which information and resources are being transmitted.

Hence, it can be argued that in order for a region to remain competitive, it needs both local buzz and global pipelines. Local buzz is beneficial to the innovation process because it generates opportunities for actors in a region to interact and form interpretative communities (Nanoka et al. 2000). Global pipelines are also necessary because they allow the integration of multiple select environments that feed the local network with knowledge residing elsewhere. Malecki (2000) sums this line of reasoning well when stating that "Some places are able to create, attract and keep economic activity... because people in those places 'make connections' with other places...".

2.2.1   Global Linkages in the Information Age

Although the importance of establishing external technological knowledge connections have been established by previous literature, it still remains a difficult task. This is because establishing a successful pipeline involves the development of a shared institutional context between partners which would enable shared learning and joint problem solving since actors are spread out in different cultural and institutional contexts (Owen-Smith and Powell 2004). In addition, building pipelines is not an automatic process and requires advance planning and investments. It is a complex and costly process especially since information about potential partners and their actual capabilities is usually incomplete (Malmgren 1961). The advancements in information and communication technologies (ICT) have lowered transport and communication costs thereby accelerating the process of knowledge creation and diffusion (Foss and Pederson 2004). In addition, ICT have made previously distant technological knowledge combinations possible (Cantwell and Santangelo 2002). Existing technological competencies have multiple uses both within and outside their primary sector of activity (Robertson and Langlois 1995).

In this current age, technology is becoming increasingly complex in character and firms must now possess a wider range of technological skills (Feldman and Audretsch 1995). As a consequence, technological interrelatedness is also rising. There is also evidence that industrialized countries are becoming more technologically specialized and differentiated from each other over time (Cantwell and Vertova 2004), thereby increasing the importance of international linkages. Because of these changes in the current information age, we expect that despite barriers to knowledge, there should be an increase in trans-local knowledge sourcing. We should therefore expect to see an increase in international connectivity between our cities with time.

We also expect to see new cities emerging as more central to the technological knowledge network with time. According to scholars in the field of urban planning and policy, a new world order exists today with a new geography of city centrality and marginality that cuts across national boundaries and the north south divide (Friedman 1986; Sassen 1991). Friedman (1995) describes the current world system as a dynamic hierarchy in which ranks and entrance criteria of cities are open. Cities that attract investment and possess more of the command control functions of the world economy will be higher up in the urban hierarchy and their ranking may change with time. Sassen (1994) also paints a similar picture of these cities today and claims that areas that were once considered core are now considered peripheral whereas peripheral areas are now joining the core city system. Our analysis will clarify whether this is true and will try and understand which cities are gradually becoming more important with time.

2.3   Data

2.3.1   Patent Data

Our main source of data is the United States Patent and Trademark Office (USPTO) database. We prefer to use the USPTO database over the patent databases such as European Patent Office (EPO) and Japanese Patent Office (JPO) because the USPTO database is user friendly because of its superior organization. From all the databases, it is known that the USPTO provides the richest information (Kim and Lee 2015). The USPTO data is rich because it provides disaggregation by cross-country, cross-firm, structural and historical dimensions (Cantwell 2006). In addition, the US patent office imposes common screening and legal procedures which provides a benchmark for comparison (Pavitt 1988). Also, since the US is the largest single market in the world, it is more likely that even international players will register for a patent there after their home country even if they are currently not selling in the market (Archibugi 1992; Cantwell 2006).

The patents in the USPTO database provide comprehensive information of the patents. This includes the patent grant date, the technological classes, information about the inventor, information about the assignees and patent citation information. For our purpose, we extracted the patent grant date, the first technological class, first inventor location and patent citation information.

We extracted patent information for the years 1976–2016 from the electronic files made available by the USPTO office. The total number of patents in our database is 5.5 million. We then cleaned and sorted the data according to metropolitan areas.

The first inventor location was used to determine the location of the patent. The reason for not using assignee location is that the assignee corresponds to the location of the headquarters of the organization rather than where the patented invention was actually developed. Hence, since we are interested in the geography of the innovations itself, it is necessary to look at the inventor locations.

We then used backward citations to identify the location of the knowledge source and the recipient. The cited patent is regarded as the knowledge source while the citing patent is the recipient. The location of the patents are determined by first inventor locations. This method of using patent citation data to identify knowledge flow has been commonly used in previous literature (Jaffe et al. 1993; Jaffe and Trajtenberg 1999; Singh 2004).

2.4   City Definitions

We conducted our study on 62 cities. We used metropolitan areas and not just the central city to determine the number of patents for every city. The reason behind this is, since we are using first inventor's address to determine the location of the patent, we have to cater for the fact that the inventor can live anywhere this is a drivable distance to the central city. We use government defined metropolitan areas to mark the boundaries of our cities.

For the European patents, we additionally used a database developed by Dr. John Cantwell while he was at the University of Reading (and used in Cantwell and Iammarino 2005) to determine the city boundaries. This database was developed by a team of expert geographers who went through the entire patent database and determined which locations should be included in the city region. In addition, we rechecked the data in the database and matched it with city boundaries defined by the European Union.

For selecting our cities, we first looked at the comprehensive list provided by GaWC (Globalization and World Cities Research Network). The GaWC chooses city on the basis of their connectivity and concentration of producer services. In these cities, the trends we want to observe will be heightened because of their characteristics. However, since we are using patent citations for our data source, we wanted to include those cities that had enough patents for us to conduct a meaningful analysis. Therefore, we set a threshold for number of patents and selected cities that were above that threshold. This threshold was lower for those cities that were from developing countries.

GaWC ranks cities into categories based on their connectivity and concentration of producer services. These categories include: alpha ++, alpha + , alpha, alpha– and beta + cities. We made sure that we included in our sample, cities from each category. We also tried to include cities from all over the world, and not just from particular regions. Therefore our sample includes cities from North America, South America, Europe, South East Asia, Asia and Oceania. In addition, we wanted to

insure that our database contained cities from different stages of development so we
included developed and emerging cities in our sample.

Our complete selection of cities is show in Table 2.1.

Table 2.1 Our selection of cities

City name	Country name	City name	Country name
North America			
Seattle	United States	Boston	United States
Austin	United States	Chicago	United States
San Diego	United States	The Bay Area	United States
Pittsburgh	United States	Miami	United States
New York City	United States	Atlanta	United States
Los Angeles	United States	Toronto	Canada
Dallas	United States	Vancouver	Canada
Houston	United States	Montreal	Canada
Mexico City	Mexico		
South America			
Buenos Aires	Argentina	Sao Paulo	Brazil
Europe			
London	UK	Paris	France
Glasgow	UK	Lyon	France
Manchester	UK	Grenoble	France
Birmingham	UK	Eindhoven	Netherlands
Berlin	Germany	Vienna	Austria
Frankfurt	Germany	Zurich	Switzerland
Munich	Germany	Basel	Switzerland
Hamburg	Germany	Stockholm	Sweden
Stuttgart	Germany	Copenhagen	Denmark
Dusseldorf	Germany	Brussels	Belgium
Madrid	Spain	Milan	Italy
Barcelona	Spain	Rome	Italy
Dublin	Ireland	Oslo	Norway
Helsinki	Finland	Moscow	Russia
Asia			
Mumbai	India	Nagoya	Japan
Delhi	India	Beijing	China
Bangalore	India	Shanghai	China
Tokyo	Japan	Guangzhou	China
Osaka	Japan		
Oceania			
Sydney	Australia	Auckland	New Zealand

2.5   Methodology

In this study, we aim to present an overall picture of the network of technological knowledge between our cities and how it has changed throughout our time period, 1981–2014. To achieve this, we calculated network statistics from the year 1981 to 2014 for the 62 cities in our study. Though we have patent data from the year 1976, the years before 1981 were excluded from the network analysis because we do not have information for a substantial number of cited patents for those years.

We divided our time period (1981–2014) into six time periods of five consecutive years each, except the last time period, 2011–2014, which consists of the four consecutive years. We calculated the average node strength and the eigenvector centrality for each of the time periods. We ranked each city in terms of average outdegree strength, indegree strength and eigenvector centrality for each of the time periods and observed how the cities changed over time relative to each other.

Using cities as nodes and unidirectional arrows representing citations from one city to another, we calculated the following network statistics:

Node Strength: This refers to the strength of ties of each node. In our case, the strength of the tie refers to the number of times a location is cited by another. Because we have a directed network, we distinguish between indegree node strength and outdegree strength. Indegree node strength of a location refers to the number of times it is cited by other locations and the outdegree node strength of a location refers to the number of times it cites other nodes.

Eigenvector Centrality: Eigenvector centrality is the measure of the influence of a node in the network. It assigns relative scores to all nodes in the network based on the concept that connections to high scoring nodes contribute more than equal connections to low scoring nodes. In our case, this implies that location connected to other high influencing locations will have a greater score than locations connected to less influencing location.

Because of the current wave of globalization and because of facilitating ICT technologies, we expect to see developing cities emerging as important contributors and recipients of technological knowledge in our network. We hope that observing the change in network over our time period will help us understand which developing cities are becoming more important to the technological network and the extent of their success. We expect that not all developing cities will succeed in becoming more central to the network and the degree of their success will also vary.

We also expect that some old technology leaders will lose their centrality in the network during our time period. Because innovation is cumulative (Pavitt 1987), it is liable to lock into a particular industrial pattern or configuration in any location and this pattern is only likely to change gradually over time since a shift to a sector in which technological opportunities are rising most rapidly might not be easy to achieve (Cantwell 1991). Therefore, we can anticipate that some cities might be still locked into the old technologies of the science paradigm which preceded this current information age and will gradually lose their importance in the network of technological knowledge.

2.6   Data Analysis

2.6.1   Outdegree Strength

We first observe how the cities change with regards to their outdegree strength throught our time period. We ranked each city according to their average outdegree strength relative to each other in each of the time periods, where rank 1 is the city with the highest outdegree strength and 62 is the one with the lowest. The outdegree strength will tell us which cities have the most trans-local (as defined by Turkina and Van Assche 2018) knowledge sources within our dataset.

We can see that New York City, Los Angeles, the Bay Area, Tokyo and Boston remain amongst the top cities with the most outdegree strength throughout our time period. Even though there is less shifting amongst the top few cities, we see developing cities such as Beijing, Taipei, Seoul and Bangalore gaining considerably in terms of outdegree strength.

We show in Table 2.2 the cities which have shown the greatest increase in rank by the end of our time period.

As expected, we also see that some cities have considerably decreased in terms of their outdegree strength.

Table 2.3 depicts which cities have decreased the most in their ranks in terms of outdegree strength by the end of our time period.

In our case, outdegree strength represents the number of citations made by the city to other cities. Therefore we can conclude from our calculation of outdegree strengths, at the end of our time period a lot of developing cities have become bigger recipients of technological knowledge compared to the beginning of our time period. Previously large recipients of technological knowledge have declined relative to other cities in our network.

Table 2.2 Cities which showed the most improvement in terms of relative outdegree strength

	Period 1	Period 2	Period 3	Period 4	Period 5	Period 6	Period 7
Seoul	56	48	31	19	14	13	13
Taipei	50	40	27	23	18	18	19
Singapore	57	57	53	45	33	31	32
Helsinki	49	43	42	37	32	29	27
Beijing		59	58	57	55	47	38
Guangzhou		62	61	62	60	57	44
Seattle	23	18	17	15	12	10	6
Austin	26	21	20	13	10	12	11
Bangalore	59	61	62	60	56	54	47
Stockholm	36	25	26	26	25	23	24

Table 2.3 Cities which showed the most decline in terms of relative outdegree strength

	Period 1	Period 2	Period 3	Period 4	Period 5	Period 6	Period 7
Frankfurt	46	38	41	46	46	50	56
Milan	25	27	25	28	30	32	35
Munich	19	19	23	25	26	30	30
Dusseldorf	15	14	16	20	23	25	28
Manchester	35	31	36	41	43	43	49
Rome	37	45	46	48	48	49	51
Mexico City	45	50	52	55	58	61	61
Oslo	28	44	45	44	45	45	45
Toronto	5	22	21	22	21	21	22
Birmingham	29	29	33	39	39	42	48
Basel	21	26	28	32	36	40	43

2.6.2   Indegree Strength

We then observe how the cities change with regards to their indegree strength throughout our time period. The indegree strength will tell us which cities are the most important sources of technological knowledge in our network. We ranked each city according to how they rank in terms of their average indegree strength in each of our time periods, where a rank of 1 indicates the highest indegree strength while 62 marks the lowest. We show in Table 2.4 the cities which have shown the greatest increase in rank by the end of our time period.

Table 2.4 Cities which showed the most improvement in terms of relative indegree strength

	Period 1	Period 2	Period 3	Period 4	Period 5	Period 6	Period 7
Seoul	55	38	19	11	13	8	9
Guangzhou		61	62	62	54	33	26
Singapore	58	53	46	33	23	23	27
Bangalore	59	62	60	59	50	37	29
Shanghai		58	57	60	53	40	30
Taipei	45	28	23	19	16	16	18
Beijing		54	56	53	51	30	31
Seattle	22	15	12	13	12	6	5
Copenhagen	47	42	38	37	37	36	33
Auckland	54	52	55	54	52	53	42
Austin	23	21	16	9	9	10	11
San Diego	19	14	11	10	7	9	7
Helsinki	48	40	36	32	31	29	37

Table 2.5 Cities which showed the most decline in terms of relative indegree strength

	Period 1	Period 2	Period 3	Period 4	Period 5	Period 6	Period 7
Basel	24	26	32	41	42	52	52
Rome	32	43	45	50	55	57	57
Dusseldorf	14	13	18	23	24	31	34
Mexico City	41	56	58	57	60	62	61
Mumbai	42	59	59	61	61	61	62
Manchester	40	39	39	45	44	51	58
Milan	27	25	26	29	34	41	45
Oslo	28	44	44	44	43	43	46
Toronto	2	20	22	20	20	19	19
Birmingham	33	33	37	40	39	47	49
Lyon	36	31	33	39	40	48	51
Sao Paulo	46	55	54	56	62	59	60

We see that in the beginning of our time period, New York City, Toronto, Los Angeles, the Bay Area, Chicago and Tokyo have the highest indegree strength in our network. However, in the later periods Osaka, Seattle and Houston also occupy the spots of the five highest indegree strength at one period or another.

Interestingly, we can see that developing cities have improved more in terms of their indegree strength as compared to their outdegree strength. As an example, we can see that Seoul is ranked 13 according to its outdegree strength but 9 according to its indegree strength. This means that these new emerging cities are better sources of knowledge than they are recipients when compared to other cities in our network.

While some cities show improvement in their relative indegree strength, we also see that some cities have considerably decrease in their relative indegree strength. Table 2.5 depicts which cities have decreased the most in their ranks in terms of indegree strength by the end of our time period.

2.6.3   Eigenvector Centrality

Lastly, we also observe how cities change with respect to their eigenvector centrality. This measure helps us determine which cites have the most influence in the network since it assigns higher weights to those cities that have more influence in the network. We ranked each city according to how they rank in terms of their average eigenvector centralities in each of our time periods, where a rank of 1 indicates the highest eigenvector centrality while 62 marks the lowest.

We see that the cities with highest eigenvector centralities change throughout the period. In our first time period, the Bay Area, New York City, Los Angeles,

Table 2.6 Cities which showed the most improvement in terms of relative eigenvector centrality

	Period 1	Period 2	Period 3	Period 4	Period 5	Period 6	Period 7
Seoul	56	46	28	21	8	19	18
Singapore	57	57	50	49	38	36	30
Austin	27	26	23	23	16	11	1
Bangalore	59	62	62	59	57	49	33
Stockholm	48	25	24	28	25	25	26
Guangzhou		61	61	62	61	53	40
Taipei	43	38	26	20	14	21	22
Seattle	23	17	10	15	17	16	4
Shanghai		58	58	58	56	50	44

Chicago and Boston have the highest eigenvector centralities. In the later periods, Tokyo and Osaka exhibit increased eigenvector centralities and rank in the top five cities with the highest eigenvector centralities in some of our periods.

We show in Table 2.6 the cities which have shown the greatest increase in rank by the end of our time period.

While some cities show improvement in their relative eigenvector centrality, we also see that some cities have considerably decrease in their relative eigenvector centrality. Table 2.7 depicts which cities have decreased the most in their ranks in terms of eigenvector centrality by the end of our time period.

As we anticipated our results show that some cities from emerging countries now play a more central role in our network of cities. On the other hand, we see some cities especially those from developed countries show considerable decline throughout our timer period.

The network of our cities in 2014 is displayed in the figure below. The size of the node represents the eigenvector centrality of each city in 2014 (Fig. 2.1).

We explore further by separating patents on the basis of their primary technological field into five broad classification fields: chemical, information and

Table 2.7 Cities which showed the most decline in terms of relative eigenvector centrality

	Period 1	Period 2	Period 3	Period 4	Period 5	Period 6	Period 7
Birmingham	25	30	37	40	39	46	50
Basel	26	28	40	32	43	42	46
Manchester	38	39	41	41	44	47	55
Mexico City	46	55	56	57	59	62	62
Frankfurt	39	41	42	45	45	43	54
Rome	37	45	44	48	49	56	52
Vienna	34	35	35	39	40	45	48

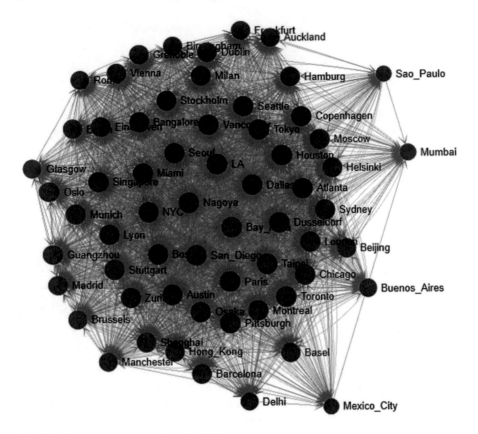

Fig. 2.1 The network of our cities in 2014

communication technologies (ICT), mechanical, other electrical equipment and transport according to their primary technological field (as done in for example Cantwell and Kosmopoulou 2002). We then analyzed the network separately for different classification fields and calculated the same network statistics outdegree strength, indegree strength and eigenvector centrality for each classification field.

2.7   Data Analysis on the Network of Chemical Patents

2.7.1   Outdegree Strength

Just like we did with the overall network of cities, we rank each city according to how they rank in terms of their average outdegree strength in each of our time periods, where rank 1 is the city with the highest outdegree strength and 62 is the one with the lowest.

Table 2.8 Cities which showed the most improvement in terms of relative outdegree strength in the chemical network

	Period 1	Period 2	Period 3	Period 4	Period 5	Period 6	Period 7
Seoul	49	48	45	31	23	21	21
Taipei	55	51	48	45	35	32	35
Seattle	28	24	19	17	15	11	10
Vancouver	42	43	36	28	28	26	24
Atlanta	30	26	21	22	17	18	15
Helsinki	46	38	41	35	27	28	32
San Diego	17	14	15	10	9	7	6

In the Chemical network, we observe that those cities that ranked highest based on outdegree strength continue to rank high throughout our time period. These cities are New York City, Tokyo and the Bay Area.

We show in Table 2.8 the cities which have shown the greatest increase in rank by the end of our time period.

We can see that in the chemical network only Seoul and Taipei have improved considerably in their outdegree strength. The rest of the cities that show improvement are developed cities.

Table 2.9 depicts the cities that have declined the most in terms of their outdegree strength.

We see that mostly developed cities, with the exception of Mexico City, have declined in terms of outdegree strength in the chemical network.

Table 2.9 Cities which showed the most decline in terms of relative outdegree strength in the chemical network

	Period 1	Period 2	Period 3	Period 4	Period 5	Period 6	Period 7
Manchester	20	22	24	27	37	35	40
Frankfurt	21	28	30	34	34	34	39
Mexico City	45	50	51	54	55	60	60
Birmingham	34	37	40	43	39	44	47
Milan	14	17	18	21	22	24	26
Brussels	27	34	34	38	38	36	38
Glasgow	41	44	46	48	49	52	52
Eindhoven	33	31	33	41	45	43	43

2.7.2 Indegree Strength

We then observe how which cities ranked the highest in terms of indegree strength. We see that cities such as New York City and the Bay Area are the only ones who are consistently amongst the top ranked cities for indegree strength. We ranked each city according to how they rank in terms of their average indegree strength in each of our time periods, where a rank of 1 indicates the highest indegree strength while 62 marks the lowest.

Cities such as Chicago which had the third highest indegree strength in the first time period had only the eighth highest at the end of our time period. Similarly Dusseldorf which had the fifth highest indregree strength in the first two time periods had only the fourteenth highest by the end of our time period and Tokyo which had the second highest indegree strength in the first two time periods had only the seventh highest at the end of our time period.

This indicates that different knowledge sources are gaining importance in the chemical network.

We show in Table 2.10 the cities which have shown the greatest increase in rank by the end of our time period.

We can see that cities such as Shanghai, Guangzhou and Beijing which did not show considerable improvement in terms of outdegree strength, still show considerable improvement in indegree strength. This implies that these cities have become important sources of knowledge for those patents that belong to the chemical classification, but have not become significant recipients of technological knowledge.

Table 2.11 depicts the cities that have declined the most in terms of their indegree strength.

We can see that there is a decline in a lot of developed cities in terms of indegree strength. This shows that their relative importance as sources of technological knowledge for chemical patents has decreased throughout our time period.

Table 2.10 Cities which showed the most improvement in terms of relative indegree strength in the chemical network

	Period 1	Period 2	Period 3	Period 4	Period 5	Period 6	Period 7
Seoul	53	46	24	21	17	13	13
Shanghai		56	59	60	59	43	26
Taipei	49	51	36	29	25	26	28
Guangzhou			61	62	62	51	41
Singapore	59	60	51	47	31	35	40
Vancouver	42	41	30	24	22	22	23
Seattle	28	22	18	15	15	12	9
Beijing		53	53	45	41	41	38
Dublin	46	43	48	48	53	55	32

Table 2.11 Cities which showed the most decline in terms of relative indegree strength in the chemical network

	Period 1	Period 2	Period 3	Period 4	Period 5	Period 6	Period 7
Frankfurt	23	28	31	38	42	36	49
Birmingham	36	42	43	46	49	62	61
Munich	21	24	33	30	44	38	44
Rome	31	39	46	42	48	52	52
Stuttgart	30	29	32	36	35	44	50
Manchester	26	26	25	39	43	46	46
Milan	14	18	21	23	24	29	34
Basel	13	10	17	22	27	28	27

2.7.3   Eigenvector Centrality

We ranked each city according to how they rank in terms of their average eigenvector centrality in each of our time periods, where a rank of 1 indicates the highest eigenvector centrality while 62 marks the lowest.

Eigenvector centrality is a measure of influence a node has on the network. We see a slight shift in the cities with the highest eigenvector centralities in the network of chemical patents. In the beginning of our time period, New York City, Tokyo and Chicago had the three highest eigenvector centralities. By the end of our time period, Boston, Los Angeles and Chicago have the three highest eigenvector centralities while Tokyo has the fifth highest eigenvector centrality and New York City has the sixth highest.

We show in Table 2.12, the cities which have shown the greatest increase in rank by the end of our time period.

We can see that developing cities such as Seoul, Shanghai, Taipei and Singapore have gained considerably more influence in our network throughout our time period.

Table 2.13 depicts the cities that have declined the most in terms of their eigenvector centrality.

Table 2.12 Cities which showed the most improvement in terms of eigenvector centrality in the chemical network

	Period 1	Period 2	Period 3	Period 4	Period 5	Period 6	Period 7
Seoul	53	47	31	26	20	16	18
Shanghai		58	59	60	56	51	31
Taipei	52	51	44	39	36	29	28
Seattle	28	24	14	18	16	13	8
Singapore	58	60	52	49	44	42	41
Vancouver	43	42	32	27	27	22	27
Toronto	29	19	20	17	19	17	15

Table 2.13 Cities which showed the most decline in terms of relative eigenvector centrality in the chemical network

	Period 1	Period 2	Period 3	Period 4	Period 5	Period 6	Period 7
Birmingham	30	41	38	44	43	48	52
Manchester	23	28	34	38	40	39	43
Rome	36	39	43	41	46	46	51
Frankfurt	25	30	33	36	37	31	39
Stockholm	20	27	30	32	30	36	34
Buenos Aires	46	54	57	54	59	59	59
Mexico City	49	52	50	57	57	61	60

We see that a lot of developed have declined considerably in terms of eigenvector centrality in the chemical network. Some developing cities such as Buenos Aires and Mexico City have also shown considerable decline.

The network of patents with chemical as their primary classification in 2014 is given in Fig. 2.2. The size of the node represents the eigenvector centrality of each city.

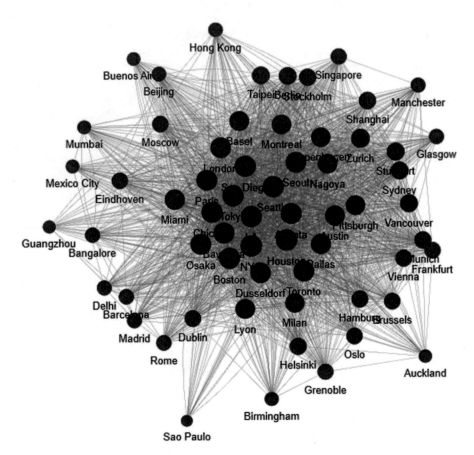

Fig. 2.2 The network of patents classified as chemical in the year 2014

2.8   Data Analysis on the Network of ICT Patents

2.8.1   Outdegree Strength

We ranked each city according to how they rank in terms of their average outdegree strength in each of our time periods, where rank 1 is the city with the highest outdegree strength and 62 is the one with the lowest.

In the ICT network, we see cities that had the highest outdegree strength, New York City, the Bay Area and Tokyo, continued to have the highest outdegree strength throughout our time period.

We show in Table 2.14 below the cities which have shown the greatest increase in rank by the end of our time period.

We see that a lot of developing cities have shown considerable improvement in terms of outdegree strength in the ICT network. This time the developing cities that show considerable improvement in outdegree strength also include cities such as Bangalore, Beijing and Guangzhou which did not show such improvement in the network of chemical patents. We also see that cities such as Guangzhou which did not have any ICT patent in the first three time period was able to catch up and move up the rankings very quickly.

Table 2.15 shows the cities which have declined the most in terms of outdegree strength in the network of ICT patents.

Table 2.14 Cities which showed the most improvement in terms of relative outdegree strength in the ICT network

	Period 1	Period 2	Period 3	Period 4	Period 5	Period 6	Period 7
Seoul	52	42	22	14	11	12	12
Singapore		56	46	34	29	27	29
Beijing		57	55	53	50	35	32
Bangalore		58	60	50	48	41	34
Sydney	42	37	39	32	33	30	19
Helsinki	43	45	41	31	27	24	22
Taipei	41	35	28	25	22	20	21
Guangzhou				61	61	55	43

Table 2.15 Cities which showed the most decline in terms of relative outdegree strength in the ICT network

	Period 1	Period 2	Period 3	Period 4	Period 5	Period 6	Period 7
Birmingham	29	32	37	41	42	46	50
Vienna	28	33	35	42	45	49	47
Brussels	34	34	36	40	44	45	52
Frankfurt	39	44	47	52	54	54	56
Hamburg	27	29	33	36	41	42	44
Manchester	31	31	30	38	37	44	48
Milan	23	24	25	28	32	33	39
Rome	35	38	40	43	47	50	51

We see that a lot of developed cities have shown considerable decline in terms of relative outdegree strength in the ICT network. We see cities such as Milan and Rome also show considerable decline in their relative outdegree strength in the ICT network although they didn't show a decline in the overall network of patents or in the network of chemical patents.

2.8.2 Indegree Strength

We ranked each city according to how they rank in terms of their average indegree strength in each of our time periods for the ICT patent network, where rank 1 is the city with the highest indegree strength and 62 is the one with the lowest.

In the ICT network, we see that cities that had the highest indegree strength, New York City, the Bay Area and Tokyo, in our first time period continue to have the highest indegree strength throughout our time periods.

We show in Table 2.16 the cities which have shown the greatest increase in rank by the end of our time period.

We see that cities such as Bangalore and Guangzhou have increased even more in terms of indegree strength than they did in terms of outdegree strength. This means that these cities have become even more important sources of technological knowledge than recipients.

Table 2.17 shows which cities have declined the most in terms of relative indegree strength in the ICT network.

A lot of developed cities have declined in terms of relative indegree strength throughout our time period. Hence we can infer that now different cities are becoming more vital sources of technological knowledge in the ICT network while older cities are declining in importance.

Table 2.16 Cities which showed the most improvement in terms of relative indegree strength in the ICT network

	Period 1	Period 2	Period 3	Period 4	Period 5	Period 6	Period 7
Seoul	46	26	11	9	10	8	8
Bangalore		57	51	51	33	29	21
Guangzhou				62	57	36	29
Sydney	41	34	31	33	26	19	13
Beijing		51	55	50	41	25	27
Taipei	40	27	24	20	17	15	16
Shanghai		54	60	61	54	40	31
Singapore		47	35	31	21	23	28
Barcelona	54	56	56	53	45	35	36

Table 2.17 Cities which showed the most decline in terms of relative indegree strength in the ICT network

	Period 1	Period 2	Period 3	Period 4	Period 5	Period 6	Period 7
Milan	25	24	26	30	32	44	50
Vienna	27	32	39	43	44	50	52
Brussels	29	43	42	48	43	55	53
Manchester	35	37	40	45	48	51	55
Eindhoven	15	15	19	21	27	31	34
Glasgow	38	44	45	46	49	54	57
Zurich	22	23	28	36	36	37	41

2.9   Eigenvector Centrality

We ranked each city according to how they rank in terms of their average eigenvector centralities in each of our time periods for the ICT patent network, where rank 1 is the city with the highest eigenvector centrality and 62 is the one with the lowest.

When calculating eigenvector centrality, we observed that the cities with the highest influence in the beginning of our time period, Tokyo, the Bay Area and New York City continued to have the highest influence till the end of our time period.

We show in Table 2.18 below the cities which have shown the greatest increase in rank by the end of our time period.

As expected, we see a lot of developing cities show considerable improvement in terms of their eigenvector centralities throughout our time period. Even Guangzhou which had no patents in the first three time periods managed to considerably increase eigenvector centrality.

Table 2.18 Cities which showed the most improvement in terms of relative eigenvector centrality in the ICT network

	Period 1	Period 2	Period 3	Period 4	Period 5	Period 6	Period 7
Seoul	49	37	17	12	9	11	12
Bangalore		59	58	49	47	34	28
Guangzhou				62	60	48	33
Beijing		55	56	51	48	32	29
Singapore		53	44	33	29	30	30
Shanghai		57	60	58	57	47	36
Taipei	39	30	27	23	20	18	18

Table 2.19 Cities which showed the most decline in terms of relative eigenvector centrality in the ICT network

	Period 1	Period 2	Period 3	Period 4	Period 5	Period 6	Period 7
Birmingham	27	33	34	37	40	51	50
Manchester	32	31	37	41	45	46	53
Hamburg	29	27	35	38	43	44	47
Vienna	30	35	36	39	38	40	48
Brussels	35	41	40	44	44	50	52
Milan	24	25	26	30	35	39	40
Glasgow	41	45	47	47	46	54	56

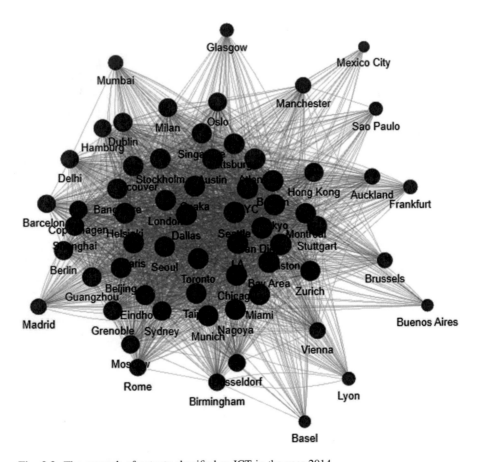

Fig. 2.3 The network of patents classified as ICT in the year 2014

Table 2.19 shows the cities which have declined considerably in terms of their eigenvector centrality in the ICT network throughout our time period.

The network of patents with ICT as their primary classification in 2014 is given in Fig. 2.3. The size of the node represents the eigenvector centrality of each city.

2.10   Data Analysis on the Network of Mechanical Patents

2.10.1   Outdegree Strength

We ranked each city according to how they rank in terms of their average outdegree strength in each of our time periods for the mechanical patent network, where rank 1 is the city with the highest outdegree strength and 62 is the one with the lowest.

We see that in the beginning of our time period, New York City, Tokyo and Los Angeles have the highest outdegree strength. However, by the end of our time period we see that the Bay Area has the highest outdegree strength while Los Angeles and New York City have second and third highest outdegree strengths respectively. Tokyo is still ranked high and has the fourth highest outdegree strength.

We show in Table 2.20 the cities which have shown the greatest increase in rank by the end of our time period.

We see that some developing cities have gradually increased in terms of relative outdegree strength in the network of mechanical patents. However, the increase is less than what we saw in the network of ICT patents. Some developed cities such as Dublin, Austin, Copenhagen and Sydney have also increased considerably in terms of outdegree strength.

Table 2.21 shows the cities which have declined considerably in terms of their outdegree strength in the mechanical network throughout our time period.

We see a lot of developed cities have decreased in terms of relative outdegree strength in the mechanical network.

Table 2.20 Cities which showed the most improvement in terms of relative outdegree strength in the mechanical network

	Period 1	Period 2	Period 3	Period 4	Period 5	Period 6	Period 7
Seoul	53	50	51	26	20	17	18
Taipei	46	37	10	21	19	19	19
Dublin	55	55	40	52	53	43	35
Austin	31	28	30	22	21	18	17
Copenhagen	38	41	44	37	33	31	24
Singapore	56	56	50	53	47	42	43
Sydney	39	32	35	30	29	23	26
Guangzhou		61	56	60	60	57	51

Table 2.21 Cities which showed the most decline in terms of relative outdegree strength in the mechanical network

	Period 1	Period 2	Period 3	Period 4	Period 5	Period 6	Period 7
Birmingham	25	25	33	33	34	40	41
Manchester	30	33	47	42	43	44	45S
Frankfurt	41	38	49	44	45	51	54
Mexico City	47	49	32	55	58	60	60
Milan	26	27	27	29	31	35	39
Basel	37	36	46	41	44	47	49
Vienna	33	40	34	39	42	41	44
Zurich	19	22	22	25	27	30	30

2.10.2   Indegree Strength

We ranked each city according to how they rank in terms of their average indegree strength in each of our time periods for the mechanical patent network, where rank 1 is the city with the highest indegree strength and 62 is the one with the lowest.

In the beginning of our time period, we see that Tokyo, New York City and Chicago have the highest indegree strength. The top three cities keep on changing throughout our time period. At the end of our time period, the Bay Area, Boston and Los Angeles are the top three cities with the highest indegree strength. This shuffling indicates that the most central sources to the network of mechanical networks changed by the end of our time period.

We show in Table 2.22 the cities which have shown the greatest increase in rank by the end of our time period.

We see that a mix of developing and developed cities have increase in terms of their indegree strength. Developing cities are ranked higher when it comes to indegree strength compared to outdegree strength. This means that these cities are more vital sources of technological knowledge for mechanical patents but are still behind when it comes to the extent of trans-local connections their mechanical patents have.

Table 2.22 Cities which showed the most improvement in terms of relative indegree strength in the mechanical network

	Period 1	Period 2	Period 3	Period 4	Period 5	Period 6	Period 7
Seoul	50	42	28	16	14	10	15
Guangzhou		59	55	59	55	35	28
Dublin	54	53	37	45	47	42	26
Shanghai		58	60	58	53	38	32
Singapore	56	52	50	42	26	31	31
Taipei	41	26	8	20	16	18	19
Copenhagen	42	41	43	32	34	30	25
Austin	28	28	22	18	19	19	12
Sydney	30	33	32	24	22	13	14

Table 2.23 Cities which showed the most decline in terms of relative indegree strength in the mechanical network

	Period 1	Period 2	Period 3	Period 4	Period 5	Period 6	Period 7
Dusseldorf	13	15	20	22	24	29	33
Manchester	38	40	48	47	44	50	56
Vienna	32	35	40	43	46	51	50
Birmingham	27	29	39	34	38	40	44
Hamburg	29	32	41	38	39	41	46
Rome	44	46	27	53	57	59	60
Stockholm	23	25	23	25	28	33	39

Table 2.23 shows the cities which have declined considerably in terms of their indegree strength in the mechanical network throughout our time period.

As expected, we see a lot of developed cities show a decline in their indegree strength throughout our time period.

2.10.3   Eigenvector Centrality

We ranked each city according to how they rank in terms of their average eigenvector centrality in each of our time periods for the mechanical patent network, where rank 1 is the city with the highest eigenvector centrality and 62 is the one with the lowest.

The cities with the highest eigenvector centrality in the beginning of our time period were New York City, Chicago and Tokyo. We see that throughout our time period the cities with three highest eigenvector centrality keeps changing. At the end of our time period, the cities with the highest eigenvector centrality are Seattle, Houston and Boston. Even though Chicago had the second highest eigenvector centrality in the beginning of our time period, it only had the sixth highest by the end.

We show in Table 2.24 the cities which have shown the greatest increase in rank by the end of our time period.

Table 2.24 Cities which showed the most improvement in terms of relative eigenvector centrality in the mechanical network

	Period 1	Period 2	Period 3	Period 4	Period 5	Period 6	Period 7
Seoul	51	49	39	22	20	19	9
Singapore	56	56	54	52	43	37	32
Austin	37	34	24	24	23	23	14
Taipei	42	36	17	21	19	17	21
Guangzhou		60	59	59	59	48	42
Seattle	16	17	15	13	12	9	1
Shanghai		58	57	58	57	49	43

Table 2.25 Cities which showed the most decline in terms of relative eigenvector centrality in the mechanical network

	Period 1	Period 2	Period 3	Period 4	Period 5	Period 6	Period 7
Basel	36	40	40	43	45	47	51
Manchester	34	37	41	44	41	43	49
Brussels	43	45	48	47	52	53	57
Moscow	30	41	51	40	42	41	44
Birmingham	26	25	31	38	38	40	39
Frankfurt	40	42	42	46	46	54	53
Mexico City	48	52	53	56	58	60	60
Rome	45	44	37	49	50	57	56

We see a mix of developed and developing cities improving in terms of their relative eigenvector centrality.

Table 2.25 shows the cities which showed the most decline in terms of their relative eigenvector centrality throughout our time period.

The network of patents with Mechanical as their primary classification is shown in Fig. 2.4. The size of the node refers to the city's eigenvector centrality.

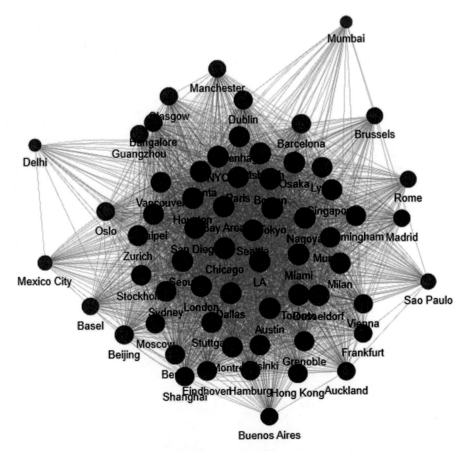

Fig. 2.4 The network of patents classified as mechanical in the year 2014

2.11   Data Analysis on the Network of Other Electrical Equipment Patents

2.11.1   Outdegree Strength

We ranked each city according to how they rank in terms of their average outdegree strength in each of our time periods for the other electrical equipment patent network, where rank 1 is the city with the highest outdegree strength and 62 is the one with the lowest.

We see that Tokyo has the highest outdegree strength throughout our time period. However, the cities with the second and third highest outdegree strength keep changing throughout our time period. At period two, even an emerging city like Taipei had the third highest outdegree strength. Other cities that in one period or another have been amongst the highest three include the Bay Area, New York City, Osaka and Boston.

We show in Table 2.26, the cities which have shown the greatest increase in rank by the end of our time period.

In terms of outdegree strength, we see that developing cities have increased considerably in their rank throughout our time period.

Table 2.27 contains cities which have shown the most decline in terms of their outdegree strength though out our time period.

Table 2.26 Cities which showed the most improvement in terms of relative outdegree strength in the other electrical equipment network

	Period 1	Period 2	Period 3	Period 4	Period 5	Period 6	Period 7
Seoul	54	47	25	11	11	7	8
Taipei	46	3	22	12	12	9	7
Guangzhou				62	51	32	26
Singapore	57	34	43	29	24	22	21
Shanghai		57	57	57	56	43	36
Beijing		58	58	51	53	48	43
Hong Kong	42	33	32	34	29	30	29
Bangalore			60	58	57	55	48

Table 2.27 Cities which showed the most decline in terms of relative outdegree strength in the other electrical equipment network

	Period 1	Period 2	Period 3	Period 4	Period 5	Period 6	Period 7
Brussels	32	41	47	47	48	47	53
Zurich	19	23	27	27	34	33	38
Frankfurt	37	43	44	45	47	56	55
Basel	39	42	46	44	45	54	56
Birmingham	26	32	35	35	38	40	41
Paris	10	12	10	15	15	18	24
Vienna	36	31	36	40	44	46	50
Manchester	34	38	39	38	40	42	46

As expected, we see that developed cities have declined considerably in terms of their outdegree strength.

2.11.2 Indegree Strength

We ranked each city according to how they rank in terms of their average indegree strength in each of our time periods for the other electrical equipment patent network, where rank 1 is the city with the highest indegree strength and 62 is the one with the lowest.

During the beginning of our time period, Tokyo had the highest indegree strength. However, at the end of our time period, the Bay Area had the highest indegree strength and Tokyo moved to the second place. Other cities that were amongst the three highest cities with the most indegree strength at some period or another include Boston, Taipei, Osaka and New York City.

We show in Table 2.28, the cities which have shown the greatest increase in rank by the end of our time period.

As we can see, developing cities have shown considerable improvement in terms of their indegree strength. Guangzhou has no patents classified as other electrical equipment but ends our time period with the 15th highest indegree strength. Other developing cities have also shown considerable improvement in indegree strength.

Table 2.29 shows the cities that have declined considerably in terms of indegree strength throughout our time period.

We see that quite a few developed cities have shown considerable decrease in their relative indegree strength during our time period. Since indegree strength refers to the number of citations to a city, this implies that these cities are no longer as vital a source of technological knowledge for other electrical equipment patents as they were in the beginning of our time period.

2.11.3 Eigenvector Centrality

We ranked each city according to how they rank in terms of their average eigenvector centrality in each of our time periods for the other electrical equipment patent

Table 2.28 Cities which showed the most improvement in terms of relative indegree strength in the other electrical equipment network

	Period 1	Period 2	Period 3	Period 4	Period 5	Period 6	Period 7
Guangzhou				62	37	17	15
Seoul	48	28	11	6	7	6	6
Taipei	43	2	13	12	8	7	7
Shanghai		52	53	50	39	28	20
Singapore	49	34	33	21	14	14	17
Bangalore			58	59	40	40	30
Beijing		54	43	55	56	39	31
Vancouver	41	39	28	31	27	24	26

Table 2.29 Cities whish showed the most decline in terms of relative indegree strength in the other electrical equipment network

	Period 1	Period 2	Period 3	Period 4	Period 5	Period 6	Period 7
London	13	16	23	24	22	30	35
Rome	40	41	44	53	57	58	61
Manchester	34	45	45	37	43	54	53
Moscow	32	48	36	36	38	43	51
Paris	10	12	15	19	24	32	29
Lyon	31	38	35	38	50	53	49
Pittsburgh	9	11	17	17	21	29	27
Basel	42	49	54	56	51	61	57
Birmingham	30	37	40	40	47	49	45
Frankfurt	39	42	42	51	46	56	54
Stockholm	27	22	31	30	32	36	42

Table 2.30 Cities which showed the most improvement in terms of relative eigenvector centrality in the other electrical equipment network

	Period 1	Period 2	Period 3	Period 4	Period 5	Period 6	Period 7
Seoul	49	38	23	12	9	6	8
Guangzhou				62	45	34	25
Taipei	46	10	19	17	12	13	9
Singapore	50	32	36	34	33	24	22
Shanghai		56	55	55	49	38	30
Bangalore			59	59	51	43	37
Beijing		55	51	52	56	47	40
Hong Kong	42	35	32	33	30	28	27
Austin	27	22	17	15	14	11	13

network, where rank 1 is the city with the highest eigenvector centrality and 62 is the one with the lowest.

Throughout our time period, either New York City or Tokyo had the highest eigenvector centrality. Other cities which are amongst the top 3 cities at some period or the other include Boston, the Bay Area, Los Angeles and Osaka.

We show in Table 2.30, the cities which have shown the greatest increase in rank by the end of our time period.

We see that developing cities have increased considerably in terms of their eigenvector centrality. This means that developing cities are gradually becoming more central to the network than older developed cities.

Cities that showed the greatest decline in terms of eigenvector centrality are given in Table 2.31.

We can see that developed cities have decreased in terms of eigenvector centrality in the network of other electrical equipment patents.

Table 2.31 Cities which showed the most decline in terms of relative eigenvector centrality in the other electrical equipment network

	Period 1	Period 2	Period 3	Period 4	Period 5	Period 6	Period 7
Brussels	34	46	49	50	48	55	54
Manchester	33	41	38	37	41	50	52
Basel	39	47	47	53	52	59	57
Zurich	17	24	27	30	34	35	35
Birmingham	25	34	35	36	40	40	41
London	12	13	15	22	19	22	28
Stockholm	22	19	26	26	27	32	38
Moscow	31	39	42	41	39	46	46
Rome	43	45	45	49	50	57	58

The network of patents with other electrical equipment as their primary classification is shown in Fig. 2.5. The size of the node refers to the city's eigenvector centrality.

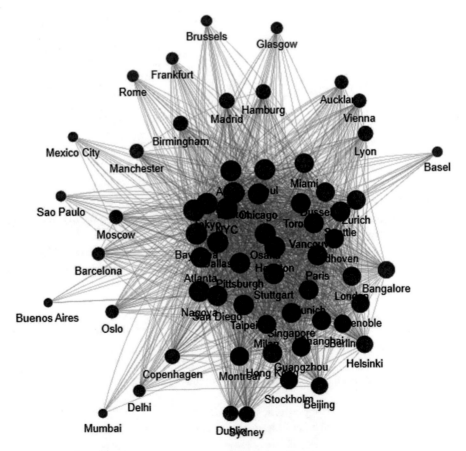

Fig. 2.5 The network of patents classified as other electrical equipment in the year 2014

2.12   Data Analysis on the Network of Transport Patents

2.12.1   Outdegree Strength

We ranked each city according to how they rank in terms of their average outdegree strength in each of our time periods for the transport patent network, where rank 1 is the city with the highest outdegree strength and 62 is the one with the lowest.

We observe that in the beginning of our time period, Nagoya, Tokyo and Stuttgart have the three highest outdegree strength. However, even though Nagoya and Tokyo maintain high rankings throughout our time period, Stuttgart decreases in rank substantially. Other cities that were ranked in the top three at one period or another include Los Angeles and New York City.

We show in the Table 2.32, the cities which have shown the greatest increase in rank by the end of our time period.

We see that with that with the exception of Seoul, Taipei and Singapore the rest of the cities that have risen in rank throughout our time period include developed cities. In the transport network, we see that cities from emerging countries have not risen in rank as much as in the other networks.

In Table 2.33, we show the cities that have shown the most decline in terms of relative outdegree strength throughout our time period.

2.12.2   Indegree Strength

We ranked each city according to how they rank in terms of their average indegree strength in each of our time periods for the transport patent network, where rank 1 is the city with the highest indegree strength and 62 is the one with the lowest.

Table 2.32 Cities which showed the most improvement in terms of relative outdegree strength in the transport network

	Period 1	Period 2	Period 3	Period 4	Period 5	Period 6	Period 7
Seoul	56	49	42	27	24	20	20
Taipei	40	35	25	22	17	17	15
Grenoble	48	39	34	28	29	30	35
Singapore	57	54	45	56	56	46	45
Frankfurt	53	40	30	42	37	44	42
Atlanta	26	24	24	16	18	16	17
Austin	30	38	33	32	28	28	21
Miami	22	16	17	15	15	13	13
Sydney	45	32	39	35	34	32	36

Table 2.33 Cities which showed the most decline in terms of relative outdegree strength in the transport network

	Period 1	Period 2	Period 3	Period 4	Period 5	Period 6	Period 7
Mexico City	39	52	54	55	57	53	60
Manchester	34	37	44	38	46	48	52
Vienna	17	17	18	31	32	35	32
Birmingham	14	15	22	23	23	27	28
Brussels	36	45	47	52	43	51	49
Barcelona	31	31	41	37	36	40	43
Lyon	25	25	26	34	35	37	37
Paris	7	9	11	12	14	15	19
London	11	12	14	19	21	22	22

Table 2.34 Cities which showed the most improvement in terms of relative indegree strength in the transport network

	Period 1	Period 2	Period 3	Period 4	Period 5	Period 6	Period 7
Seoul	48	36	27	18	18	16	9
Madrid	53	44	50	34	49	35	31
Guangzhou			58	60	40	36	38
Hong Kong	46	55	43	40	36	32	29
Bangalore						47	32
Berlin	49	47	48	36	32	29	34
Taipei	38	29	22	22	17	19	23
Sydney	42	34	36	32	28	24	28

Just as the with the indegree strength, Nagoya, Tokyo and Stuttgart have the three highest indegree strength. However, even though Nagoya and Tokyo maintain high rankings throughout our time period, Stuttgart decreases in rank substantially. Other cities that were ranked in the top three at one period or another include Osaka, Los Angeles and Boston.

We show in Table 2.34, the cities which have shown the greatest increase in rank by the end of our time period.

Even though we did not see a lot of developing cities improving in terms of outdegree strength, we see that is not the case with indegree strength. Cities like Guangzhou and Bangalore also show considerable improvement in terms of indegree strength. This means that even though patents classified as transport within these cities may not use trans-local links as much as other cities, they still are becoming important sources of technological knowledge in this network.

Table 2.35 shows the cities which have declined the most in terms of relative indegree strength in the transport network.

As expected, we see a lot of previously developed cities decline in terms of relative indegree strength.

Table 2.35 Cities which showed the most decline in terms of relative indegree strength in the transport network

	Period 1	Period 2	Period 3	Period 4	Period 5	Period 6	Period 7
Helsinki	32	41	41	48	39	44	61
Vienna	15	16	21	35	35	34	42
Birmingham	20	21	23	27	27	30	43
Rome	37	39	35	54	50	56	58
Barcelona	34	32	45	46	58	45	54
Lyon	25	30	31	39	43	41	44
Manchester	35	45	54	49	41	48	53
Moscow	41	56	46	33	33	55	56
London	12	14	20	21	20	23	26

2.12.3   Eigenvector Centrality

We ranked each city according to how they rank in terms of their average eigenvector centrality in each of our time periods for the transport patent network, where rank 1 is the city with the highest eigenvector centrality and 62 is the one with the lowest.

We see that almost throughout our time period, Tokyo has the highest eigenvector centrality. However, at the very end of our time period, Los Angles has the highest eigenvector centrality. Other cities that were amongst the top three at one period or the other include New York City and Boston.

We show in Table 2.36, the cities which have shown the greatest increase in rank by the end of our time period.

Surprisingly, we see that developing cities have risen substantially in terms of eigenvector centrality in the transport network. This is despite the fact that Beijing, Guangzhou and Bangalore did not rise considerably in terms of their outdegree strength.

In Table 2.37, we display cities that show considerable decline in terms of eigenvector centrality throughout our time period.

Table 2.36 Cities which showed the most improvement in terms of relative eigenvector centrality in the transport network

	Period 1	Period 2	Period 3	Period 4	Period 5	Period 6	Period 7
Seoul	51	40	33	23	23	16	13
Bangalore						62	37
Taipei	32	30	21	21	19	19	15
Guangzhou			57	61	54	42	41
Beijing		57		56	55	53	44
Hong Kong	44	51	46	37	39	33	32
Grenoble	46	33	34	30	33	32	36

Table 2.37 Cities which showed the most decline in terms of relative eigenvector centrality in the transport network

	Period 1	Period 2	Period 3	Period 4	Period 5	Period 6	Period 7
Mexico City	38	48	56	57	57	58	60
Barcelona	36	38	41	38	47	39	52
Birmingham	16	22	22	24	26	27	31
Rome	39	44	38	44	48	54	54
London	11	12	16	20	16	22	20
Lyon	25	29	31	34	35	36	34
Manchester	40	35	49	40	41	49	49
Pittsburgh	12	18	17	19	20	21	21
Vienna	29	26	32	35	36	41	38

As expected, we see a lot of developed cities declining considerably in terms of relative eigenvector centrality throughout our time period.

A snapshot of what the network looked like in 2014 is shown in Fig. 2.6. The size of the node represents the eigenvector centrality of the city.

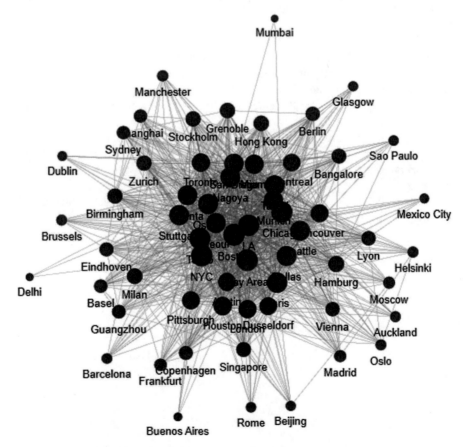

Fig. 2.6 The network of patents classified as mechanical in the year 2014

We can see from the figure above that the network of patents classified as transport is not as dense as the other networks displayed before.

2.13   Some Conclusions

Overall, we see that the technological knowledge network is getting denser and more dispersed. Technological knowledge connections to already connected cities are becoming stronger, i.e. more technological knowledge is being shared between different cities than before. As expected, we see that this new information age is characterized by more technological knowledge sharing than previous time periods.

At the same time, more locations are joining the technological knowledge network. Some of these emerging locations have become strong sources as well as receivers of knowledge. On the other hand, some cities that were once very central to the network are gradually decreasing in centrality over our time period. This falls in line with the observations from scholars who stated that even though for most of the twentieth century, the United States housed most of the world's innovative regions with a few scattered in Europe and Japan, this is no longer true (Florida 2005). New innovative urban areas are starting to emerge, while some former rich, middle sized cities are starting to decline (Storper and Manville 2006).

We see many developing cities showing considerable improvements in their rankings in terms of outdegree strength, indegree strength and eigenvector centrality. Some cities, such as Seoul and Taipei stand out from the rest of the developing cities, because they show a tremendous increase in all three of our measures. Singapore and Hong Kong also show a lot of improvement across the board, but to a lesser extent. Some cities such as Shanghai, Beijing, Guangzhou and Bangalore are also improving in their outdegree strength, indegree strength and eigenvector centrality but not as much as Singapore and Hong Kong.

When we look at networks divided by specialization we see that cities like Guangzhou, Beijing and Bangalore have shown the most improvement in all three measurements in the network of ICT patents. In addition, Bangalore and Guangzhou also shows consider improvement in the network of 'other electrical equipment' patents. It is interesting to see that newer cities are more central to networks of newer technologies rather than the old. This is expected since innovation is cumulative (Pavitt 1987), and a location might get locked in with a particular industrial pattern or configuration. This pattern is only likely to change gradually over time, and a quick shift to a sector in which technological opportunities are rising most rapidly might not be easy to achieve (Cantwell 1991). In contrast, newer emerging locations are able to develop new technologies quicker (Awate and Mudambi 2017). These new technologies are also more central to the network as compared to older technologies, therefore emerging cities are likely to grow more central as well (Awate and Mudambi 2017).

There are cities which continue to remain central to the network throughout the time period. These include The Bay Area, New York City and Tokyo. Therefore, even though there is a lot of movement and shift in the importance of cities, some cities have managed to stay important sources and recipients of technological knowledge. Interestingly, when we separate the networks according to their technological knowledge classification, we see that these cities have relatively high centrality in all networks.

Even though a lot of emerging cities have gained in centrality during our time period, some cities consistently rank low. These include Mexico City, Buenos Aires, Moscow and Sao Paulo.

We also observe that there is more change in the cities in terms of indegree strengths when compared to outdegree strengths. This indicates that emerging cities are gradually becoming more important sources of knowledge, but they have not started receiving technological knowledge at the same rate. This may be because their innovative rates are over all low, and therefore they utilize their external technological knowledge connections less than the developed cities.

References

Archibugi, D. (1992). Patenting as an indicator of technological innovation: A review. *Science and public policy, 19*(6), 357–368.

Audretsch, D. B., & Feldman, M. P. (1996). R&D spillovers and the geography of innovation and production. *The American Economic Review, 86*(3), 630–640.

Awate, S., & Mudambi, R. (2017). On the geography of emerging industry technological networks: The breadth and depth of patented innovations. *Journal of Economic Geography, 18* (2), 391–419.

Bathelt, H. (2007). Buzz-and-Pipeline dynamics: Towards a knowledge-based multiplier model of clusters. *Geography Compass, 1*(6), 1282–1298.

Bathelt, H., & Glückler, J. (2011). *The relational economy: Geographies of knowing and learning.* Oxford University Press.

Bathelt, H., Malmberg, A., & Maskell, P. (2004). Clusters and knowledge: Local buzz, global pipelines and the process of knowledge creation. *Progress in Human Geography, 28*(1), 31–56.

Bathelt, H. (2001). Regional competence and economic recovery: Divergent growth paths in Boston's high technology economy. *Entrepreneurship & Regional Development, 13*(4), 287–314.

Bramanti, A., & Ratti, R. (1997). The multi-faced dimensions of local development. In R. Ratti, A. Bramanti, & R. Gordon (Eds.), *The dynamics of innovative regions: The GREMI approach* (pp. 3–45). Aldershot, UK: Ashgate.

Bresnahan, T., Gambardella, A., & Saxenian, A. (2001). 'Old economy' inputs for 'new economy' outcomes: Cluster formation in the New Silicon Valleys. *Industrial and Corporate Change, 10* (4), 835–860.

Burt, R. S. (1992). *Structural holes: The social structure of competition.* Harvard university press.

Cairncross, F. (1997). *The death of distance: How the communications revolution will change our lives.* Cambridge: Harvard Business School Press.

Cantwell, J. A., & Kosmopoulou, E. (2002). What Determines the Internationalisation of Corporate Technology? In M. Forsgren, H. Hakanson, & V. Havila (Eds.), *Critical perspectives on internationalisation,* (pp. 305–334). Pergamon: Oxford.

Cantwell, J. A., & Iammarino, S. (2005). *Multinational corporations and European regional systems of innovation.* London: Routledge.

Cantwell, J. A., & Santangelo, G. (2002). The new geography of corporate research in information and communications technology (ICT). *Journal of Evolutionary Economics, 12*(1), 163–197.

Cantwell, J. A., & Vertova, G. (2004). Historical Evolution of Technological Diversification. *Research Policy, 33*(3), 511–529.

Cantwell, J. A., & Zaman, S. (2018). Connecting global and technological knowledge sourcing. *Competitiveness Review, 28*(3), 277–294.

Cantwell, J. A. (1991). Historical trends in international patterns of technological innovation. In J. Foreman-Peck (Ed.), *New perspectives on the late victorian economy: Essays in quantitative economic history, 1860–1914* (pp. 37–72). Cambridge: Cambridge University Press.

Cantwell, J. A. (2006). *The economics of patents*. Edward Elgar Publishing.

Chesbrough, H. (2003). The logic of open innovation: Managing intellectual property. *California Management Review, 45*(3), 33–58.

Florida, R. (2005). The world is spiky globalization has changed the economic playing field, but hasn't leveled it. *Atlantic Monthly, 296*(3), 48.

Florida, R. (2002). The Rise of the Creative Class. and how it's Transforming Work, Leisure and Everyday Life.

Forstall, R. L., & Fitzsimmons, J. D. (1993). Metropolitan growth and expansion in the 1980s. US Department of Commerce, Economics and Statistics Administration, Bureau of the Census.

Foss, N., & Pedersen, T. (2004). Organizing knowledge processes in the multinational corporation: An introduction. *Journal of International Business Studies, 35*(5), 340–349.

Frey, W. H., & Alden Jr, S. (1988). *Regional and metropolitan growth and decline in the US*. Russell Sage Foundation.

Friedman, T. L. (2005). *The world is flat: A brief history of the twenty-first century*. Macmillan.

Friedmann, J. (1995). Where we stand: A decade of world city research. *World cities in a world system*: 21–47.

Friedmann, J. (1986). The world city hypothesis. *Development and Change, 17*(1), 69–83.

Gertler, M. S. (2001). Best practice? Geography, learning and the institutional limits to strong convergence. *Journal of Economic Geography, 1*(1), 5–26.

Goerzen, A., Asmussen, C., & Nielson, B. (2013). Global cities and multinational enterprise location strategy. *Journal of International Business Studies, 44*, 427–450.

Grabher, G. (2002). Cool projects, boring institutions: Temporary collaboration in social context. *Regional Studies, 36*(3), 205–214.

Hall, P. (2001). Global city-regions in the twenty-first century. In A. Scott (Ed.), *Global city— regions: Trends. Theory, policy*. Oxford: Oxford University Press.

Hotz-Hart, B. (2000). Innovation networks, regions and globalization. In *The Oxford handbook of economic geography* (PP. 432–450). Oxford: Oxford University Press.

Jacobs, J. (1969). *The economy of cities*. New York: Vintage.

Jaffe, A., Trajtenberg, M., & Henderson, R. (1993). Geographic localization of knowledge spillovers as evidenced by patent citations. *The Quarterly Journal of Economics*, 577–598.

Jaffe, A. B. (1989). Real effects of academic research. *The American Economic Review*, 957–970.

Jaffe, A. B., & Trajtenberg, M. (1999). International knowledge flows: Evidence from patent citations. *Economics of Innovation and New Technology, 8*(1–2), 105–136.

Kern, H. (1996). Vertrauensverlust Und Blindes Vertrauen: Integrationsprobleme Im Ökonomischen Handeln Handeln (Loss of Trust and Blind Confidence in Economic Action). *SOFI Soziologisches Forschungsinstitut Göttingen, 24*, 7–14.

Kim, J., & Lee, S. (2015). Patent databases for innovation studies: A comparative analysis of USPTO, EPO, JPO and KIPO. *Technological Forecasting and Social Change, 92*, 332–345.

Leamer, E. E., & Storper, M. (2001). The economic geography of the internet age. *Journal of International Business Studies, 32*(4), 641–666.

Lorenz, E. (1999). Trust, contract and economic cooperation. *Cambridge Journal of Economics, 23*(3), 301–315.

Lucas, R. (1988). On the mechanics of economic development. *Journal of Monetary Economics, 22*(1), 3–42.

Maillat, D. (1998). Interactions between urban systems and localized productive systems: An approach to endogenous regional development in terms of innovative milieu. *European Planning Studies, 6*(2), 117–129.

Malecki, E. J. (2000). Knowledge and regional competitiveness (Wissen Und Regionale Wettbewerbsfähigkeit). *Erdkunde,* 334–351.

Malecki, E. J., & Oinas, P. (1999). *Making connections: Technological learning and regional economic change.* Ashgate Publishing Company.

Malmgren, H. B. (1961). Information, expectations and the theory of the firm. *The Quarterly Journal of Economics, 75*(3), 399–421.

Marshall, A. (1890). Principles of economics, 8th Edn (1920). London: Mcmillan.

Nonaka, I., Toyama, R., & Nagata, A. (2000). A firm as a knowledge-creating entity: A new perspective on the theory of the firm. *Industrial and Corporate Change, 9*(1), 1–20.

Oinas, P. (1999). Activity-specificity in organizational learning: Implications for analysing the role of proximity. *GeoJournal, 49*(4), 363–372.

Owen-Smith, J., & Powell, W. (2004). Knowledge networks as channels and conduits: The effects of spillovers in the boston biotechnology community. *Organization Science, 15*(1), 5–21.

Pavitt, K. (1987). The objectives of technology policy. *Science and Public Policy.*

Pavitt, K. (1988). Uses and abuses of patent statistics. In *Handbook of quantitative studies of science and technology* (PP. 509–536). Elsevier.

Robertson, P., & Langlois, R. (1995). Innovation, networks, and vertical integration. *Research Policy, 24*(4), 543–562.

Sassen, S. (1991). *The global city: New York, London and Tokyo.* Princeton, NJ: Princeton University Press.

Sassen, S. (1994). The urban complex in a world economy. *International Social Science Journal, 46,* 43–62.

Sassen, S. (2012). Cities: A window into larger and smaller worlds. *European Educational Research Journal, 11*(1), 1–10.

Saxenian, A. (1994). *Regional networks: Industrial adaptation in silicon valley and route 128.* Cambridge, MA: Harvard University Press.

Scott, A. (1998). *Regions and the world economy: The coming shape of global production, competition, and political order.* Oxford: Oxford University Press.

Scott, A. J. (2001). Industrial revitalization in the ABC municipalities, Sao Paulo: Diagnostic analysis and strategic recommendations for a new economy and a new regionalism. *Regional Development Studies, 7*(2001), 1–32.

Scott, A., & Storper, M. (2003). Regions, globalization, development. *Regional Studies, 37*(6–7), 579–593.

Singh, J. (2004). *Multinational firms and knowledge diffusion: Evidence using patent citation data, 2004*(1), 1543–8643.

Storper, M. (2013). *Keys to the city: How economics, institutions, social interaction, and politics shape development.* Princeton University Press.

Storper, M., & Manville, M. (2006). Behaviour, preferences and cities: Urban theory and urban resurgence. *Urban Studies, 43*(8), 1247–1274.

Storper, M., & Venables, A. J. (2004). Buzz: Face-to-face contact and the urban economy. *Journal of Economic Geography, 4*(4), 351–370.

Summers, A. A., Cheshire, P. C., & Senn, L. (1993). *Urban change in the United States and Western Europe: Comparative analysis and policy.* The Urban Insitute.

Turkina, E., & Van Assche, A. (2018). Global connectedness and local innovation in industrial clusters. *Journal of International Business Studies,* 1–23.

Uzzi, B. (1997). Social structure and competition in interfirm networks: The paradox of embeddedness. *Administrative Science Quarterly,* 35–67.

Uzzi, B. (1996). The sources and consequences of embeddedness for the economic performance of organizations: The network effect. *American Sociological Review,* 674–698.

Vatne, E. (2001). Local versus extra-local relations: The importance of ties to information and the institutional and territorial structure of technological systems.

Veltz, P. (1996). Mondialisation, Villes Et Territoires. In *L'économie d'archipel*. Paris: Presses Universitaires de France.
Wall, R. S., & Van der Knaap, G. (2011). Sectoral differentiation and network structure within contemporary worldwide corporate networks. *Economic Geography, 87*(3), 267–308.

Salma Zaman (Ph.D., Rutgers University) is Assistant Professor at the Lahore University of Management Sciences (LUMS) in Lahore, Pakistan. She completed her Ph.D. in International Business from Rutgers Business School in Newark, New Jersey under the supervision of Dr. John Cantwell. Her previous degrees include an MBA and an undergraduate degree in Computer Science. Her most recent work, entitled "Connecting local and global technological knowledge sourcing" was published in *Competitiveness Review* in 2018. Her research interests include exploring the changing patterns of technological knowledge flows across cities and the possible causes and implications of these changing patterns.

Chapter 3
World-Wide Dispersion of Research and Development (R&D) Capabilities

Takabumi Hayashi and Atsuho Nakayama

Abstract This paper examines to what extent research and development (R&D) capabilities have geographically disseminated and dispersed worldwide, analyzing scientific papers and patents as R&D outputs. As a result, global dispersion was revealed using Lorenz curve and Gini Coefficient, in the main field of scientific journals from INSPEC; Physics, Electrical Engineering, Electronics, computer and control engineering including Information Technology, and Mechanical engineering. In the case that we examine foreign invented US patents excluding domestically invented patents in the US, we can see geographical dispersion among nationalities since 1990, among others, much more than all the US patents including domestically invented patents in the US. Herfindahl-Hirschman (HH) indicator also shows the decline in the degree of the concentration ratio. The results obtained in this paper is that the number of nationalities of authors' affiliation and that of nationalities of affiliations of U.S. patent inventors have increased and diversified. Especially, R&D capabilities measured by nationalities of authors' affiliation remarks that point, rather than by those of US patent inventors, which means that more and more countries have improved R&D capabilities, "R" in particular.

Keywords Research and development capability · Global dispersion of R&D capabilities · Scientific paper · US patent · Lorenz curve · Gini coefficient · Herfindahl-Hirschman index

T. Hayashi (✉)
Rikkyo University, Tokyo, Japan
e-mail: takabumi@rikkyo.ac.jp

A. Nakayama
Marketing Sience, Tokyo Metropolitan University, Tokyo, Japan
e-mail: atsuho@tmu.ac.jp

© Springer Nature Singapore Pte Ltd. 2019
J. Cantwell and T. Hayashi (eds.), *Paradigm Shift in Technologies and Innovation Systems*, https://doi.org/10.1007/978-981-32-9350-2_3

3.1 Introduction

3.1.1 Research Objectives

Since the end of 1980s, digital networking of industrial and social activities with the advance in information technology has not only accelerated the speed of knowledge transfer and dispersion with the cross-border human migration, but also helped research and development (= R&D) capabilities disperse globally ever than before. Globalization of multinational enterprises (MNEs)' R&D activities has also given impetus to this phenomenon.

It should be noted that some firms and nationalities which have strategically adapted to global dispersion of R&D capabilities could leverage these dispersed R&D resources.

R&D capability has captured much attention from various viewpoints. The paper refers to "international technology and knowledge transfer" (Hayashi 1995; Komoda 1987; Jeremy 1991; Kotabe et al. 2007; Sorenson and Fleming 2004), "national innovation system including R&D expenses, human resources in R&D, industrial cluster" (Badaracco 1990; Lundvall 1992, 2007; Nelson 1993; Porter 1998; Saxenian 1996, 2006; Akashi and Ueda 1995; Nonaka 1995; Gotoh and Kodama 2006; Hayashi and Komoda 1993), and "product life cycle model" (Vernon 1979; Cantwell 1995). The globalization of MNEs' R&D activities has also been discussed since the 1970s. The question naturally arises, to what extent R&D capabilities are decentralized if the globalization of R&D activities proceeds. Although it has been analyzed by Freeman and Hagedoorn (1995), Patel and Pavitt (1998), Hayashi (2004a, b), a focus on the recent trend since the 2000s has not been verified.

This paper, therefore, examines to what extent R&D capabilities have geographically disseminated and dispersed worldwide by the methodology discussed below.

3.1.2 Methodology

In this paper, we examine global dispersion of R&D capabilities, analyzing scientific papers and patents as R&D outputs, not inputs such as R&D expenses or human resources in R&D. We collect data of papers from INSPEC. INSPEC offers information referring to scientific papers in the fields of computer, control engineering, and mechanical engineering including physics, electrical engineering, and information technology, excluding chemistry, pharmaceuticals, and biotechnology. Thus, this paper analyzes, setting the target at these fields mentioned above. Also, in order to reduce noise, we search scientific papers published in any of the United States, Britain, and Netherlands where major scientific papers are published. In the paper, we do not consider citation index, which assess the degree of importance. We validate the trend of dispersion by looking across 45 year transition, to mitigate such noise.

As to technological capabilities, we examine US patents from open data source of USPTO.

In order to measure how capabilities are decentralized, we scrutinized the number of nationals of author's affiliation of papers and the number of nationals of inventor's affiliation of US patents. Therefore, data we processed in this paper refers to nationalities of the location of affiliation, not individual passport nationality. Lorenz curve, Gini coefficient, and Herfindahl-Hirschman Index are used in this paper as statistical methods.

How are research and development capabilities distinguished and interrelated? In other words, in what ways should we make distinctions and interrelations between each other? This paper elucidates how much scientific papers contribute to the technological invention, considering technological development capabilities of firms or industries from the perspective of patented technologies. It is recognized that it gives an insight into the process in which scientific papers lead to invention by construing cited references in the statement for patent application of newly developed technologies (Jaffe and Trajtenberg 2002; Nagaoka 2011).

In that case, to what degree scientific papers pertain to the invention of patented technology? According to the US and Japanese inventor's survey which contributed to patent applications to the United States, Europe, and Japan as well as domestic application (Nagaoka 2011), the ratio of respondent who answered that scientific references gave important ideas is on average of US and Japan 31% in chemistry, 51% in pharmaceuticals, and 51% in biotechnology respectively. It is 15% each in hardware and software of the computer, 17% in data storage, 19% in communication technology, 22% in semiconductor device. Scientific references affect positively invention processes in these fields. Furthermore, the same methodological survey taken from 2013 through 2014 shows that 57.9 inventions were made, including those not applied, and 12.2 papers were published on academic journals on average by 823 respondents. That survey from 2003 to 2005 demonstrated that 56.2 inventions and 10.9 papers, respectively (Nagaoka and Yamauchi 2014, p. 8). Essentially, which do engineers and researchers prioritize, publications of academic papers or invention of patented technologies? As is obvious, private companies require them to prioritize the invention of patented technologies because their bosses might underscore the rights of exclusive use and commercialization, given ROE or stock price. As above, the number of patents tends to exceed that of papers.

The point that should be paid attention is that inventors do not always accomplish their R&D activities with the help of cited papers. When they present or publish newly developed technologies, they are adjured to ask for patent applications from their affiliations, especially on technologies which would bring strategic benefits to the firm. It means that a certain degree of their papers refers to their patentable technological achievement. That is to say, engineers or researchers open their technological result to the public after taking measures to prevent patent infringement, while gaining an insight from some scientific documents. It is denoted that the number of patents and that of papers will move in an interrelated way each other. In other words, both of them as indicators of R&D capabilities complementally change each other.

3.2 Global Dispersion of Research and Development Capability

In this section, we examine the global trend of scientific and technological capabilities since 1970s by looking across the nationalities of author's affiliation of papers from INSPEC database. INSPEC offers the field of computer, control engineering, and mechanical engineering including physics, electrical engineering, and information technology, excluding chemistry, pharmaceuticals, and biotechnology. Thus, this paper analyzes, targeting at these fields mentioned above.

Table 3.1 shows the transition in the number of nationalities of authors' affiliation appeared on journals in the United States, Britain and Netherlands from 1970 to 2015, dividing into four layers; more than one, ten, a hundred, and a thousand.

In 1970, out of journals in INSPEC, 86 nationalities were involved on at least one paper in journals published in these countries, 47 on more than ten, 20 on more than a hundred, and 11 on more than a thousand. This figure has augmented and in 2015, 187 nationalities appeared on at least one paper, 131, 79, and 59 in that order. Provided that more than one thousand papers, nationalities incremented by 48 from 1970 to 2015. Whereas online publication has become more and more common thanks to the rapid evolution of digital technology, the number of nationalities of authors' affiliation has dramatically increased, considering that we limit the publication to main three nationalities. Inevitably, the ratio that the affiliation from G7

Table 3.1 The transition in the number of nationalities of authors' affiliation

Number of papers/Year	1970	1980	1990	2000	2010	2015
Over 1 paper	86	110	124	136	176	187
Over 10 papers	47	56	72	87	114	131
Over 100 papers	20	36	43	52	73	79
Over 1,000 papers	11	16	27	35	50	59
Share of G7 nationalities	80.4%	72.0%	68.1%	62.1%	48.7%	39.7%
Share of top ten nationalities	90.8%	87.8%	83.3%	72.6%	66.7%	64.8%

Note
(1) Publication countries are limited to US, UK and Netherland
(2) G7 countries are Canada, France, Italy, Germany, Japan, UK, and US
(3) The number of authors whose names are appeared on papers is generally plural. In that case, the number of nationalities of institutions to which authors belong is also plural. For example, when the number of authors of the same paper is five, 1 Japanese, 2 Chinese, 2 US, the number of nationalities are counted 1 for Japanese, 2 for Chinese, and 2 for US in Table 1. Accordingly, total number of the paper is counted as 3. The total number of papers including overlapped ones is 42,036 in 1970, 98,822 in 1980, 133,517 in 1990, 169,625 in 2000, 339,243, in 2010, and 530,843 in 2015
(4) Percentage share of G7 is that of the total number of papers including overlapped ones
(5) Language which is used in papers published in three countries mentioned above is mainly English, accounting for 99.98% as of 2015
Source Compiled from INSPEC search

nationalities account for in major journals, reaching 80% before 2000, has dwindled even to less than 40% in 2015. It means that while the number of nationalities of authors' affiliation, having enough R&D capabilities to complete a certain level of scientific papers, has increased in these 45 years, the number of human resources and institutions which have acquired high standard of R&D capabilities, have dispersed globally over these years.

As Lorenz curve, shown in Fig. 3.1, helps express the gap appeared on the data, the extent to this curve, the degree of inequality is shown by how far the Lorenz curve appears above or drops below the line of perfect equality. The Lorenz curve, here, can usually be represented by a function L(F), where F, the cumulative portion of the number of nationality, is represented by the horizontal axis, and L, the cumulative portion of the total papers, is represented by the vertical axis. Lorentz curve L(F) for continuous distribution can be defined as follows by expressing the expectation value of the entire group as μ.

$$L(F) = \frac{\int_0^F x(F')dF'}{\mu}$$

As it got closer to the 1970 curve, the curve got closer to 1.00 apart from the even distribution line, and the concentration in the higher ranked nationalities has increased.

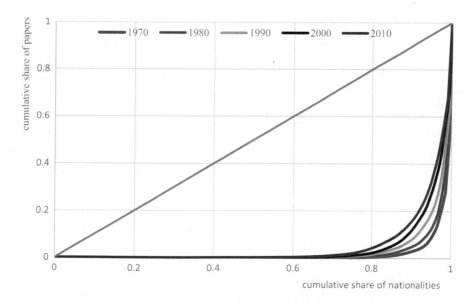

Fig. 3.1 Lorenz curve, showing the number of papers by nationalities of author's affiliations. *Note* (1) The way of counting the number of papers by nationalities is shown in the note 3 of Table 3.1. (2) Each year is the average of every 5 years. For example, 1970 is the average of 1970 and 1975. *Source* Compiled from ISPEC database search

Conversely, as approaching the curve in 2015, it approaches the even distribution line and moves away from 1.00, and the degree of dispersion is increasing. As the area between the Lorenz curve and diagonal gets larger, the Gini Coefficient rises to reflect greater inequality, defined as follows. The higher the number, the greater the degree of papers.

$$G = \frac{\sum_{i=1}^{n}\sum_{j=1}^{n}|Y_i - Y_j|}{2\overline{Y}n^2}$$

In this formula, n means the number of papers, and Yi means the number of nationalities, which are ranked i (i = 1 ... n) ordered by nationalities of author's affiliation. $\overline{Y}$ is the average number of papers by nationalities. A Gini Coefficient of zero expresses perfect equality, and one, where all the nationalities' paper is earned by a single nationality. Thus, Gini Coefficient for a Lorenz curve is defined as the ratio of the area between the Lorenz curve and the diagonal to the triangle below the diagonal line.

Figure 3.2 illustrates the change in Gini Coefficient during this research period. As is shown, since 1970, Gini Coefficient has decreased much with no exception. In other words, the number of the nationalities of author's affiliations who issue scientific papers increases and the concentration ratio on specific nationalities decreases in recent years, showing that the more dispersion has been accelerated, the closer to 2015.

Next, we processed the data to reveal the tendency of dispersion using Herfindahl-Hirschman Index (= HHI). HHI is one of indices that show the degree of the market concentration by companies in some industries. HHI is calculated by the sum of the squares of market shares of each company belonging to the industry. The maximum index score is 1 if the sum of the market shares is 1. The maximum

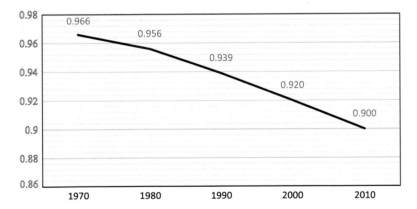

Fig. 3.2 The transitive change in Gini coefficient. *Note* Each year is the average of every 5 years. For example, 1970 is the average of 1970 and 1975. *Source* Compiled from INSPEC database search

index score is 10,000 if we assume the sum of the market shares as percentage. It shows the case of a monopoly. HHI is defined as follows

$$\text{HHI} = \sum_{i=1}^{n} S_i^2$$

In this formula, **n** means the number of papers (and patents), and S_i means the ratio of nationality i.

The concentration ratio of papers by nationalities of authors' affiliation can be revealed by HHI obtained from the sum of the squares of the ratio of papers by nationalities of authors' affiliation.

Table 3.2 shows the changes in the concentration ratio of papers by nationalities of authors' affiliation. Even though the proportion which top nationality's papers accounts for is the same, HHI gets lower figure if the proportion which other nationalities' papers is fewer. That is, the more papers are dispersed, the smaller HHI is and vice versa.

Even though the proportion which top nationality's papers accounts for is the same, HHI gets lower figure if the proportion which other nationalities' papers is fewer. That is, the more papers are dispersed, the smaller HHI is and vice versa. Herein we recognize that HHI gets higher when the proportion of the top nationality's papers account for is larger than that of other nationalities, which implies that it is not dispersed. As shown in the same table, the concentration ratio of the number of scientific papers is decreasing consistently. In that sense, the tendency of global dispersion cannot be denied. In other words, as long as we verify the number of papers by nationalities of author's affiliation, global dispersion on R&D capabilities has steadily been advanced.

It should be noted that while the total share of G7 has declined by more than 40 points, that of top ten nationalities has not so much declined as G7. Let us look at the transition of top ten nationalities by twenty years.

Table 3.3 discloses that the top ten nationalities have been replaced in these 45 years. As of 2015, China and India have been ranked. To conclude, geographically worldwide dispersion has proceeded during 45 years, seen from the perspective of the ability to complete the concerned scientific papers. Also, the

Table 3.2 The changes in the concentration ratio by HHI

	HHI
1970	2699.05
1980	2062.56
1990	1600.69
2000	1055.17
2010	716.92

Source Compiled from INSPEC database search

Table 3.3 The transition of top ten nationalities of authors' affiliations

		1970		1990		2010		2015
1	US	54.8	US	38.8	US	20.4	CN	17.7
2	GB	11.9	SU	10.9	CN	11.7	US	15.3
3	SU	6.5	JP	7.2	DE	6.0	DE	5.2
4	CA	4.1	GB	7.0	GB	5.7	GB	5.0
5	DE	3.0	DE	5.4	JP	5.2	IN	4.5
6	JP	2.7	FR	3.8	FR	4.8	FR	4.2
7	FR	2.5	CA	3.5	IT	3.5	JP	3.9
8	IN	2.4	CN	2.4	CA	3.2	IT	3.3
9	AU	1.5	IT	2.4	IN	3.2	KP	2.9
10	NL	1.5	IN	2.0	KP	3.1	CA	2.7
Total share of top ten		90.8		83.3		66.7		64.8

Note Nationality code is as follows: US (United States), GB (UK), SU (Soviet Union), CA (Canada), DE (Germany), JP (Japan), FR (France), IN (India), AU (Australia), NL (Netherland), CN (China), IT (Italy), KP (Korea)
Source Same to Fig. 3.1

concentration ratio of top nationalities represented by top ten nationalities has been decreasing, not so much as that of G7. What is notable here is that the composition of top nationalities has been dramatically changed for 45 years, and emerging nationalities such as China or India have rapidly risen.

The next section examines whether the ability of technological development capabilities, the capabilities needed when research outcome is applied to techno-logical development at the commercialization stage, has globally dispersed alike.

3.3 Global Dispersion of Technological Development Capabilities

When it comes to comparing the technological development capabilities among nationalities, the number of patents is widely used, which guarantee exclusive license to use newly invented technology for firms. Specifically, the nationality of firms and individuals to whom patents were granted in the U.S. should be scruti-nized, which has many firms with competitive R&D capabilities and a huge market. Herein we examine the ratio of the number of U.S. patent granted to foreign firms or individuals.

In 1965, the ratio of overseas patents other than the U.S. was 19.9%, but it reached 40% in 1980s, and finally more than half in 2010s. Interestingly enough, the number of nationalities of firms or individuals to whom US patents were granted, has increased, to more than 100 after 2010, and 123 as of 2015 (See Table 3.4 [3]). In more detail, the number in every range shows a tendency to

Table 3.4 The ratio and the number of US patents granted to foreign nationals

		1965	1970	1980	1990	2000	2010	2015
[1]	Number of US patents granted	100.0	100.0	100.0	100.0	100.0	100.0	100.0
[2]	Overseas patents: %	19.9	26.9	39.6	47.6	46.0	50.9	52.8
[3]	Number of nationalities other than the US (over 1 patent)	66	71	69	72	100	102	123
(3-1)	Number of nationalities over 20	22	26	28	32	35	45	57
(3-2)	Number of nationalities over 100	12	15	16	21	27	30	35
(3-3)	Number of nationalities over 500	8	9	9	10	16	20	23
(3-4)	Number of nationalities over 1,000	3	6	6	7	11	16	19
	Share of top ten nationalities (%)	21.6	27.4	36.8	40.9	38.6	44.2	44.7

Note Country of patent origin by USPTO is determined by the residence of the first-named patent assignee
Source USPTO, Calendar Year Patent Statistics

increase. It means that more and more nationalities acquired U.S. patent by newly invented technologies in these 50 years and improved technological capabilities.

The structure diagrammed in Fig. 3.3 is Lorenz curve, showing the number of U.S. patents granted to each nationality of the residents of affiliations to which first-named assignee belong from 1965 to 2015, every five years. The ratio of US patents, granted to firms and individuals in developing or emerging nationalities where industrial infrastructure is not in place, is smaller than that of scientific papers, since a large amount of expense is incurred to invent patented technology. In contrast, as shown in Fig. 3.3, the number of patents, granted to U.S. firms or individuals, exceed that of any other nationalities. It is because these patent applications are regarded as domestic. As a result, the concentration ratio on U.S. nationalities gets higher and that on top nationalities is on the rise.

The closer to the curve of 1965, the higher the degree of concentration to the upper level is. However, the closer to the curve of 2015, the smaller the area surrounded by the even distribution line and the Lorenz curve. To that extent, the degree of decentralization is getting higher.

What is noteworthy is that, as already mentioned, the ratio of the top ten nationalities (excluding the United States) also rises so that the ratio of overseas nationality patents is rising sharply.

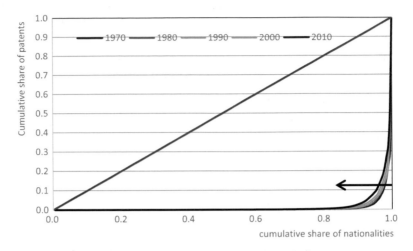

Fig. 3.3 Lorenz curve of the number of UP patents of overseas nationalities (including US nationalities). *Source* Same to Table 3.4

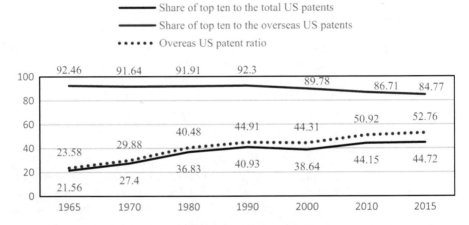

Fig. 3.4 The ratio of the top ten nationalities to the number of overseas nationality in US patents. *Note* Top ten nationalities other than the US. *Source* Same to Table 3.4

However, the ratio of the top ten nationalities to the number of overseas nationality patents tends to decrease gradually. In that sense, the dispersion of technological capabilities to many other nationalities seems to proceed worldwide (See Fig. 3.4).

Next, we again address Lorenz curve of the number of US patents of overseas nationalities other than those of the U.S. nationalities (Fig. 3.5).

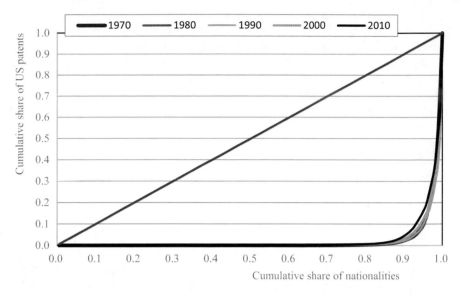

Fig. 3.5 Lorenz curve of the number of US patents of overseas nationalities (excluding US nationalities). *Source* Same to Table 3.4

As are shown in Figs. 3.4 and 3.5, it is difficult to conclude that the dispersion trend is going on unlike what Fig. 3.1 shows clearly. We draw a comparison with the transition of Gini Coefficient (G). In this formula, n means the number of U.S. patents, and Yi means the number of nationalities, which are ranked i (i = 1 … n) ordered by nationalities of inventor's affiliation. $\bar{Y}$ is the average number of patents by nationalities. A Gini Coefficient of zero expresses perfect equality, and one, where the gap is the largest. Thus, Gini Coefficient for a Lorenz curve is defined as the ratio of the area between the Lorenz curve and the diagonal to the triangle below the diagonal line. Table 3.5 shows that Gini Coefficient of patents including whose nationality is the U.S. is declining.

Table 3.5 Gini Coefficient of US patents by the nationality of the residence of the first-named assignee

	Gini coefficient (including US)	Gini coefficient (excluding US)
1970	**0.980**	0.958
1980	0.977	0.962
1990	0.976	**0.964**
2000	0.973	0.958
2010	*0.965*	*0.949*

Note Figure of each year is the average of every 5 years. For example, 1970 is the average of 1970 and 1975

Source Same to Table 3.4

In contrast, Gini Coefficient of patents except whose nationality is the U.S., rose until 1990 and turned into drops after that. In 2010, it fell down from the figure that include all patents (0.965) to 0.949, the average between 2010 and 2015. In other words, it means that the number of nationalities other than the U.S. has also been globally dispersed, but not so much as that of scientific papers.

We scrutinized HHI like as the number of nationalities of papers. Table 3.6 shows the result. The figure has been consistently decreasing since 1970 when we include patents of U.S. nationality, whereas the concentration ratio of patents other than U.S. nationality, had increased until 1990, and dropped rapidly. The reason why HHI increased after 1980s is as follows. In 1970s, while there was not so much difference between the proportion of U.S. nationalities and that of other than the U.S. In 1980s, the number of patents whose nationalities other than the U.S., which was ranked top, increased and the deviation of the ratio occurred. The trend that HHI has decreased from 2000s means that global dispersion is going on, because the gap between the number of patents of top nationalities other than the U.S. and other nationalities has shrunk. The top three nationalities from 1970s to 2010s excluding the U.S. is Japan, Germany, and Great Britain, respectively. Each percentage share was 0.253, 0.211, 0.133 in 1970, 0.351, 0.221, 0.087 in 1980s, 0.466, 0.160, and 0.064 in 1990s. However, since 2000, other nationalities ranked below Japan increased the number of patents. As a result, the gap of the number of patents has been closer between upper and lower ranked nationalities towards the international diversification and divergence of technological capabilities.

Next, we ascertain that which nationalities have been ranked to date. As is shown in Table 3.7, while European nationalities such as Germany and Great Britain, and Japan, which consist of G7 nationalities, accounted for a large portion in 1965, Taiwan emerged in top ten in 1990s, and South Korea, Taiwan, China, Israel and India ranked within top ten nationalities in 2015.

It implicates that conventional world system composed of nationalities represented by G7 which used to hold competitive R&D capabilities, has dramatically changed and globally dispersed by the rise of emerging nationalities in the last fifty years.

Table 3.6 HHI: The trend of concentration of US patents by nationality

	HHI (including US)	HHI (excluding US)
1970	**4868.05**	1460.38
1980	3665.98	1928.10
1990	3438.47	**2559.02**
2000	3304.75	2242.17
2010	*2761.27*	*1684.92*

Note Figure of each year is the average of every 5 years. For example, 1970 is the average of 1970 and 1975
Source Same to Table 3.4

Table 3.7 Top ten overseas nationalities of US patents and the ratios to the total US patents

	1965		1990		2015	
	US	80.1%	US	52.4%	US	47.2%
	Other the US	19.9%	Other the US	47.6%	Other the US	52.8%
1	DE	(5.3)	JP	(21.6)	JP	(17.6)
2	UK	(4.1)	DE	(8.4)	KP	(6.0)
3	FR	(2.2)	UK	(3.1)	DE	(5.5)
4	JP	(1.5)	FR	(3.2)	TW	(3.9)
5	CH	(1.4)	CA	(2.1)	CN	(2.7)
6	CA	(1.4)	IT	(1.4)	CA	(2.3)
7	SW	(0.9)	CH	(1.4)	FR	(2.2)
8	IT	(0.7)	NL	(1.1)	UK	(2.2)
9	NL	(0.8)	SW	(0.8)	IS	(1.2)
10	BE	(0.3)	TW	(0.8)	IN	(1.1)
Share of top ten other than US	21.6		40.9		44.7	
Share of G7	95.5		92.2		77.0	

Note Nationality code is as follows: US (United States), DE (Germany), FR (France), JP (Japan), CH (Switzerland), CA (Canada), SW (Sweden), IT (Italy), IN (India), NL (Netherland), BE (Belgium), TW (Taiwan), KP (Korea), CN (China), IS (Israel)
Source Same to Table 3.4

3.4   Conclusion

The basic research question in this chapter is as follows. Firms need now to explore and exploit R&D resources globally in order to improve these capabilities that take a fundamental role for new products or services, in line with the recent globalization of business activities and innovation of information technologies. However, if firms rely on overseas R&D resources excessively, their capabilities at home would become hollowlose. At the same time, if firms construct a comprehensive system that enables them to exploit overseas R&D resources effectively and create synergistic effects, R&D capabilities would be globally intensified. It means that global dispersion and concentration on specific nationalities of R&D capabilities could simultaneously progress in parallel.

Thus, this paper has tried to clarify the global trend of R&D capabilities from 1960s to 2015, examining scientific papers and US patents. As a result, global dispersion was revealed using Lorenz curve and Gini Coefficient, in the main field of scientific journals from INSPEC; Physics, Electrical Engineering, Electronics, computer and control engineering including Information Technology, and Mechanical engineering.

In the case that we examine foreign invented US patents excluding US patents invented in the US, we can see geographical dispersion among nationalities since

1990, much more than all the US patents including US patents invented in the US. HHI indicator also shows the decline in the degree of the concentration ratio.

The results obtained in this paper is that the number of nationalities of authors' affiliation and that of nationalities of affiliations of the first-named U.S. patent assignees have increased and diversified. Especially, R&D capabilities measured by nationalities of authors' affiliation remarks that point, rather than by those of US patent inventors, which means that more and more countries have improved R&D capabilities, "R" in particular.

A limitation of this study is that we cannot conclude whether and to what degree research and development capabilities would complementarily progress in step with each other or in parallel. We need further in-depth research study by specific companies, industries and technologies.

References

Akashi, Y., & Ueda, H. (1995). *Nihon kigyou no kenkyu-kaihatsu sisutemu (R&D system of Japanese Companies)*, Tokyo: Tokyo University Press.

Archibugi, D., & Michie, J. (Eds.). (1997). *Technology, globalization and economic performance.* Cambridge: Cambridge University Press.

Asakawa, K. (2011). *Gurobaru R&D manejimento (global R&D management)*. Tokyo: Keio University Press.

Badaracco, J, Jr. (1990). *The knowledge link: How firms compete through strategic alliances.*

Cantwell, J. A. (1995). The globalization of technology: What remains of the product cycle model? *Cambridge Journal of Economics, 19*(1), 155–174.

Freeman, C., & Hagedoorn, J. (1995). Convergence and divergence in the internationalization of technology. In J. Hagedoorn (Ed.), *Technical change and the world economy* (pp. 34–57). Vermont: Edward Elgar.

Gotoh, A., & Kodama, T. (Eds.). (2006). *Nihon no inobeishon sisutemu (Innovation system of Japan)*. Tokyo: Tokyo University Press.

Hayashi, T. (1995). Higashi Ajia ni okeru gijutsu chikuseki to nihongata gijutsu iten sisutemu, (Technological accumulation in East Asia and the Japanese-style technology transfer system). In B. Chin & T. Hayashi (Eds.), *Ajia ni okeru gijutsu hatten to gijutsu iten (Technological development and technology transfer in Asia)*, Chap. 2. Tokyo: Bunshindo.

Hayashi, T. (2004a). Globalization and networking of R&D activities by 19 electronics MNCs. In M. Serapio & T. Hayashi (Eds.), *Internationalization of research and development and the emergence of global R&D networks* (pp. 85–112). Oxford: Elesvier.

Hayashi, T. (2004b). Kenkyu kaihatsu nouryoku no kokusaiteki bunsanka to shuchuka (International dispersion and concentration of R&D capabilities). *Rikkyo Keizaigaku-Kenkyu, 57*(3), 63–87.

Hayashi, T., & Komoda, F. (1993). *Gendai No Sekaikeizai To Gijutsu Kakusshin (Contemporary world economy and technological innovation)*. Minerva Shobo: Tokyo.

Jaffe, A., & Trajtenberg, M. (Eds.). (2002). Patents, citations, and innovations: A window on the knowledge economy. In *Library of Congress Cataloging-in-Publication Data.* MIT.

Jeremy, D. J. (1991). *International technology transfer.* Edward Elgar: Aldershot.

Komoda, F. (1987). *Kokusai gijutsu iten no riron (Theory of international technology transfer).* Yuhikaku: Tokyo.

Kotabe, M., et al. (2007). Determinants of cross-national knowledge transfer and its effect on firm innovation. *Journal of International Business Studies, 38,* 259–282.

Lundvall, B.-A. (Ed.). (1992). *National systems of innovation: Towards a theory of innovation and interactive learning*. London: Pinter.
Lundvall, B.-A. (2007). National innovation systems: analytical concept and development tool. *Industry and Innovation, 14*(1), 95–119.
Nagaoka, S. (2011). Amerika to nihon no inobeishon purosesu (Innovation process of the US and Japan). In S. Fujita, & S. Nagaoka (Eds.), *Seisansei to inobeishon sisutemu (Productivity and innovation system)* (Chap. 4, pp. 147–190). Tokyo: Nippon Hyouronsha.
Nagaoka, S., & Yamauchi, Y. (2014). Hatsumei no kagakuteki gensen (Scientific sources of inventions). In *RIETI Discussion Paper Series* 14-J-038 (pp. 1–74).
Nelson, R. (Ed.). (1993). *National innovation systems*. Oxford: Oxford University Press.
Nonaka, I. (1995). *Nihon gata inobeishon sisutemu (Japanese style innovation system)*. Hakutoshobo: Tokyo.
Patel, P., & Pavitt, K. (1998). Uneven technological accumulation among advanced countries. In G. Dosi, D. J. Teece, & J. Chytry (Eds.), *Technology, organization, and competitiveness* (pp. 289–317). NY: Oxford University Press.
Porter, M. E. (1998). *The competitive advantage of nations*. NY.: The Free Press.
Saxenian, A. (1996). *Regional advantage: culture and competition in Silicon Valley and Route 128*. Cambridge: Harvard University Press.
Saxenian, A. (2006). *The new argonauts: regional advantage in a global economy*. Cambridge: Harvard University Press.
Sorenson, O., & Fleming, L. (2004). Science and the diffusion of knowledge. *Research Policy, 33*, 1615–1634.
Vernon, R. (1979). The product cycle hypothesis in a new international environment. *Oxford Bulletin of Economics and Statistics, 41*(4), 255–267.

Takabumi Hayashi (Ph.D. in Economics, Rikkyo University) is Professor Emeritus of Rikkyo University, Tokyo. He successfully filled the position of senior lecturer at Fukuoka University, associate professor and professor of International Business at Rikkyo University, and Professor at Kokushikan University, Tokyo. His recent research arears are innovation systems and R&D management, focusing on knowledge creation and diversity management. His works have been widely published in books and journals. His book "Multinational Enterprises and Intellectual Property Rights" (in Japanese; Moriyama Shoten, Tokyo, 1989.)" is widely cited, and "Characteristics of Markets in Emerging Countries and New BOP Strategies" (in Japanese; Bunshindo, Tokyo, 2016) received the award from Japan Scholarly Association of Asian Management (JSAAM) in 2018. He has been sitting on the editorial board of several academic journals.

Atsuho Nakayama (Ph.D. Rikkyo University) is Associate Professor of Marketing Science at Graduate School of Management, Tokyo Metropolitan University. His recent research interests include the visualization and reduction of large and complex data about consumer behavior. A huge quantity of information is often available but standard statistical techniques are usually not well suited to managing these kinds of data. He has studied image and text data analysis using machine learning techniques and deep learning techniques. He plans to use consumers' uploading habits on the internet for marketing purposes. He has published many papers about behaviormetrics and multivariate analysis.

Chapter 4
International Standardization of the New Technology Paradigm: A Strategy for Royalty-Free Intellectual Property

Yasuro Uchida

Abstract Since the establishment of the World Trade Organization in 1995, international standardization of intellectual property (IP) has become more important as multinational enterprise form their international competitive advantages. This activity means "the internationalization of IP." In addition, Internet of Things (IoT) has become important in recent years, and the activities of related technologies are being aggressively promoted. What is important in IoT is the development of an environment that can utilize this technology regardless of industry type. In other words, it is "inter-industrialization of IP." Today's technological development competition is inevitably the "inter-industrialization of IP" while at the same time the "internationalization of IP." Moreover, when internationalization and inter-industrialization are promoted at the same time, it turns out that there is a big change in the role of IP. In the past, one of the roles of IP was to generate a source of revenues from royalties, but what is now increasing in IoT-related business fields is actually royalty-free. Thus, a paradigm shift in IoT field has occurred. Now there are more cases where patent holders have made their IP royalty-free. Consequently, it is royalty-free IP that holds the key to the paradigm shift in the IoT field. Why royalty-free cases are increasing is examined in detail along with the phenomenon's background in this chapter. Furthermore, this chapter aims to clarify what kind of technology strategy is required for MNEs promoting technology development in such a complex business environment. At the same time, a new "viewpoint of inter-industrial business studies" is added to existing international business studies.

Keywords Royalty-free · Intellectual property · International standard · Inter-industrialization · International business

Y. Uchida (✉)
Graduate School of Business, University of Hyogo, Kobe, Hyogo, Japan
e-mail: y-uchida@mba.u-hyogo.ac.jp

© Springer Nature Singapore Pte Ltd. 2019
J. Cantwell and T. Hayashi (eds.), *Paradigm Shift in Technologies and Innovation Systems*, https://doi.org/10.1007/978-981-32-9350-2_4

4.1   Introduction

The purpose of this chapter is to clarify the mechanism of royalty-free intellectual property (IP) that can be recognized as a new trend of innovation. As companies aim to innovate and spread new technology to the global market through international standardization required by the World Trade Organization (WTO), at the same time they hope to earn huge profits through collecting royalties. However, in recent years, although IP has been a consistent a source of such revenue, a movement to make it free of charge is rapidly growing, particularly in new technical fields as seen in the Internet of Things (IoT). This is different from the consumer-driven free innovation that Hippel talks of (Hippel 2016)—is the phenomenon that technology developers themselves are making products free of charge.

What is common to these phenomena is they are occurring in a business field where internationalization and inter-industrialization progress simultaneously. Indeed, the assertion that "international business" and "inter-industrial business" are simultaneous can be confirmed in today's IoT business. In current research on international business, what is developing is more than "synchronization between internationalization and inter-industrialization". In this environment and as this phenomenon progresses, royalty-free IP will be the standard.

For example, Fig. 4.1 shows how Toyota Motor Corporation, one of Japan's leading companies, has progressed in development projects with other companies since 1980. As shown in this figure, Toyota's alliance strategy is classified into four types from I to IV. Two points stand out markedly in the early 1990s. At this time, the US political press was calling for the Japanese market to open up. In Japan, domestic automobile manufacturers, including Toyota, had shown rapid internationalization of the domestic market through alliances with US automobile manufacturers. The second feature is that, as shown in type IV, Toyota has been increasing its partnership with "different industry firms in other countries" since 2000. It has promoted alliances with leading overseas IT companies such as Microsoft and Google, and companies involved with electric vehicle (EV) charging technology. After 2010, "Type II" type trends also increase. This means that internationalization and inter-industrialization were progressing at the same time. Under such circumstances, in January 2015 Toyota announced that it would make the technology related to fuel cell vehicles (FCV) royalty-free.

In traditional technology strategies, IP of key technologies has been utilized significantly in proprietary licenses in business activities, and firms gained licensing income through spread of the technology. However, in a business environment where internationalization and inter-industrialization are synchronized, the role of IP has changed dramatically. It can be seen that the change of role of IP means that the paradigm shift has been occurred. Why did such changes come emerge?

Here, the research question of this chapter is proposed: What is the strategic meaning of IP that holds the key to "synchronization between internationalization and inter-industrialization"? As mentioned above, in international business studies,

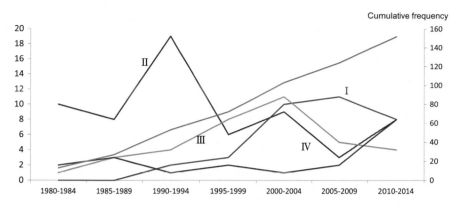

Fig. 4.1 Type of Toyota's alliance strategy (1980–2014). *Notes* This table shows following four types of alliances by Toyota. I: the same industry enterprises in Japan. II: the same industry enterprises in other countries. III: the different industry enterprises in Japan. IV: the different industry enterprises in other countries. *Source* Nihon Keizai Shimbun (Nikkei), January 1, 1980–December 31, 2014. Search words are "Toyota" and "Alliance", 152 cases out of 3,304 cases. Subsidiary cases are excluded. Duplicate articles were put together as one case

IP has been regarded as a source of overwhelming competitive advantage or as a source of patent revenue income. The competitive advantages of the alliance between Microsoft and Intel, once called "Wintel", were captured in such a context. Also, as firms group in consortia for technology development, the patent pool is important in distributing income from patent fees to the participating companies.

However, in the current field of technology development where internationalization and inter-industrialization progress simultaneously, a different type of IP has been emerging. In recent years, consortia to develop technology related to IoT to spread the developed technology at once, and make essential patents royalty-free. Cases like Toyota illustrated earlier are becoming the norm in IoT (Uchida 2016).

In this chapter, the next section examines the research question, positioned against the background of previous research. Next, progress of royalty-free in the IoT field is confirmed based on data of from Telecommunication Technology Committee (TTC), an organization affiliated with the Ministry of Internal Affairs and Communications of Japan, which is investigating the development of information technology around the world. In a TTC survey published in 2017 (TTC 2017), about 60 organizations were listed as active in the IoT consortia. This research reveals that in consortia founded after 2008, approximately 70% of the consortia make their IP royalty-free. This chapter will discuss the trend and background of that development to clarify the mechanism of the paradigm shift in the IoT field.

In addition, the third section introduces two kinds of cases regarding to above argument. The first one is the Open Connectivity Foundation (OCF), one of the representative agencies among those consortia, and the activities of Qualcomm, the core member of the OCF. Although patent revenue accounted for about one-third of total sales, the OCF is helping firms make their technology free of royalties. The meaning of the strategy will be explored. And the second one is the case of GS1 EPCglobal which is one of the biggest standard developing organization. Through introducing the process of standardization, the meaning of royalty-free IP will be discussed.

4.2 Transition of International Business Studies

Until the 1990s, much of the research related to innovation was limited to specified industries. Discussion on innovation focused on firms seeking self-sufficiency through internal resources and R&D systems. However, such "closed innovation" had gradually begun to show limitations. Also, companies with existing businesses could have an advantage when sustaining innovation, but it became clear that there was a tendency to lag behind disruptive innovation (Christensen 1997).

Attention shifted to the importance of "open innovation" (Chesbrough 2003). It became necessary for companies to transform their internal business model to incorporate external resources and technologies (Chesbrough 2006). Thus, in the 2000s, the direction of research on innovation began to change, and as a result many remarkable discoveries were made. Cantwell presented a new perspective on traditional MNEs theory by combining research on corporate innovation management with research on national innovation systems (Cantwell and Iammarino 2003; Cantwell and Molero 2003; Armann and Cantwell 2012). In particular, it is worth noting that firms in emerging countries did not limit innovation management to specific areas (Armann and Cantwell 2012); innovation not only transcended national boundaries, but also spread beyond industries and businesses, including information and telecommunications, electric machinery, and even the automobile industry (Armann and Cantwell 2012). In other words, the viewpoint of inter-industrialization has been added. The phenomenon that such internationalization and inter-industrialization are advancing is consistent with the trend of Toyota as mentioned above.

Meanwhile, Chesbrough discussed the relationship between innovation management and the theory of dynamic capability (Chesbrough 2001). Dynamic capability is the ability of a company to integrate, build, and reconfigure internal and external resources and capabilities to deal with rapidly changing business environments (Teece 2009). Armann and Cantwell (2012) also analyzed the evolution mechanism of innovation management by companies in emerging markets, introducing perspectives of dynamic capabilities and open innovation.

Cantwell also showed an important perspective, co-authoring with Dunning: multinational enterprises respond to complex uncertainties by shifting to a flexible and open business network structure in order to adapt to the changes of the environment. He pointed out that MNEs have come up with solutions to acquire decentralized knowledge and recombine it. The cause was found in the transition to an open business network structure (Cantwell et al. 2010). This means that multinational enterprises must deal with inter-industrialization. MNEs are now responding to the call of inter-industrialization, crossing boundaries across industries. For example, it is well known that GE is advancing service sales, leasing power per hour, as well as manufacturing aircraft engines.

Today, as MNEs promote both internationalization and inter-industrialization, we see that internationalization and inter-industrialization are part of the innovation process. Next, previous research on the innovation process is outlined.

4.3   Changes in Research on Innovation Process

As the business activities of MNEs became inter-industrialized, innovation processes became part of inter-industrialization. In the background, there has been an increase in the influence of the WTO in inter-industrialization and innovation processes.

4.3.1   Influence of WTO

Of the over 160 WTO member countries, 23 are advancing toward affiliation. Since most major countries have joined the WTO, the organization's rules have a noteworthy influence on international business as a whole.

When the WTO was established in 1995, all of the WTO members agreed to observe an agreement on technical barriers to trade (TBT). According to the TBT agreement, in international business, when companies need new standards, the WTO requires the members to base them on international standards, like those of the International Standardization Organization (ISO), the International Electrotechnical Commission (IEC), etc. already published in principle. As such, using an original standard that obstructs free trade is forbidden. A duty to conform to the standards that define international business, that is, the de jure standard, is imposed.

Unfortunately, a company with its own defined standard, unless it is the international standard, cannot export its products.

In the early 2000s, Japan's cell phone manufacturers failed to export their products because Japanese communication standards were not accepted as internationally standardized. At that time, a phenomenon known "Galapagos Syndrome" occurred. This describes a situation in which a superior fails to spread globally and is forced to continue its evolution in only the country it was created in. The term is named after organisms living on the Galapagos Islands, i.e. organisms that evolve independently in a closed environment.

It goes without saying that communication technology is of upmost importance in the mobile phone industry. Currently, preparations are being made toward the practical use of the fifth generation mobile phone technology, however, originally this industry had been promoted mainly in Europe. In particular, Nokia had played the central role in this industry from the first generation. Gower and Cusumano's investigation (2002) is helpful in elaborating on Nokia's strategic behavior at the time.

In the first generation, Nokia actively promoted the development of Nordic Mobile Telephone (NMT), which was a common communications system in Northern Europe, where their company was located. However, in Europe, communication standards had been muffled because each country had constructed its own communication system.

Nokia failed to disseminate the NMT system as each European country had its own standards, like TACS in the UK, Radiocom 2000 in France, C-NET in Germany and RTMI/RTMS in Italy. However, Nokia learned how important an international standard in communication technology was through this experience. This lesson was put to use in the second generation. Nokia promoted the development of GSM (Global System for Mobile Telecommunications) with other mobile phone manufacturers like Motorola. GSM is based on NMT, the system actively promoted by Nokia. They proposed this communication system to the European standardization body, the European Conference of Postal and Telecommunications Administrations (CEPT), and since the European Commission also supported the GSM, 18 European countries decided to adopt it in 1991. GSM became an international standard in this way and it has been actively introduced in other countries and regions.

In contrast, Japan had developed its own communication system of second-generation communication technology, called PDC (Personal Digital Cellular). It was developed mainly by NTT DoCoMo. PDC became the standard in Japan's mobile phone industry, but it did not conform to GSM, which was being advanced in most countries and regions of the world. Thus, the phenomenon of distinctive evolution, the aforementioned "Galapagos Syndrome," progressed.

PDC was standardized only in Japan, and thus was not considered an international standard, even though the technical evaluation of PDC was very positive and Japan fell behind in the international standardization process based in Europe. In the end, as GSM had become the standard, a unique domestic industrial structure was built up in Japan.

By this example, it is clear that international standardization is the key to the spread of technology. International standardization has the possibility to create a strategic position for standard-setting companies.

4.3.2   Changes in the Standardization Process

Naturally, WTO regulation is not the only reason behind international standards becoming important in international business; it is also because international standards directly connect to international competitive advantages. Since the standardized products can be mass-produced for international standards regardless of differences in regions and countries, it acts as a driving force for the manufacturing industry and produces great economic effects (Takeda 1998).

In addition, network externalities work with standards that are popular enough to surpass competing standards in the market. Network externalities bring about a phenomenon called "positive feedback" that makes the dominant standards stronger and stronger, giving a standard an overwhelming competitive advantage (Arthur 1994, 1996; Schilling 2002; Burg and Kenney 2003; Suarez and Lanzolla 2005), if it becomes the de facto standard. Therefore, the de facto standard also has a remarkable meaning in global business (Katz and Shapiro 1985; Farrell and Saloner 1986; Christensen et al. 1998).

As a result of such competition, the "cx-ante standard", which standardizes before competition, has become general in recent years from the "ex-post standard" where standardization advances.

The biggest reason for this is to avoid the risk of competition (Yamada 2004). As mentioned above, companies with dominant standards have enough momentum to monopolize the market based on their technology, and it is extremely difficult for inferior companies to reverse it. In areas where huge investments in development are required, a company must avoid losing competition over standardization. For this reason, ex-ante standards are selected to promote standardization, while firms negotiate with companies that have competing standards before putting their products on the market. In the past, for example, the "DVD Consortium", formed in 1995 by companies with technology licenses, standardized DVDs before the products were released to the market.

Therefore, in recent years, the process of forming a consortium before market introduction, developing standardization within it, and submitting it to international standardization organizations like the ISO and IEC has become generalized. DVDs have also been certified by ISO/IEC in 1999 through the DVD Consortium ("DVD Forum" since September 1997) through Ecma International (European Computer Manufacturer Association).

In the case of the ex-ante standard, unlike the de facto standard, standard development companies can avoid competition for standardization, but in order to standardize, consensus must be obtained within the consortia. It is difficult to agree on standardization at the consortium level based on the coordination needed among companies to reach a consensus.

Thus, the method with which to advance the standardization is clearly different from the process of the ex-post standard in terms of whether competition in the market is necessary, or adjustment to obtain an agreement in the consortium is necessary (see Fig. 4.2). Recently, standards following these processes have come to be called "Consensus-Based Standards". Because these consensus-based standards cannot promote standardization with only one company, standardization must be promoted in cooperation with other companies at a consortium. Therefore, it is assumed that such standards should be shared with members of the consortium. So, in the case of the standardization of such processes, a business model that promotes both cooperation and competition is essential (Krechmer 2006; Shintaku and Eto 2008; Uchida 2008). Various verifications have been promoted from the

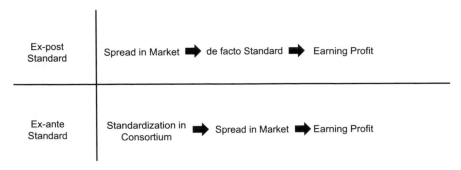

Fig. 4.2 Difference between Ex-ante Standard and Ex-post Standard

relationship between standards and IP in international business (Sakakibara 2005; Sakakibara and Kayama 2006). For example, one a business model embeds IP in the standard itself and thus obtains license revenue based on it.

Even if finished products become commoditized and lose their competitive advantage, it is possible to anticipate IP income if the company's technology is adopted as a core technology of products that are spread globally. Currently, a business model combining international standards and IP is seen as a powerful strategy for the commoditization of innovation, when consensus-based standards have become common.

In this way, although the standardization process has greatly differed from the past, it can be confirmed that the technology standard also plays an important role for international business. International standards are becoming more meaningful as part of a competitive strategy aimed at earnings based on IP.

4.3.3 Changes in Innovation Process

Figure 4.3 is a summary of the areas in which technology is utilized. The horizontal axis shows the regional classification where the developed technology is utilized. It is a classification according to whether technology is utilized only within a specific country or region, or whether it is based on an international situation. For example, when the technology utilized in "B" is also utilized in "C", "international standardization of technology" will be pushed forward and as such, that strategy also becomes necessary.

On the other hand, the vertical axis is categorized according to the business area where the developed technology is utilized. This axis separates cases where technology is used only in a specific industries and cases where technology is used beyond the industry. When the technology utilized in "B" is also utilized in another industry, it is necessary to make an "inter-industrial standardization of technology." Companies' business activities are, in general, similar. After the company's

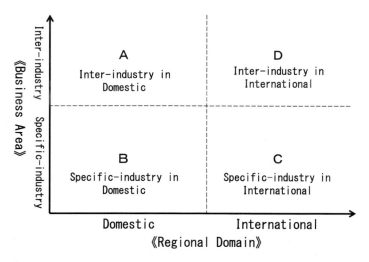

Fig. 4.3 Application area of technology

activities shift from B to C in this figure, international business studies will also become active. In the 1960s, various innovation theories were published since Hymer, which cited "removal of conflict" as a reason for companies to work across borders. For example, an "internalization theory" describes a relationship between overseas subsidiaries and headquarters, global competitive advantage, and open innovation. However, most of these are theories which analyze the movement from B to C or the activities in C from various perspectives. Until today, much of the research focused on the existence of the border as a matter of course.

Research on technical standards is summarized as follows. Up to now, studies on technical standards were often analyzed based on the situation of information technology and the PC software industry that focused on economic effects, such as network externality, the bandwagon effect, the lock-in effect, switching costs, etc. (Farrell and Saloner 1986; Cargill 1989; Basen and Farrell 1994; Shapiro and Varian 1998; Shapiro 2000; Jakobs 2000, 2005).

In addition, there have been many studies on competitive advantage that focused on cost reduction due to economies of scale created by de facto standardization in certain specific technologies and products (Yamada 1993; Shintaku et al. 2000; Doi 2001; Takeda Uchida Kajiura 2001; Gawer and Cusumano 2002). These are the research contents included in "B" and "C" of Fig. 4.3 shown earlier.

Also, there was research on the strategic behavior of companies based on proprietary technology (Wegberg 2006; Warne 2005; Ogawa 2009; Arai and Uchida 2012; Uchida 2015) and the appropriability of IP (Simcoe 2006; Kajiura 2010, 2012, 2013); these studies are based on the premise of a process in which the business environment shifts from "B" to "C" in Fig. 4.3.

Meanwhile, since the latter half of the 2000s, research on consensus based standards based on "inter-industrial alliances" promoted by specific countries and regions had also appeared (Krechmer 2006; Shintaku and Eto 2008; Tatsumoto 2011, 2017; Kajiura 2013). This corresponds to "A" in Fig. 4.3.

Of course, technology crossing borders and being used beyond industries had already existed (Yasumoro and Manabe 2017). However, as described above, in previous research (in particular existing research on technical standards), the viewpoint that new technology was developed in sync with "internationalization" and "inter-industrialization" was inadequate. In this research, we focus on this part, namely "D" in Fig. 4.3.

Next, the specific cases seen in each cell of Fig. 4.3 will be described.

4.4 Specific Cases in the Innovation Process of Each Cell

Here, cases seen in cells A to D are introduced. Through those cases, the cause of royalty-free IP which has become confirmed in Cell D will be discussed.

4.4.1 Innovation Process in Cell A

This area is not dealt with in international business, because it describes cross-industry collaboration in the domestic field. However, innovation also occurs in these areas. For example, a domestic small company diversifies to expand its business opportunities.

In China, Alibaba, one of the leading e-commerce companies, is taking on the challenge of a brick and mortar business as well. They have developed a shopping mall called Hema. In Hema, customers can not only shop at real stores but also can order on the Internet. Customers who come to the physical location can actually leave with fresh vegetables and fish in hand, and eat at the dining area as well. To customers who order on the Internet, Hema's staff deploys services to deliver to customers' houses within 60 min. In both cases, "Alipay" is used for payment. We can understand Alibaba have spread their business from EC to different fields recent years.

These cases are by no means representative of international business. However, such changes seen in Cell A have steadily increased in recent years.

4.4.2 Innovation Process in Cell B

This cell shows a pattern in which technology used in a specific industry is developed within a specific country. General cases so far may be included in this area.

For example, the case of home video cassettes is one which serves as a great starting point. VHS, released in 1976, was developed by JVC in Japan. A year before, Sony developed a standard called Betamax, but TV broadcasts could be recorded only for one hour. However, VHS was able to record for up to two hours. This meant VHS was advantageous for recording movies broadcast on TV, as movies broadcast in Japan are often edited down to fit two hours. In other words, it was a technology developed by a company of a specific industry to adapt to special circumstances of a specific country, Japan.

Later, this technology was internationalized and its use increased, but at the beginning of development it was technology advanced in cell B.

This case also made companies around the world learn the advantages of de facto standards. It demonstrated the possibility of bringing enormous profits to companies that succeeded in setting standards in specific industries (Takeda 1998).

One of the most trend-setting companies surely is Sony. After defeating the competition with VHS technology, Sony developed a multitude of technologies that became de facto standards. However, in the case of DVDs, Sony chose to collaborate with competitors in the same industry in order to avoid the risk of aiming at a de facto standard with only one company, hence the DVD Consortium was created in 1995. At the same time, however, Sony experienced difficulty in coordination within the consortium. The consortium, created by 10 companies, conflicted over the allocation method of patent fees and divided into two groups: one led by Sony the other with Toshiba as a leader. The division has not yet been resolved (Uchida 2007).

4.4.3 Innovation Process in Cell C

Here, it is assumed that internationalization of technology commonly used in a specific industry is common. The QR code developed by DENSO Corporation is a typical example. DENSO is a Japanese company that is one of the biggest suppliers for automobile industry. They develop advanced automotive technologies, systems, and components for major automakers. QR codes were generally utilized as automatic identification technology in production settings or physical distribution. The codes incorporate the signboards Kanban used at auto manufacturing sites; they are a piece of technology that can be read correctly and quickly. Since it has a feature that can also recognize a 2-byte language, Chinese characters can also be recognized. Therefore, it had greater utility compared to the somewhat similar barcode. As a result, after starting in car production in Japan, it gradually permeated the industry.

In fact, there was agreement on an international standardization of a similar code, the "data matrix," developed by ID Matrix in the US (see Fig. 4.4). Global companies used the data matrix in international business, however, since this code cannot effectively render Chinese characters, Japanese companies are not convinced to use it. In addition, differences in codes used between Japanese firms and overseas production subsidiaries have resulted in a deterioration of production efficiency.

Fig. 4.4 QR code and
datamatrix

QR code   Datamatrix

As a result, DENSO decided to send QR code to the organization for standardization. This meant DENSO decided to aim at the international standardization of QR codes. Having taken charge of international standards in automatic identification technology proposed to the Joint Technical Committee 1 (JTC1, a joint committee of ISO and IEC), international standardization was formally carried out in June 2000.

Now, we can understand that DENSO's strategy lead to the creation of their position. Formal standardization had been adapted for the customers' accessibility, especially with regard to global companies, abiding the rule of the WTO.

4.4.4 Innovation Process in Cell D

Here is the cell we want to pay attention to in this chapter; it describes a process that advances while simultaneously aiming at internationalization and inter-industrialization.

Until now, there have been cases in which many go toward inter-industrialization through the process of internationalization. As the DVD and QR code examples showed earlier, when the technologies were first developed, the use of the technology was initially specific to one industry.

However, in recent years, innovation on the premise of internationalization and inter-industrialization has come to be seen from the beginning (Armann and Cantwell 2012). In other words, innovation is advanced in sync with both internationalization and inter-industrialization. In satisfying this situation, in many cases it is "Royalty-Free IP" which most appropriately fits the bill. This is clearly different from other innovation processes.

The development of RFID (Radio Frequency Identification) is a good example. Initially, RFID standardization was advanced with a standardization organization called EPCglobal, which was created as a subordinate to GS1. EPCglobal was organized by collecting 550 companies from around the world centered on Intermec, Wal-Mart, and the US Department of Defense, which made tags used in RFID in 2003. Then, technology standardized by EPCglobal was proposed to JTC1.

For RFID technology, users requested correspondence not only from internationalization but also from inter-industrialization as well. In other words, rather than inter-industrializing a technology specified for one industry, it was intended to be utilized in various industries from the very beginning. In the process of standardization, the basic policy for RFID was royalty-free, standardized technology. This standardized technology means essential patents would be applied to many industries.

At one time, Hitachi developed tags with technology superior to the technology developed at EPCglobal, called the "Hibiki tag." Hitachi aimed for international standardization, arguing to JTC 1 that the Hibiki tag was superior to EPCglobal's tag in terms of security. If Hibiki tags became internationally standardized, they would become popular both internationally and inter-industry, and eventually lead to the possibility of obtaining huge royalty revenues.

However, the EPCglobal side fought back with fierce opposition, and JTC1 supported the proposal by EPCglobal (who has many members) rather than Hitachi's proposal (proposed by only one company).

EPCglobal, and its upper organization GS1, had another reason it wanted to advance royalty-free. In order to standardize more reliably, it needed an international and inter-industrial consortium, and members could probably not be collected without having essential patents royalty-free. Today, the standardization promoted in GS1 is basically all royalty-free. They were promoting standardization in the 2000s according to these ideas.

According to Porter, a smart product is made up of the following three elements; (1) physical elements such as mechanical parts and electrical parts, (2) smart elements such as sensors, data storage and software, and (3) connection functions such as antennas, ports and protocols connected to the Internet (Porter 2014). These functions are being utilized in many products, including home appliances.

In the case of such a product, it is extremely difficult for one specific MNE to develop technology corresponding to various uses. So today, innovation on IoT technology is generalized and conducted within a consortium composed of companies across borders and industry barriers. To further elaborate, the next section discusses the process of innovation by which internationalization and inter-industrialization progress at the same time.

4.4.5 Difference Between Each Cell

Figure 4.5 shows the difference between each cell. As this shows apparently, there is a big difference between cell A to C and D. In the cell A to C, business activities are being conducted within some boundaries. The boundaries are created by borders or industrial barrier.

The other hand, cell D is completely different from them. In the cell D, the situation which internationalization and inter-industrialization are progressed simultaneously is the premise. Therefore, the direction of innovation is also

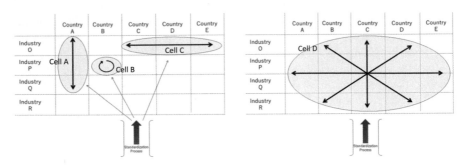

Fig. 4.5 Difference between each cell

different from cell A to B. The important thing is to cover a wide range of user needs. In order to this requirement, Wide Innovation, so to speak, is needed in cell D from the beginning regardless to some boundaries. The consortium is often used under that situation.

Now, how innovation process in the consortium is discussed in detail in the next section.

4.5 Actual Situations of Royalty-Free IP

In this section, we pay attention to the consortia that are advancing the standardization of technology, while considering both internationalization and inter-industrialization. The survey on the actual situations will be summarized.

4.5.1 Overview of IPR Survey

This study adopts a survey report of the Telecommunication Technology Committee of Japan. The TTC is an external organization of the Ministry of Internal Affairs and Communications, an organization that promotes the standardization of information and communication technology, which is engaged in surveys of consortia around the world. Though there are few surveys of active situations, TTC's surveys are excellent.

The TTC published a list of actively operating consortia in 2017. Most of the consortia are developing technologies that are compatible with the elements of smart products, which Porter describes.

Based on that list, Table 4.1 shows the results of investigating how all consortia deal with IP. Recently, to avoid conflicts over rights or license fees, IP Rights (IPR) policy has been approached through the consortia.

Table 4.1 IPR policy survey in IoT consortia

	Consortium	Full name	Foundation	No. of members	RF	RAND
1	TMForum	TMForum	1988	850	Closed	
2	OMG	Object Management Group	1989	264	◎	
3	ITS America	The Intelligent Transportation Society of America	1991	370	Closed	
4	OASIS	Organization for the Advancement of Structured Information Standards	1993	279	◎	
5	BBF	Broadband Forum	1994	153		◎
6	IMTC	International Multimedia Telecommunication Consortium	1994	30	◎	
7	LONMARK	LonMark International	1994	118		◎
8	W3C	World Wide Web Consortium	1994	398	◎	
9	FSAN	Full Service Access Network	1995	73	Closed	
10	TOG	The Open Group	1996	521	○	○
11	ECHONET	ECHONET Consortium	1997	266	○	○
12	OIF	Optical Internetworking Forum	1998	99	○	○
13	Bluetooth.SIG	Bluetooth.SIG	1998	8000	◎	
14	GCF	Global Certification Forum	1999	284	Closed	
15	FCIA	Fibre Channel Industry Association	1999	24	Closed	
16	ITS Forum	ITS Info-communications Forum	1999	94	Closed	
17	OSGi	OSGi Alliance	1999	144	◎	
18	HPA	HomePlug Alliance	2000	39	Closed	
19	SIP Forum	SIP Forum	2000	29	◎	
20	MEF	Metro Ethernet Forum	2001	206	◎	
21	IIC(ITS)	Internet ITS Consortium	2002	86	Closed	
22	OMA	Open Mobile Alliance	2002	70	◎	

(continued)

Table 4.1 (continued)

	Consortium	Full name	Foundation	No. of members	RF	RAND
23	ZigBee	ZigBee Alliance	2002	408		◎
24	DLNA	Digital Living Network Alliance	2003	175	Closed	
25	EPC Global	EPC Global (GS1)	2003	1500	◎	
26	MoCA	Multimedia over Coax Alliance	2004	45		◎
27	NFC Forum	Near Field Communication Forum	2004	167	O	O
28	Ethernet Alliance	Ethernet Alliance	2005	87	◎	
29	Continua Health Alliance	Continua Health Alliance	2006	Closed	Closed	
30	NGMN	NGMN Alliance	2006	96		◎
31	OGF	Open Grid Forum	2006	19	O	O
32	Hadoop	Apache Hadoop Project	2008	57	◎	
33	HbbTV	HbbTV Association	2008	82	◎	
34	HomeGrid Forum	HomeGrid Forum	2008	58	◎	
35	IPTVFJ	IPTV Forum Japan	2008	109	◎	
36	Kantara	Kantara Initiative	2009	45	◎	
37	SGIP	Smart Grid Interoperability Panel	2009	144	◎	
38	JSCA	Japan Smart Community Alliance	2010	272	Closed	Closed
39	OpenADR	OpenADR Alliance	2010	127		◎
40	JSSEC	Japan Smartphone Security Association	2011	145	Closed	Closed
41	OCP	Open Compute Project	2011	97	◎	
42	ONF	Open Networking Foundation	2011	141	◎	
43	OPEN Alliance SIG	OPEN Alliance special Interest Group	2011	311	Closed	Closed
44	Wi-SUN	Wi-SUN Alliance	2012	91	◎	◎

(continued)

Table 4.1 (continued)

	Consortium	Full name	Foundation	No. of members	RF	RAND
45	FIDO	Fast Identity Online alliance	2012	262	◎	
46	OCC	Open Cloud Connect	2013	18	Closed	
47	AllSeen	AllSeen Alliance	2013	169	◎	
48	OpenDaylight	OpenDaylight Project	2013	50	◎	
49	IIC	Industrial Internet Consortium	2014	238	◎	
50	THREAD	THREAD GROUP	2014	216	◎	
51	OPNFV	Open Platform for NFV	2014	58	◎	
52	AOM	Alliance for Open Media	2015	16	◎	
53	UHD	UHD Alliance	2015	42	Closed	
54	OpenFog	Open Fog Consortium	2015	32	◎	
55	MulteFire	MulteFire Alliance	2015	17	Closed	
56	LoRa	LoRa Alliance	2015	246	◎	
57	WIoTF	Wireless IoT Forum	2015	6	Closed	
58	Hyperledger	Hyperledger Project	2016	81	◎	
59	OCF	Open Connectivity Foundation	2016	179	◎	O
60	TIP	Telecom Infra Project	2016	40	◎	
61	DMTF	Distributed Management Task Force	Dissolution	162	O	O

Notes "◎" demonstrates a clear choice of either RF or RAND, and "O" means choice of both. Though OCF is permitted RAND as well, RF is highly recommended in their IPR policy. *Source* Forum report of Telecommunication Technology Committee (2017)

It must be noted in this survey whether the consortia make licenses for standard essential patents paid or free. In the event that a licensor seeks appropriate royalties, the licensor grants patent use that is RAND (reasonable and non-discrimination) or FRAND (fair, reasonable, and non-discriminatory). IPR policy specifies that patent management be performed by RAND or FRAND. In the DVD Consortium example shown earlier, FRAND is specified.

On the other hand, if the licensor has to give up royalties, it is specified as royalty-free in the IPR policy.

4.5.2   The Actual Situation of IPR Policy

As shown in Table 4.1, the consortium surveyed consisted of 61 institutions. Table 4.1 is organized in chronological order, according to the time when each consortium was created. All IPR policies were subject to the survey. The consortia are solicited through websites used to recruit members, and many consortia have IPR policies on their websites to ensure the new members understand the policy before joining.

In this survey, 44 institutions out of 61 institutions had released IPR policies. Of those 44 institutions, in the case of obviously onerous contracts, or obviously royalty-free, "◎" was attached. However, some consortia had accepted both onerous contracts and royalty-free. In that case, "○" was included to distinguish it from the previous. The Table 4.2 organized these contents.

What we can see from this table is that 31 institutions are royalty-free; that figure is much greater than the seven institutions of RAND. More than half of the 44 organizations that publish IPR policies promote royalty-free.

This trend appears even stronger when limited to the last 10 years. The trend since 2008 is particularly surprising. Since 2008, 20 of 23 organizations (about 87%) that publish IPR policies promote royalty-free.

In other words, we can understand that a paradigm shift of the innovation of IoT related technology had been occurred in the early 2000s. Figure 4.6 shows that there is a significant difference between before 2007 and after 2008 in royalty-free and RAND. This figure shows that royalty free has risen sharply, and the other hand RAND has fallen instantly.

Today, at consortia that promote internationalization and industrialization at the same time, we need to realize that RF is generalized in the innovation process.

Why have such phenomena increased? The cause is considered next.

Table 4.2 Comparison of RF and RAND (1988–2016)

	1988–2016		2008–	
RF	31	70.50%	20	87.00%
RAND	7	15.90%	2	8.70%
RF or RAND	6	13.60%	1	4.30%
Total※	44	100.00%	23	100.00%

Notes The total number excluded N/A

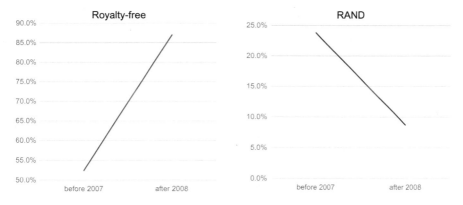

Fig. 4.6 Difference between before 2007 and after 2008

4.6   Onerous Contract or Royalty Free?

It has long been understood that, as a part of a competitive strategy, companies earn royalty income. In many cases, companies held IP included in the technology as a proprietary license. Additionally, in strategies related to technology, great significance was found in licensing revenues, which increase with the spread of a technology. Therefore, in order to further such a spread, it is necessary to expand the use of the technology and for the technology to be used in various fields. As DVDs have been used for a variety of purposes, MNEs first internationalized the technology and then chose a process to make it inter-industrialized.

However, this chapter discusses technological development in simultaneously-realized internationalization and inter-industrialization. In other words, rather than inter-industrializing a technology for a specified use, it is assumed that it will be used by in various industries from the very beginning. In this case, as aforementioned, it is extremely difficult to develop technology for various applications within one MNE. Therefore, standardization of such technology should occur in consortia composed of companies across borders and industries.

The Fig. 4.7 is constructed based on the contents discussed above in this chapter. It shows who led the initiative and how to lead the company in responding to the synchronization of internationalization and inter-industrialization. In other words, it can be said that this figure explains cell D in Fig. 4.3. The horizontal axis shows the subject leading the standardization, and the vertical axis shows whether it is RAND/FRAND or royalty-free.

Therefore, if the licensors do not renounce royalty income, it is positioned as Type I. Type I, in which it is allowable to grant licenses with RAND/FRAND is the most common so far.

From the IPR policy survey, the existence of Type II and III in this figure has become known, the majority being TypeII. Interestingly, Type III can also be seen, in part. In Type II, licensors play a leading role in the operation of a consortium;

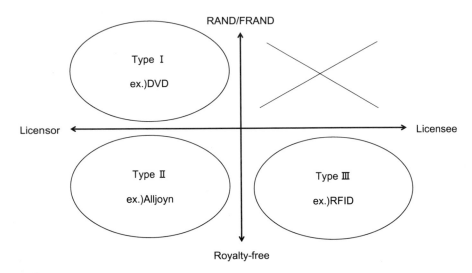

Fig. 4.7 Typology of standardized process

at the same time, they make essential patents royalty-free as part of their strategy. In Type III, the licensee plays the role of actively managing the consortium. That is, the licensor's license is made royalty-free by the licensees. Since Type IV only exists theoretically for the time being, it is not shown in this figure.

Thus, it is necessary to consider the backgrounds of Type I–Type III innovations.

4.6.1 Type I

After 2008, only two consortia, Open ADR Alliance and Wi-SUN Alliance, are classified as Type I. Both are consortia building a mechanism to optimally control the balance of electricity supply and demand. IP management is based on RAND, as members of these two consortia have to rely on the technology of licensors.

In the case of Type I, the most important feature is that the licensor secures the appropriability of IP. In these consortia, it is recognized that standardization is underway on the premise of simultaneous internationalization and inter-industrialization but the scope of the technical application after standardization will be limited.

4.6.2   Type II

In Type II, the licensor proactively advances royalty-free. This is different from the pattern in which the licensor aims at royalty revenue through possession of IP. It is also different from open innovation. Open innovation is a strategy to actively utilize external technologies to revitalize internal technologies. Open innovation theory, however, does not include the scope of utilization of the technology after innovation.

In Type II, it is premised that standard, essential patents are utilized both internationally and inter-industrially. For this reason, members of this type of consortia are gathered from various countries and from various industries. That is why many consortia are created in advance, recognizable as being open and wide in innovation practices.

When considering such a situation, the recent case of Qualcomm serves as great reference. That is because that Qualcomm has begun to choose a new strategy which is different from before.

Qualcomm was founded in San Diego, California in 1985. In the latter half of the 1980s, the FCC (Federal Communications Commission) entrusted technical development of the mobile telecommunication system concept to Qualcomm and it proposed the CDMA (Code Division Multiple Access) method. Although this concept ended up failing, Qualcomm promoted CDMA development for the mobile phone system on the ground.

In the communication standard 3G, the W-CDMA system and the CDMA 2000 system coexisted. However, there was also a plan to integrate with W-CDMA in 3G. Qualcomm held essential patents of each method. Since CDMA 2000 was compatible with CDMA One, if CDMA 2000 disappeared, it would have meant that Qualcomm would lose these royalty revenues at the same time. Therefore, Qualcomm insisted on the coexistence of both standards in 3G, and announced that it would not license essential patents of their W-CDMA if they could not be accepted. This was known as "Qualcomm shock" at the time.

In recent years, Qualcomm's main product is its chip "Snapdragon" which is included in most commercially available smartphones. As such, royalties from smartphone makers have become a huge profit source.

In January 2017, Apple filed a lawsuit against Qualcomm because the royalties were too expensive and stopped royalty payments to Qualcomm, a move that became the subject of much discussion.

In this way, Qualcomm followed the business model that focused on earning extensive profits based on their IP. However, in recent years, Qualcomm has begun to trend downward. In comparison with the previous year, it has had a continued declining trend since 2011. The Fig. 4.8 shows this situation.

However, under such circumstances, Qualcomm is promoting another business model for IoT business. With regard to IoT business, they have changed their core technology to royalty-free. That core technology has created what is called "AllJoyn" software. AllJoyn is a framework developed to make the IoT's

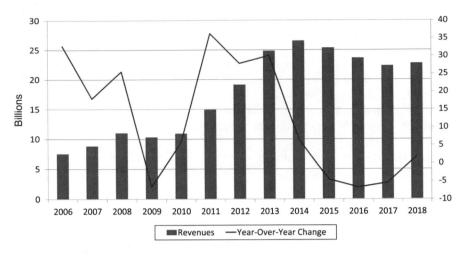

Fig. 4.8 Qualcomm's revenue from fiscal year 2006 to 2018. *Source* Qualcomm's annual report

environment more convenient. AllJoyn, which is currently in widespread use, was announced in December 2016.

Increasing the number of devices that can be connected to the network is said to benefit IoT at home. However, on the other hand, an increase in IoT devices not only brings convenience but also may cause confusion. When home appliances such as a rice cookers, refrigerators, air conditioners, air purifiers, and water heaters support IoT, people need to use different communication methods, operating systems, and types of application software.

In order to handle IoT devices, users must select and operate each one, each time through special smartphone applications, PC software, or operation panels installed at home. Due to the inherent confusion caused by having to remember such a range of information, it is not easy for ordinary people to operate multiple pieces of IoT equipment. On the other hand, there are also many specifications commonly required for IoT devices in different devices.

With this in mind, Qualcomm has provided AllJoyn as an IoT framework to link such common parts (see Fig. 4.9). Qualcomm prepared a library of functions common to various devices, and licensees can develop IoT devices and applications easily through a royalty-free license.

AllJoyn's core technology was developed by Qualcomm. Qualcomm has made this technology free and promoted standardization with a consortium called the OCF (Open Connectivity Foundation). Board members of this consortium include not only US companies like Qualcomm, Intel, Microsoft, and Cisco Systems, but also Samsung, Electrolux, Haier, LG Electronics, and many others. Today, over 300 companies participate from all over the world. This consortium is made up of the leading companies in various countries and industries. This consortium is promoting both internationalization and inter-industrialization at the same time. To tell the truth, the IPR policy of the OCF lists both royalty-free and RAND. There is

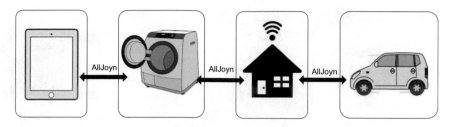

Fig. 4.9 AllJoyn framework. *Note* This figure has been partially revised by author. *Source* "An Open Source project building the framework for the Internet of Things", AllSeen Alliance, https://allseenalliance.org/sites/default/files/pages/files/intro_to_alliance_10.14.15_0.pdf

freedom to admit RAND to a licensor, however, the OCF is basically proceeding with royalty-free.

AllJoyn has three major features: (A) communication without going through the cloud, (B) devices that connect to each other to provide functions, and (C) the ability to connect everything regardless of manufacturer or product differences. Since in general, IoT devices rely on specific clouds, operating systems, and communication systems, the versatility of AllJoyn is revolutionary.

And AllJoyn is composed of three layers, the core of which is the "AllJoyn Core Library." Its function is to find and connect equipment to realize the functionality of AllJoyn, and provide functions such as access control and encryption as an API. Above that layer, there is a function called "AllJoyn Service Framework," which provides AllJoyn's functionality to the devices. Various companies belonging to the Allseen Alliance have played a role in enhancing this service function[1]. Finally, the top layer has a function called the "AllJoyn Application Layer," which defines the user interface (Fig. 4.10).

AllJoyn makes it easy to collaborate and connect AllJoyn compatible devices in an average. It also is helping to disseminate domestic IoT devices as a mechanism to provide services in cooperation with other IoT devices.

Products such as AllJoyn compatible TVs, air purifiers, wireless speakers, and others are appearing one after another. All editions of Windows 10 already work with AllJoyn. The possibility of widespread adoption seems to be all but inevitable.

Qualcomm also developed the "DragonBoard 410c" in 2016, as a tool to develop IoT equipment that can cooperate with AllJoyn. This is an extension board with Qualcomm's Snapdragon 410 processor. With DragonBoard 410c, customers can easily create an environment that supports AllJoyn. In addition, Qualcomm started embedding Snapdragon, which had been included mostly in smartphones, in various devices for AllJoyn.

This Snapdragon was not royalty-free, so the sale was directly linked to Qualcomm's revenue. In other words, Qualcomm made AllJoyn software

---

[1] AllJoyn was standardized at Allseen alliance at first. Later, the consortium was absorbed by the OCF.

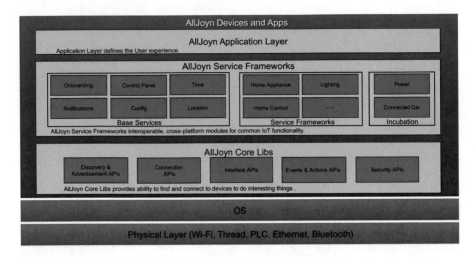

Fig. 4.10 AllJoyn Framework. *Note* This figure has been partially revised by author. *Source* "An Open Source project building the framework for the Internet of Things", AllSeen Alliance, https:// allseenalliance.org/sites/default/files/pages/files/intro_to_alliance_10.14.15_0.pdf

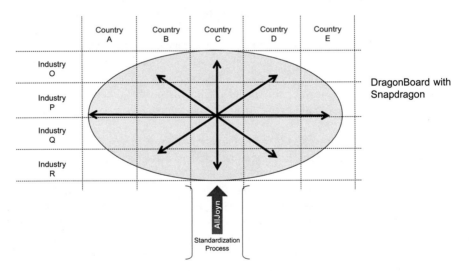

Fig. 4.11 Application Area of AllJoyn

royalty-free to make the IoT environment more convenient to develop international and inter-international standards at the OCF consortium, but it also created a business model to earn revenue from the Snapdragon hardware that was required for the IoT environment. In short, since AllJoyn had been standardized widely, Qualcomm can do their business beyond the border and the industrial barrier with Snapdragon (Fig. 4.11).

This example with Qualcomm demonstrates just how important royalty-free IP is. While royalty-free IP gathers companies from various countries and industries, standardization is promoted at a specific consortium.

This phenomenon is not mentioned in conventional international business studies. It is a new revelation that royalty-free IP can bring new business opportunities in IoT business.

4.6.3   Type III

Despite the fact that there are few Type III cases, in which technology is made royalty-free by the licensee, they are surely increasing (Uchida 2012, 2013). The most familiar example is the Internet communication protocol "http." This was standardized in a consortium called the W3C (World Wide Web Consortium) created in 1994. Looking back at the aforementioned RFID—a newer case than W3C—the standardization of RFID was done by a consortium called GS1 and EPCglobal was once a subordinate organization of GS1. Subsequently, EPCglobal was advancing standardization led by licensees. EPCglobal was created in 2003, and it consisted of about 500 companies from around the world. It was an organization where both internationalization and inter-industrialization progressed simultaneously. Intermec, a company with an important license, was one of the members. Intermec was acquired by Honeywell in 2012.

EPCglobal had adopted an royalty-free policy with respect to essential patents from its inception. Therefore, any member of WG who developed standardization in EPCglobal was able to use the technology free of charge.

However, even if a company became a member of EPCglobal, not everyone could become a member of the WG. In order to participate in the WG, a firm was required to further sign a consent form related to the IPR policy[2]. EPCglobal's most focused standardization was the technology called "ISO/IEC 18000-6 Type C," which was internationally standardized in 2006, hereinafter referred to as an "EPC tag."

Deliberation on the international standardization of RFID was advanced by ISO/IEC JTC 1 SC 31 (Subcommittee 31). EPCglobal proposed the EPC tag to SC 31 in December 2004. At the same time, Japan's Ministry of Economy, Trade and Industry aimed for international standardization of RFID based on Hitachi's technology. Japan suggested Hibiki tags to SC 31 in March 2005, four months after the proposal of the EPC tag. As a result, the Hibiki tag competed completely with the EPC tag. Unfortunately for the Japanese side, members of SC 31 at that time contained many EPCglobal board members.

---

[2]Although it was called "IP policy," to be exact, it is consistent with the content described in the W3C above and is referred to as "patent policy" here. Furthermore, now that it is "GS 1 EPCglobal," agreement with GS 1 IP policy is required.

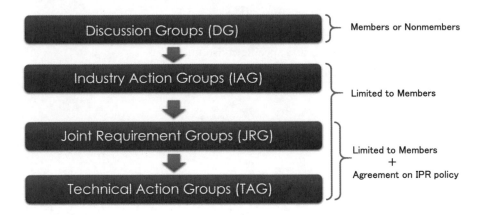

Fig. 4.12 Standardization process of GS1 EPCglobal. *Note* Based on the information provided by EPCglobal and the contents of the interview survey

Moreover, because the deliberation of the EPC tag was already underway in SC 31, the reaction of the committee was extremely cold, as testified by a member who proposed the Hibiki tag at that meeting. In the end, the Hibiki tags could not be endorsed at this conference and never had its potential realized in the same way as the EPC tags. Thus, the Hibiki tag project failed.

The standardization process in EPCglobal is shown in Fig. 4.12. All standardization work is first started with Discussion Groups (DG). These DGs are positioned as a place where volunteers from various industries, as potential future users, gathered and called for standardization.

Interestingly, EPCglobal allows to seek standardization of the new technology as a user, even though it is not a member at this stage. On the other hand, there is also an agreement to exclude allegations from the licensor side at this point. Activities aim at a developing a business model through future standardization, based on existing technologies, has been regulated.

As a result of deliberations at a DG, after the request is approved by EPCglobal, Industry Action Groups (IAG) are created. From this point, participants must be official members of EPCglobal, but not yet required to sign an agreement of the IPR policy. This is because the purpose of IAGs is to let users summarize the requirement specifications approved by a DG. Therefore, work at the WG is not yet advanced.

Until this stage, requirement specifications are summarized for each industry. However, not many requirement specifications are very similar. Therefore, in the Joint Requirement Groups (JRG), concrete discussion on specifications is advanced so that they could be arranged and unified, and further utilized beyond industry. Therefore, from the stage of JRG, members need to agree to IPR policy. After that, Technical Action Groups (TAGs) develop the required specifications.

From this example of the standardization process at EPCglobal, we see that many requests on the licensee side are adopted. In this sense, it closely resembles

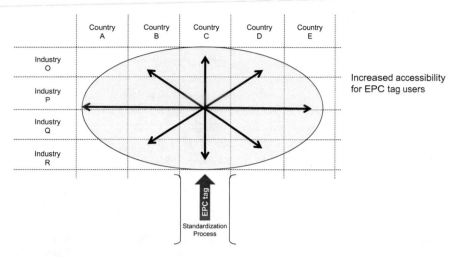

Fig. 4.13 Application area of EPC tag

the W3C. When standard development activities proceed that reflect the licensee's requests, it can be confirmed that it is a licensee-led standardization process (Fig. 4.13).

4.7 How About Technology Areas for the Future?

In this section, we looked at consortia that have standardized technology based on both internationalization and inter-industrialization.

At consortia that promote simultaneous internationalization and inter-industrialization, it is found that royalty-free is overwhelmingly selected instead of RAND/FRAND. Type II and Type III in the Fig. 4.7 correspond to this.

Regardless of which type is selected, in order to disseminate a standardized technology, it is vital to acquire many adopters from a wide range of international and inter-industrial environments. At first, may be difficult to specify where a standardized technology will be utilized and how it will be utilized; therefore, a royalty-free IP strategy is indispensable. For open and wide innovation, there are many cases where IP is free.

In that sense, Type I is a different innovation process from Type II and III. This is because the location and use of developed technology in Type I are within a range assumed by the licensor beforehand.

Based on the above considerations, it is worth considering the business areas where further expansion of the market is expected. One promising field in the future is the next generation of cars: currently, automatic driving technology is being developed all over the world and companies are competing to realize the "CASE" functions possessed by future cars. CASE stands for Connectivity, Autonomy,

Shared, and Electric. This word itself was coined in the medium-term management plan announced by Daimler's CEO Dieter Zetsche at the 2016 Paris Motor Show.

In the development process of these standard essential patents for new fields of the future, we cannot confirm which will lead to royalty-free IP at the present time. For example, a consortium called the CHAdeMO Council was created in 2010 to set battery standard of electric vehicles (EVs), and it possesses technologies certified as international standards by IEC in 2014. This is one of the big consortia on EVs composed of members from many countries and various industries and both RAND and royalty-free are accepted in its IPR policy. If both RAND and royalty-free are accepted, it is easy to imagine that few companies will choose royalty-free. This indicates that the CHAdeMO Council is qualitatively different from Type II and III innovators.

This type of collaboration applies to the automotive semiconductor market, as well. In this market, Intel and NVIDIA are developing technologies, aiming at de facto standards, but there is no movement to make their IP free. Even in the development of a new technology, it depends on whether the post-development technology will correspond to both internationalization and inter-industrialization. Advanced technologies are being developed in next-generation automobiles at the moment; however, the uses of these various technologies seem to be limited to the automobile market. In other words, if the scope of the application range after standardization is limited to some extent, it is difficult to make IP free.

4.8 Conclusion

In IoT business strategy, there are clear differences among strategies related to international business. In existing research, technical standards are recognized as a strategic tool to establish a competitive advantage for a specified industry. However, it turns out that a technology embedded in a standard should not be exclusive; it should be utilized as technology that can used for any purpose in IoT business.

Technology standards in the IoT field have been positioned as ways to create new markets. In order to create new markets for IoT, consortia to promote standardization have opened their entrance widely to deal with the myriad possibilities and securely acquire users by guaranteeing royalty-free to members belonging to the consortia.

Despite the fact that IP has been regarded as a powerful revenue source in the past for international businesses, setting another revenue source, and aiming for an international standard on IoT is increasingly occurring. In other words, instead of aiming to earn from a particular technology, constructing a business model that earns revenue in cooperation with a licensee is required (Ogawa 2015; Tatsumoto 2017). Such a strategy has been rapidly generated since the late 2000s. In this chapter's view, the paradigm shift of IoT field was occurred at that time. And, in the competition related to IoT, those strategies will become more widespread in the future.

In particular, in the case of technology standards required for IoT, we need to recognize the apparent difference in the innovation process between Type C and Type D, as shown in Fig. 4.3 above. Although not a traditional international business strategy, an international business consortium of inter-industrial business will be required. The key to international business in the future is to recognize the essence of inter-industrial business. Also, the most important factor for MNEs in the near future is to deal with "inter-industrialization of international business".

Acknowledgements   This work was supported by JSPS KAKENHI Grant Numbers JP17K03973, JP18K01833, JP18K00883.

References

Arai, M., & Uchida, Y. (2012). Pitfall of the international standardization process: The consensus-based standard in the Japanese manufacturing industry. *International Journal of Business Research, Academy of International Business and Economics, 12*(2), 23–43.

Armann, E., & Cantwell, J. (Eds.). (2012). *Innovative firms in emerging market countries*. Oxford. Oxford University Press.

Arthur, B. (1994). *Increasing Returns and Path Dependence in the Economy*. Ann Arbor. University of Michigan Press.

Arthur, B. (1996, July–August). Increasing returns and the new world of business. *Harvard Business Review*, 100–109.

Bartlett, C. A., & Ghoshal, S. (1989). Managing across borders: The transnational solution. Harvard Business School Press.

Besen, S. M., & Farrell, J. (1994). Choosing how to compete: Strategies and tactics in standardization. *Journal of Economic Perspectives, 8*(2), 117–131.

Burg, U., & Kenney, M. (2003). Sponsors, communities and standards: Ethernet vs. token ring in the local area networking business. *Industry and Innovation, 10*(4), 351–374.

Cantwell, J., & Iammarino, S. (2003). *Multinational corporations and european regional systems of innovation*. London. Routledge.

Cantwell, J., & Molero, J. (Eds.). (2003). *Multinational enterprises, Innovative Strategies and Systems of Innovation*. Edward Elgar.

Cantwell, J., Dunning, J., & Lundan, S. (2010). An Evolutionary Approach to Understanding International Business Activity: The Co-evolution of MNEs and the Institutional Environment, *Journal of International Business Studies*, 41.

Cargill, C. F. (1989). *Information technology standardization: Theory, Process and Organizations*. Digital Press Newton.

Chesbrough, H. (2001). Assembling The Elephant: A review of empirical studies on the impact of technical change upon incumbent firm. In R. A. Burgelman & H. Chesbrough (Eds.), *Comparative studies of technological evolution*. Emerald Publishing Limited.

Chesbrough, H. (2003). *Open innovation: The new imperative for creating and profiting from technology*. Boston. Harvard Business Press.

Chesbrough, H. (2006). *Open business models*. Boston. Harvard Business School Publishing Corp.

Christensen, C. (1997). *The Innovator's Dilemma: When New Technologies Cause Great Firms to Fail*, Harvard Business Review Press.

Christensen, C., Suarez, F., & Utterback, M. (1998). Strategies for survival in fast-changing industries. *Management Science, 44*(12-part-2), 207–220.

Doi, N. (2001) Gijutsu Hyojun to Kyoso (Technology Standard and Competition), Nihonkeizaihyoronsya. (in Japanese).

Farrell, J., & Saloner, G. (1986). Standardization, compatibility, and innovation. *RAND Journal of Economics, 16,* 70–83.

Gawer, A., & Cusumano, M. (2002). *Platform leadership: How intel, microsoft, and cisco drive industry innovation.* Harvard Business School Press.

Hippel, E. V. (2016). *Free Innovation,* The MIT Press.

Jakobs, K. (Ed.). (2000). *Information technology standards and standardization: A global perspective.* Idea Group Publishing.

Jakobs, K. (Ed.). (2005). *Advanced topics in information technology standards and standardization research* (Vol. 1). Idea Group Publishing.

Kajiura, M. (2010). The strategic consortia movement in standardization. *International Journal of Manufacturing and Management, 21*(3/4), 324–339.

Kajiura, M. (2012). Open innovation of consensus standard: cases of business model creation in ICT. *International Journal of Enterprise Network Management, 5*(2), 126–143.

Kajiura, M. (2013). *ICT Consensus based standard Bunshindo.* (in Japanese).

Katz, M. L., & Shapiro, C. (1985). Network externalities, competition, and compatibility. *The American Economic Review, 75,* 424–440.

Krechmer, K. (2006). Open standards requirements. In J. Kai (Ed.). *Advanced topics in information technology standards and standardization research,* 27–48.

Ogawa, K. (2009). *International & business strategy.* Hakutosyobou. (in Japanese).

Ogawa, K. (2015). *Open & close strategy.* Shoueisya (in Japanese).

Porter, M., & Heppelmann, J. (2014, November). How smart, connected products are transforming competition. *Harvard Business Review, 92*(11), 64–88.

Sakakibara, K. (2005). Innovation no Shuekika (Profit Process of Innovation), Yuhikaku. (in Japanese).

Sakakibara, K., & Koyama (Eds.). (2006). Innovation to Kyoso Yui (Innovation and Competitive Advantage), NTT Shuppan. (in Japanese).

Shapiro, C., & Varian, H. R. (1998). *Information rules: A strategic guide to the network economy.* Boston. Harvard Business School Press.

Shapiro, C. (2000). Navigating the patent thicket: Cross licenses, patent pools, and standard-setting. *Working Paper No CPC00–11,* University of California at Berkeley.

Schilling, M. A. (2002). Technology Success and Failure in Winner-Take-All Markets: The Impact of Learning Orientation, Timing, and Network Externalities, *Academy of Management Journal, 45.* 387–398.

Shintaku, J., Konomi, Y., & Shibata, T. (Eds.). (2000). de facto standard no honshitsu (Nature of de facto standard), Yuhikaku. (in Japanese).

Shintaku, J., & Eto, M. (Eds.). (2008). Consensus Hyojun senryaku (Consensus BassedStarategy), Nihonkeizaishinbunsya.

Simcoe, T. S. (2006). Open standard and intellectual property rights. In H. Chesbrough, W. Venheaverbeke, & J. West (Eds.), *Open innovation, researching a new paradigm.* Oxford University Press.

Suarez, F., & Lanzolla, G. (2005). The half-truth of first mover advantage. *Harvard Business Review, 83*(4), 121–127.

Takeda, S. (1998). Takokusekikigyo to senryakuteikei (MNEs and Staretegic Alliance), Bunshindo. (in Japanese).

Takeda, S., Uchida, Y., & Kajiura, M. (2001). Kokusaihyojun to senryakuteikei (International Standard and Strategic Alliance), Chuokeizaisha. (in Japanese).

Tatsumoto, H. (2011). Global standard, consensus hyojunka to kokusai bungyo (Global standard, consensus based standard and international division of labor). *Kokusai Business Kenkyu, 3*(2), 81–97. (in Japanese).

Tatsumoto, H. (2017). Platform kigyo no global senryaku (Global Strategy of Platform Companies), Yuhikaku (in Japanese).

Teece, D.J.(2009). *Dynamic Capabilities and Strategic Management: Organizing for Innovation and Growth.* Oxford. Oxford University Press.

TTC. (2017). Johotsushinkankei no forum katsudo ni kansuru chosahokokusyo (Survey report on information and communication related forum activities). *Telecommunication Technology Committee.* http://www.ttc.or.jp/files/8114/8592/9731/forum_report_v23_2017-1.pdf.

Uchida, Y. (2007). Hyojun no ruikeika to open policy ni motoduku Hyojunka no senryaku (Standardization strategy based on open policy and typology of standard). In M. Kajiura (Eds.), *Kokusai Hyojun to Gijutsu Hyojun* (pp. 52–93). (in Japanese).

Uchida, Y. (2008). de jure standard no tsuikyu to senryakuteki kadai (Investigation of de jure standard and strategic issue). *Sekai Keizai Hyoron, 52*(8), 22–32 (in Japanese).

Uchida, Y. (2012). User shudo no hyoujunka process to royalty free: kokusai hyoujunka ni muketa aratana process ga motarasu senryakuteki imi (User-driven standardization process and royalty free: Strategic implications of the new process towards international standardization). *Kokusai Business Kenkyu, 4*(2), 99–113. (in Japanese).

Uchida, Y. (2013). The Process of International Standardization and Royalty Free. *Journal of International Business and Economics, Academy of International Business and Economics, 13* (2), 153–160.

Uchida, Y. (2015). The relationship between technology and diffusion process. *International Journal of Business and Economics, 15*(2), 87–94.

Uchida, Y. (2016). IoT no Shintento Kokusai Business no Kankei ni tsuite: Gijutsu Hyojun no Gyosaika eno torikumi wo chushin ni, *Working Paper, 364*, pp.1–13, University of Toyama. (in Japanese).

Yamada, H. (1993). Kyoso Yui no Kikaku Senryaku (Competitive Advantage of de facto standard), Daiamondsha. (in Japanese).

Yamada, H. (2004). Gyakuten no Kyoso Senryaku (Reverse Competitive Strategy), Seisanseishuppan. (in Japanese).

Yasumoto, M., & Manabe, S. (2017). Openka senryaku: Kyokai wo koeru Innovation (Open Strategy: Innovation Across Boundaries), Yuhikaku. (in Japanese).

Warne, A. G. (2005). Block alliances and the formation of standards in the ITC industry. In J. Kai (Ed.), *Advanced topics in information technology standards and standardization research* (pp. 50–70). Idea Group Publishing.

Wegberg, M. V. (2006). Standardization and competing consortia: The trade-off between speed and compatibility. *Journal of IT Standards & Standardization Research, 2*(2), 18–33.

Yasuro Uchida (Ph.D, Yokohama National University) is Professor of Strategic Management at Graduate School of Business, University of Hyogo, Japan and Professor Emeritus of University of Toyama. He was a member of technology standard council of Ministry of Economy, Trade and Industry of Japan. His research interest is competitive strategy, and the international standardization of technology. His book "International Standard and Strategic Alliance" (in Japanese; Chuokeizaisha, Tokyo 2001) received the award from Japan Academy for International Trade and Business (JAFTAB) in 2001.

Chapter 5
New Roles for Japanese Companies at the Knowledge-Based Economy

Adaptation to Newly Emerging Techno-Economic Paradigm

Fumio Komoda

Abstract Information technology (IT) lies at the heart of today's techno-economic paradigm. In the latter half of the twentieth century, Japanese companies succeeded in achieving high international competitiveness in the home appliance and industrial electronics markets by making technological progress in various hardware, including microfabrication technologies. However, as symbolized by the emergence of the term "knowledge-based economy," the relative importance of software, rather than hardware, in IT began to increase during the transition from the twentieth to 21st century. Successful adaptation to this change in techno-economic paradigm is essential to maintaining international competitiveness. However, Japanese companies did not succeed in shifting their R&D focus to software. This failure to adapt in a timely manner is considered one of the important reasons for Japanese companies' prolonged stagnation. This paper draws on patent and magazine article data to verify that the core of techno-economic paradigm is changing from hardware to software and that Japanese companies failed to adequately adapt to this change.

Keywords Techno-economic paradigm · Software · Knowledge based economy · Software driven economy · Japanese company

5.1 Introduction

Throughout history, technological innovation has served as the engine of economic growth. The emergence of new technologies over the past 200 years, such as steam engines, textile machinery, steel, inorganic chemistry, internal combustion engines, power generation, communications, petrochemicals, semiconductors, and computers,

F. Komoda (✉)
Honorary Professor, Saitama University, Saitama, Japan
e-mail: techtra@ae.auone-net.jp

© Springer Nature Singapore Pte Ltd. 2019
J. Cantwell and T. Hayashi (eds.), *Paradigm Shift in Technologies and Innovation Systems*, https://doi.org/10.1007/978-981-32-9350-2_5

has been a source of capitalism's dynamism. Undeniably, the timing of occurrence of technological innovations biased in time is related to the business cycle (long-term) as well as a shift in the central country driving technological innovation and thus to a change in hegemon countries in the world political economy. One of the most powerful models explaining the relationship between the economy and technology is the techno-economic paradigm (TEP) theory.

Information Technology (IT) has undoubtedly lain at the core of TEP since the last quarter of the twentieth century. Its penetration into almost all businesses activities, households, and social infrastructure offers a paradigm for economic growth. For example, the US led the development of computer, electronics, semiconductors, and other technologies and enjoyed a position at the center of the world economy; similarly, Japanese companies that focused on developing these technologies enjoyed favorable results.

IT has advanced rapidly, leading to ongoing changes to its nature. In particular, the emergence and diffusion of distributed processing system called the Internet, where computers are used to form and access networks rather than stand-alone workstations, has dramatically changed the character of IT. Advances in IT are still supported by progress in electronics technology as hardware, but as networks evolve, the relative importance of software has increased. Moreover, the maturity of industrialized countries' economies has seen a leveling off of consumer expenditure on products; the increasing emphasis on services rather than products will further increase the importance of software in the future. These trends mean that software now accounts for a larger share of IT than hardware, and technological progress in software will be faster than in hardware. This has increased productivity of enterprises, created new customer needs, and offered a new source of economic growth. Moreover, the country/region where this innovation occurs will be the hegemon of the world's political economy. The strong performance of US companies stemmed from the fact that they successfully adapted to new realities spawned by the evolution of IT technologies. Conversely, the sluggishness of Japanese companies is largely attributable to their failure to grasp the importance of software.

This paper aims to discuss the process that placed IT, particularly software, at the heart of today's TEP. Also, the author will argue that insufficient adaptation to this new TEP was one cause of Japanese companies' stagnation. For this purpose, a variety of databases have been consulted.

5.2   Evolution of TEP Theory

5.2.1   TEP Theory

J. Schumpeter highlighted the link between technology and the economy, emphasizing that the development of new technology and its dissemination throughout society will promote economic growth (Schumpeter 1912). Neo-Schumpeterians took a more elaborate view of the relationship between technology and economic growth, introducing the concept of TEP, where uneven occurrence of technological innovations appears both temporally and geographically.

TEP was proposed by Dosi (1982) and others, inspired by the "science paradigm" proposed by Kuhn (1962). Kuhn stated that scientific progress is achieved within the scientific research paradigm prevailing in each era (Kuhn 1962). Dosi also described technology in the context of a dominant paradigm in each era, which he named the "technological paradigm." Generally, incremental innovation occurs along the technological trajectory within a technological paradigm. In contrast, radical and innovative technologies that are not within the frame of technological paradigm emerge accidentally, resulting in a completely new technical paradigm. Thus, Dosi argues that technological progress proceeds both continuously and discontinuously.

Many researchers adopted the technological paradigm concept, refining it into the concept of TEP. This reflects a perception that the dominant paradigm related to innovation is also affected by various economic and social conditions.

Dosi's pioneering argument was theoretical, not an empirical study on the current TEP. In response, many researchers, including Dosi himself, sought to elucidate the characteristics of today's TEP. One such study was carried out by Perez (1985), who argued that pattern of economic growth was transitioning from a mass-production system based on low-cost oil to a flexible production system based on low-cost electronics and digital telecommunication. Thus, cheap microelectronic chips can be said to have enabled the new TEP. Perez further argued that the IC chips as hardware led to a shift from a small-range, high-volume production system to an information-intensive, high-mix, low-volume production system.

Perez mainly focuses on IT hardware. Of course, Perez also refers to the importance of software, describing how "in product engineering, there would be a tendency to redesign existing goods to make them smaller, less energy consuming, with less moving parts, more electronics and more software" (Perez 1985, p. 447), but the main focus is on IC chips and digital communication systems as hardware. This reflects the fact hardware was at the core of the technical system at that time.

5.2.2 Knowledge-Based Economy, Data-Driven Economy

As early papers including Dosi and Perez argue, there is no doubt that IT is at the core of today's TEP. In companies, factory automation started around 1960, followed by automation of routine clerical work such as accounting. Computers have continued to further penetrating work practices even in office work including sales, which is difficult to computerize given its complicated and atypical nature. Furthermore, "potential fusion between formerly unrelated technologies through ICT" has begun to provide new opportunity for innovation (Cantwell and Santangelo 2000, p. 132). Households also are filled with various electronics products including televisions, audio equipment, refrigerators, and air conditioners. The management of social infrastructure such as public transportation, electric power, urban buildings etc., is also no longer possible without IT (Hayashi 1995, 2007). Many TEP theories appeared to reflect such changes (Drechsler et al. 2009; Freeman 1987, 2009; Perez 2004).

However, the actual situation began to change from the twentieth to 21st century, with software becoming increasingly important compared with hardware. This is reflected in the emergence of new concepts such as "knowledge-based economy" and "data-driven economy."

The knowledge-based economy concept was developed by the OECD and other researchers. A few such researches, such as those of D. Bell and A. Toffler, which stress the importance of new technologies and knowledge obtained from R&D activities in economic growth, have already gained significant influence. The concept of a knowledge-based economy has emerged as an extension of this line of research. For example, by claiming that "OECD economies are more strongly dependent on the production, distribution, and use of knowledge than before" (p. 9), OECD (1996) argues that technologies embodied in the human capital play an important role in the economic growth of modern industrialized countries. As a result, economic theory, which had formerly focused on capital and labor as factors of production, is necessarily increasingly emphasizing the role played by technology. Therefore, the OECD states that government's policy should focus on (1) enhancing knowledge diffusion, (2) upgrading human capital, and (3) promoting organization change.

The knowledge-based economy concept emerged in part due to the economic stagnation of European countries. The OECD came up with the recognition that the slowdown in the European economy was caused by the slowdown in R&D.

However, the explosive expansion of the Internet since the beginning of the 21st century is further changing the meaning of knowledge-based economy. The digitization of text, image, and audio data has created and accumulated an unprecedented amount of data, known as "big data." Analyzing these data makes it possible to obtain valuable knowledge that could not be obtained previously. These new vast amounts of data comprise the infrastructure of a knowledge-based economy. Reflecting such a reality, new concepts such as "data-driven economy" and "data-driven innovation" have emerged. For example, OECD (2014) emphasizes the critical importance of data, stating that "in the current context of a weak global recovery, with lingering high unemployment in major advanced economies, governments are looking for *new sources of growth* to boost the productivity and competitiveness of their economies and industries, to generate jobs and promote the well-being of their citizens" (p. 9). The OECD further states that "techniques and technologies for processing and analyzing large volumes of data, which are commonly known as "big data," are becoming an important resource that lead to new knowledge, drive value creation, and foster new products, processes and markets" (p. 4).

In this way, the idea of a data-driven economy emphasizing the fact that big data is creating the foundation for future economic development is beginning to be widely accepted. For example, Cavanillas et al. (2016) assert that analyzing big data will promote economic growth and competitiveness, arguing that "data has become a new factor of production, in the same way as hard assets and human capital"

(p. 4). Hardware serves as a sensor to gather this data in large quantities in real time or non-real time. More important, however, to analyze and gain valuable knowledge from these data, software is required. For example, according to Lyko et al. (2016), the large amount of big data collected is analyzed by big data analytics algorithms.

The fact that the center of TEP is shifting from hardware such as IC chips to software is understood by looking at the top 10 global market capitalization. In 1992, GE, NTT, and AT & T essentially hardware-oriented firms, were among the top 10. In 2006, the top 10 included GE, Microsoft, and AT & T. By 2017, the top 10 was mainly composed of software companies such as Apple, Alphabet, Microsoft, Amazon, and FaceBook (the top 5), along with China's Tencent and Alibaba within the top ten. It can be seen that TEP has shifted from an energy- and materials industry-led to a processing and assembly industry-led type, and has also shifted to an IT-led type, with software companies emerging as leaders in IT.

This review has demonstrated how software has become an important, profitable growth area.

5.2.3 Increasing Importance of Software in Economic Growth

The emergence of new concepts such as knowledge-based economy and data-driven economy reflects the fact that the core of IT is shifting from hardware to software.

Although the source of corporate profits remains new knowledge and technology, the relationship between knowledge/technology as the source of earnings and products has gradually weakened. In the past, knowledge/technology were directly connected with products, typically shown in quality and functional improvements. For example, not only is the importance of knowledge in industries like financial services unrelated to the "monozukuri" (manufacturing) industry increasing, but also the competitiveness of products/apparatuses is increasingly determined by added value such as design, sensibility value, after-sales services, etc., which are gained from knowledge rather than functions or quality. Consequently, as Fig. 5.1 shows, the growth of both companies and macro economies has become dependent on whether or not the data generates knowledge and in turn harnessing that knowledge to generate data.

Software plays an important role in this cycle. Not only data analysis software, as shown in Fig. 5.1, but also software for data collection and communications can be classified as software indispensable for goods circulation. This is attained by the development of artificial intelligence (AI) including machine learning.

This cycle requires sophisticated software to process vast amounts of data. AI will further change the nature of the software-centered TEP and have a major impact on the pattern of economic growth. It has frequently been predicted that AI

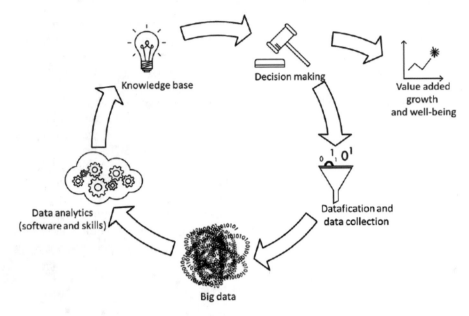

Knowledge base

Decision making

Value added growth and well-being

Datafication and data collection

Big data

Data analytics (software and skills)

Fig. 5.1 The data value cycle *Source* OECD (2014), p. 23

will spread. AI has generally considered to have experienced three booms. The first boom in the late 1950s and 1960s centered on the possibility of machine translation based on the establishment of "search" and "inference" technologies, but petered out due to the great difficulty in this task. The second boom in the 1980s stemmed from advances in knowledge-based algorithms, but the difficulty in teaching computers to understand human knowledge led to another downturn. The third boom began in 2000, when the advent of "machine learning" created the possibility that computers could generate knowledge themselves. In particular, in the 2010s, the potential of "deep learning" technologies increased dramatically, and the boom is further expanding now.

There are two conflicting views as to whether artificial intelligence will evolve rapidly in the future, will permeate industrial activities, civic life, social infrastructure, medical care and the like, and will significantly change society. In order to understand the future trends of AI, the number of articles on Nikkei BP and patents published by Japan Patent Office that include the term "AI" has been examined. These results are shown in Table 5.1. The fact that almost no such articles and patents were found before 2000 is consistent with the fact the disappearance of the first and second booms, and the fact that they began to increase around 2000 is consistent with the beginning of the third boom. Since then, the number of Nikkei BP articles temporarily declined around 2010 before increasing again around 2013. This is consistent with the fact that the significance of deep learning technology become widely known around that time, and AI began to have a

Table 5.1 Trends in the number of journal articles and patents on AI

	Before 1990	91	92	93	94	95	96	97	98	99	2000	01	02	03	04	05	06	07	08	09	10	11	12	13	14	15	16	17
Nikkei BP	1	0	0	0	0	0	0	0	0	0	26	48	28	26	29	39	21	12	15	16	10	18	21	67	99	303	772	555
Patent	1	9	17	20	11	4	6	7	8	12	19	22	29	22	29	31	37	26	22	17	9	14	10	23	25	45	55	28

Source Nikkei BP, Japan Patent Office

full-fledged impact on society. Moreover, the gradual increase in the number of patents since around 2010 seems to indicate a technical backing for the third boom toward the commercialization of AI.

The impact of AIs can largely be attributed to their versatility. According to Cockburn et al. (2017), deep learning, which began to evolve rapidly around 2009, differs from conventional intelligent software. Most conventional intelligent software was developed and utilized for specific narrow applications such as industrial robots. In contrast, deep learning can be used for a range of applications in a variety of dynamic and evolving fields, including prediction of potential compounds for new drugs and automated driving systems. That is, deep learning can be considered "general-purpose technology."

In order to understand the general-purpose nature of AI, companies that have introduced AI in Japan are examined by industry. Of these, the "living-related service industry/entertainment industry," the industry that most commonly introduced AI, accounts for 13.6%. This is followed by "finance and insurance" (7.9%), "information and communications" (4.7%), "electricity, gas, heat supply and water" (4.2%), and "manufacturing" (3.5%) (Center for Global Communication, International University of Japan 2018, p. 116). However, this only shows the companies that had adopted the system at the time of the survey. Future trends in AI and its impact on the economy as a whole need to be taken into account. Frequency of publication in journals serves as one of the key indicators for understanding the industry sectors in which AI is expected to be introduced in the future.

Accordingly, the full text of AI-related articles contained in Nikkei BP, a database provided by the Nihon Keizai Shimbun (a Japanese financial newspaper), and AI-related patent gazettes contained in the patent database provided by Japan Patent Office are examined to find industrial fields and applications introducing AI using the appearance frequency of terms as a criterion. The results are shown in Tables 5.2 and 5.3.

According to this survey of mentions, the number of robot-related mentions is large in both tables. It can be seen that the robot business is regarded as the largest application field of AI. Automated driving is the second most common. In addition, the term is often mentioned in relation to "energy" and "electricity" as social infrastructure, "factory" as manufacturing, "electrical," "home appliances" and "information communication system" as IT, and "education." Although many cases concerning medical-related terms are found, these are characterized by a large diversity of terms rather than that. This may show that multifaceted use in the medical field is expected.

Compared with Nikkei BP magazines, the difference in frequency of occurrence between robotics and other fields of use is not large for patent publications. It is interesting to note that Nikkei BP has a large number of "automatic driving" but not many in the patent gazette, with only three in the latest 2011–2017. There is no doubt that the rapid development of automated driving technology is proceeding, but the analytical result the automatic operations are mentioned much more frequently in the Nikkei BP seems to exaggerate reality.

Table 5.2   Application of AI (Number of Cases in Nikkei BP)

Robot		Power saving 27		Clinical information **2**	
Robot	696	Manufacturing		Service	
Industrial robot	90	Manufacturing	205	Information service	25
Humanoid robot	65	Factory	303	Big data	467
Care robot	11	Factory automation	3	Secretary	18
Service robot	25	Production line	71	Business efficiency	41
Communication robot	17	Electric, IT		Sales, market	
Conversational robot	S	Consular electronics	189	Customer	566
Hospice robot	5	Information telecommunication	23	Marketing	246
Self-supporting robot	1	Satellite communication	1	Needs	284
Automobile		Crowd	324	Customer needs	19
Automatic driving	376	Medical		Consumer needs	8
Driving assistance	59	Medical	213	Customer satisfaction	3
Transportation, logistics		Diagnosis	81	Product development	80
Transportation	19	Care	83	Education	
Traffic	76	Health	87	Education	164
Railway	48	Biological	19	Education industry	3
Logistics	81	Patient information	7	Finance	
Electronic commerce	72	Clinical	13	FinTech	6
Warehouse management	5	Regenerative medic	11	Securities	58
Social infrastructure		Medical equipment	55	Stock price	80
Social infrastructure	40	Electronic medical records	18	Mutual fund	24
Energy	134	Health care	13		
Electricity	127	Medical image	8		

Note Terms are translated from Japanese into English
Source Nikkei BP

This survey of publication mentions further underscores that, AI can be considered a general-purpose technology that will further change the character of today's TEP.

5.2.4   Software Positioning in TEP

As shown above, software is becoming increasingly important in numerous spheres. For example, as the term "bioinformatics" implies, progress in biotechnology has become impossible without IT.

Table 5.3 Application of AI (Number of cases in Japan Patent Office)

	1991–2000	2001–2010	2011–2017	Total		1991–2000	2001–2010	2011–2017	Total
Robot					Crowd	0	0	40	40
Robot	8	26	51	85	Medical				
Industrial robot	0	4	1	5	Medical	5	20	25	50
Conversational robot	0	0	3	3	Biological	1	19	17	37
Humanoid robot	0	1	1	2	Health	4	14	19	37
Automobile					Clinical	1	14	12	27
Automatic driving	1	3	3	7	Health management	2	3	6	11
Driving assistance	4	0	1	5	Patient information	1	2	5	8
Transportation, logistics					Medical equipment	0	2	7	9
Transportation	3	12	6	21	Clinical information	1	1	3	5
Traffic	5	8	9	22	Care	0	2	5	7
Railway	3	2	2	7	Diagnosis	15	23	30	68
Logistics	1	2	0	3	Electronic medical records	0	0	9	9
Electronic commerce	1	4	2	7	Service				
Warehouse management	0	0	1	1	Big data	0	0	9	9
Social infrastructure					Secretary	0	0		2
Social infrastructure	0	1	2	3	Sales, market				
Electricity	8	30	30	68	Customer	12	28	15	65
Energy	8	26	36	70	Marketing	1	5	3	9
Power saving	0	2	2	4	Advertising	4	19	14	37
Manufacturing					Needs	5	9	24	38
Manufacturing	0	2	0	2	Customer needs	0	1	0	1
Factory	10	8	7	25	Education				
Production line	2	0	1	3	Education	3	151	15	33
Electric, IT					Finance				
Consumer electronics	1	11	7	19	Securities	1	3	0	4
Information telecommunication	6	26	19	51	Stock price	0	1	3	4
Satellite communication	1	14	3	18	Mutual fund	0	0	7	7

Note Terms are translated from Japanese into English
Source Japan Patent Office

Needless to say, today's technology architecture does not consist solely of IT. New state-of-the-art technologies, such as biotechnology and nanotechnology, are creating new industries and will support economic development in future. Biotechnologies underpinned by constant advances in genomics including genome editing are spreading in earnest into medicine, agriculture, and other industries.

Moreover, although IT does not unidirectionally support other technological advances, advances in IT have been made possible by a number of technological advances, including in material chemistry. For example, nano-level fabrication technologies have enabled miniaturization of IC chips, enable quantum devices, and pave the way for future quantum computers. Conversely, nano-level microfabrication is impossible without computer control technology. Thus, it can be argued that IT advances cannot be made without advances in nanotechnology.

These facts show that not all technologies exist in isolation from other technologies. Rather, all technologies are interdependent and complementary. Moreover, this fusion continues to evolve constantly. For example, using R&D expenditure statistics, Kodama (1992) analyzed the actual state of fusion among technologies, clarifying that all technologies are mutually complementary and their mutual relationships are getting becoming increasingly closer.

Thus, technology is interrelated. However, IT plays the most important role in connecting to all technical fields and in changing the technology architecture as a whole. IT is at the center of technological linkages, and the importance of software within IT is growing, while IT is at the heart of today's TEP.

5.3 Position of Software in IT

5.3.1 Definitions of Software and IT Classifications

As described above, software is becoming increasingly important. However, it is not easy to define exactly what software comprises. It is clear, however, that hardware lies outside the software domain. However, even if software is viewed as a non-physical intangible source code, it often appears as a physical tangible object when used or bought or sold. Thus, source code is often viewed as hardware. In fact, statistically, the sales value of software is often expressed as sales of devices such as hard disks and servers. In addition, the CPU/MPU, commonly considered as hardware, contains a sophisticated algorithm, and most of the value of a CPU/MPU is attributed to this algorithm. Because of the difficulty in developing this algorithm, Intel dominates the MPU market for PCs and Qualcom dominates the MPU market for smart phones and other mobile devices. Therefore, CPU/MPUs may not be simply viewed as hardware.

In addition, content such as music data is often discussed as software, but of course is not included within the scope of software in the context of this article. Also, in keeping with the increasingly "data-driven economy," big data are analyzed by software to generate tremendous amounts of knowledge and added value. Data are subject to analysis by software, just like computer contents. Nonetheless, it is also true that data are not separated from software because high quantities of high-quality data are needed to obtain the necessary knowledge. In the sense that program intelligence is made possible by repeating analysis of data, software and data are inseparable. Furthermore, in today's computer systems, which are based on the principle of the Turing machine, programs are also designed as data and so can be seen both data and program depending on point of view.

Thus, it is in fact difficult to clearly distinguish between software and hardware, and it is also difficult to separate the software from the data. Therefore, as shown in Fig. 5.1, this paper considers "data," "narrowly defined software (software in the true meanings)," "technologies in fusion area of software and hardware," and "narrowly defined hardware." Komoda et al. (1996, 1997), Komoda (2000) "Technologies in the fusion area of hardware and software" include technology with strong software properties as well as technology with strong hardware properties. In general, when discussing software, "narrowly defined software" is the focus, and in many cases "technologies in the fusion area of hardware and software" and "narrowly defined hardware" are considered hardware. Narrowly defined software technology is represented by algorithms, which are defined as procedures for solving problems. To resolve a problem, the procedure must be unambiguous and clearly defined. Of course, there is not only one procedure, but a range, among which the most efficient one is adopted and serves as the source of corporate profits.

These algorithms are depicted as programs. For this reason, programming languages have been created, and high-level languages such as C and JAVA have appeared. Computer languages have become more sophisticated with the evolution of computer programs and are expressed in a way that is intuitive to human understanding. However, this has led to a greater distance from machine languages, and these languages must therefore be translated into a machine language. Therefore, an interpreter or an assembler is also included in software.

When looking at software in a hierarchical manner from the point of contact with people, software can be divided into application software and OS. Furthermore, computers have become networks instead of a stand-alone units. As a result, software can be grasped as an application, OS, middleware, network OS, or the like.

This basic algorithm for modern computer systems was proposed by von Neumann. Over the years, it has continually become more sophisticated, and research on neuroalgorithms, parallel calculations, and so on has progressed. Today a rapid pace of machine learning is exemplified, for example, by the progress in deep learning, a trend is leading to the further advancement of AI.

Narrowly defined software is increasingly becoming to be associated with devices, such as IC chips and hard disks. The core of technology in the fusion area is a logic circuit as a physical medium for reading and processing machine languages. Logic circuits range from CPUs, MPUs, and ASICs, which account for a large proportion of algorithms, to memories and dedicated processors with a relatively small role of software. Although the role played by software differs in importance among circuits, logic circuits can be considered products in a fusion of software and hardware in the sense that it has properties of both software and hardware.

The IC chip reads a high-level language, which is translated into a machine language processable by the machine. In firmware, the program translated into the machine language is implemented as an unchangeable circuit in the IC chip. The logic circuit includes an arithmetic circuit, a memory circuit, etc., and it is classified into general-purpose CPU/MPU, specialized ASIC for special purpose, for example.

A typical technique among narrowly defined hardware is microfabrication techniques, which enable improved performance of logic circuits. Sensors, which can be narrowly defined as hardware, are also important.

IT started with advances in hardware. The development of transistors by AT&T in 1947, the development of IC manufacturing technologies by Fairchild and Intel in the late 1950s, and the subsequent development of microfabrication technologies that supported the advancement of IC to LSI and LSI to VLSI as indicated by Moore's Law has led to dramatic improvements in IT abilities. IT has increased its information processing capability, and its application fields have been steadily widening.

However, progress is beginning to hit physical limitations. Needless to say, new technological developments are taking place, such as miniaturization of transistors, crystal technology, exposure technology, and thin-film manufacturing technology, among others. However, such progress has been slower than before and its limits are fast approaching. Promoting miniaturization of IC is not as easy as previously, and there is little scope for significant progress relative to software.

Advances in nanotechnology are widely expected to lead to (1) the creation of quantum devices with new operating principles and (2) the creation of self-organizing chip. These developments will enable a major breakthrough but are likely to take time.

In terms of data, relational databases have a high affinity as a conventional computer system. Today, however, new data, such as image data, text data, and voice data, which have a low affinity with a relational database, have been generated, creating big data. Cloud computing is indispensable for storing big data.

5.3.2 Advantages/Disadvantages of Hardware and Software

As mentioned in the previous section, IT advances born in the latter half of the twentieth century occurred in hardware, software, and in their fusion, but until the end of the twentieth century, technological advances in narrowly defined hardware and the fusion area of software and hardware had been ahead of developments in software. In particular, miniaturization from ICs to LSIs and VLSIs has greatly contributed to the improvement of information processing capabilities. Advances in hardware technology mean that today's personal computers have far more information processing capabilities than the large computers from the 1970 and 1980s.

However, this progress has finally met the limitations of current technology. Of course, even though the limit of miniaturization has been forecast many times, miniaturization has managed to advance by overcoming these limits through various technical improvements. However, further miniaturization is not as easy today, and the potential for technological progress is shrinking. To overcome this, new architecture based on the latest advanced fabrication techniques such as quantum devices will eventually be realized, but not for a long time.

In contrast, the potential of software has grown, supported by the introduction of intelligent technology, and has begun to evolve even more rapidly in the 21st century. As part of this, IT advances are changing from hardware-driven to software-driven.

However, this shift in the core of IT from hardware to software cannot be attributed only to the difference in the potential of technology. Economic and social environmental changes also are factors, as can be seen by comparing the advantages/disadvantages of hardware and software.

Software and hardware have their own advantages and disadvantages, respectively, in information processing. Hardware's advantage is its speed of information processing. For example, comparing a dedicated processor specializing in image processing and processing with software using a general-purpose processor, the former is far faster than the latter. The general processor's need to read the program or data necessary for processing necessarily extends the processing time. Hardware is therefore more suitable for repeatedly processing the same or similar types of data in large quantities using the same program.

The disadvantage of hardware, on the other hand, is its high cost. Today, the cost of building a new manufacturing plant for IC chips exceeds $10 billion. It is thus also difficult for hardware to respond to changing market needs because of the substantial investment costs associated with frequent modifications in specifications. In today's consumer society, customer needs are characterized by uncertainties, fluctuations, and unpredictability. Responding to such changes in customer needs via narrowly defined hardware or technologies in the fusion area of software and hardware is disadvantageous in terms of cost.

In addition, hardware technology has become more difficult to monopolize and keep proprietary. Even a latecomer is able to manufacture highly-qualified IC chips that are not inferior to that of innovators simply by introducing the latest semiconductor manufacturing equipment. In other words, technology transfer to competitors has become easier than before, meaning that innovators are findings it increasingly difficult to gain monopoly profits based on R&D investment. If semiconductor technology can easily be transferred from innovators to latecomers, products would be put on the market in large quantities, prices would fall, and innovator profits would be lost.

The advantages and disadvantages of hardware are, conversely, the disadvantages and advantages of software, respectively.

Software specifications can be changed at low cost simply by rewriting the program. Therefore, despite the slower speed, it is preferable to implement the same function in software rather than in hardware when customer needs are constantly changing and the future prospects are uncertain.

This also holds true for the advantages of microprocessors compared with dedicated circuits. Microprocessors with computing and control circuits perform a variety of functions by replacing narrowly defined software. In other words, because a range of information processing duties can be performed simply by replacing programs unincorporated with the circuit on chips, the use of computers has expanded. One of the reasons why PCs have become widely accepted is not their ability to perform predetermined calculations, such as calculators, but ability to perform a variety of data processing tasks simply by switching programs. Even a single circuit specification can be used as a system for a variety of different applications by changing the computing program.

Software's flexibility and versatility makes it possible to respond quickly, accurately, and cheaply to diversifying and ever-changing social needs. As a result, diversification of services and applications will become basic trends in modern consumerism.

As can be seen from the above, technical and social factors both have contributed to the shift in the technology at the heart of today's TEP from hardware to software. In other words, hardware microfabrication technology faces technical limitations, and no dramatic progress can be expected. In addition, it is advantageous to respond with software rather than hardware because of the ease of responding to changing needs. For these reasons, software is becoming more important.

5.4 Verification of Transition to a Software-Driven Economy by Number of Patents/Business Articles

As described in the previous section, the core of today's TEP is becoming software-centric and the importance of intelligent software is increasing. This section examines this transformation based on changes in the number of patents and business articles focused on these topics, respectively. The analysis starts by extracting important technologies from the "narrowly defined software," "technologies in the fusion area of hardware and software," and "narrowly defined hardware," as shown in Table 5.4, and then looking at time-series changes in the frequency of occurrence in terms, how the center of technological progress is changing. For this purpose, patent publication filed with Japan Patent Office and EU Patent Office are analyzed.

Table 5.4 Classification of IT

Classification	Technology *emerging technology*
Data	Data big data database relational database, cloud database
Narrowly defined software (software as a true meaning)	Software architecture algorithm *neural computing* *artificial intelligence,* *machine learning,* *deep learning,* *ultra parallel processing,* *non von Neumann algorithm* program (programming) programming language high-level language interpreter, compiler, assembler application software middleware operating system network operating system
Technologies in fusion area of software and hardware	Machine language firmware embedded software microprogramming circuit design neural circuit, non linear circuit tensor processing unit ASIC, graphic processor CPU, MPU memory
Narrowly defined hardware	Microfabrication technology transistor, crystal structure thin film technology exposure technology NP structure sensor *nanotechnology* *quantum device* *superconducting devices,* *molecular electronic devices* *bio device* *self-organizing material*

Source Created by author

5.4.1 Verification Based on Japan Patent Office Data

The Japan Patent Office's Digital Library contains most patents published since 1971. The library also contains information on patents published by some overseas patent offices, including the full text data from patents published by five offices (WIPO, US Patent Office, EU Patent Office, Chinese Patent Office, and Korea Patent Office). However, the technical reasons mean that all overseas patents are not included in the Library's database, and older patents in particular are more likely to be missing. As a result, when looking further back in time, the more likely it is that patents obtained by Japanese companies are over-represented in comparison with foreign ones, making it impossible to infer the competitiveness of Japanese companies and foreigners based on the time-series change in the number of patents published by the Japan Patent Office and foreign patent offices. Nonetheless, careful analysis can provide important insights.

Naturally, Japanese patent data are written in Japanese, and overseas patents are translated into Japanese using machine translation technology. Therefore, in this paper, patent publications written in Japanese are mined, followed by the translating Japanese into English. The results are shown in Table 5.5.

Paying attention first to the number of cases including the terms "software" and "hardware," the number of cases with "software" is higher than that of "hardware." For example, in 2013–2017, 727,599 instances of "software" were found, compared with 591,403 for "hardware." However, from 1993–1997 to 2013–2017, the number of mentions of "software" grew 2.3-fold, whereas mentions of "hardware" increased 2.8-fold. This result is the opposite of the basic understanding that TEP is shifting from hardware to software. This could possibly be attributed to the fact that the terms "software" or "hardware" are used in a broader sense than in this article.

Accordingly, author extracts representative terms from Table 5.4 and considers the number of occurrences. This table shows that occurrences of terms related to "narrowly defined hardware" is high, and terms related to "narrowly defined software" are mentioned the least often. In 1993–1997, terms related to "narrowly defined hardware" accounted for 60.4% of the total. On the other hand, terms referring to "technologies in the fusion of software and hardware" accounted for 32.2%, whereas those related to "narrowly defined software" accounted for only 7.5%. The high number of mentions of "narrowly defined hardware" reflects the frequent mentions of "sensors," but the number of cases referring to "narrowly defined hardware" is the highest even with the exception of "sensor" mentions. Since then, however, the percentage of "narrowly defined hardware" has declined to 52.1% in 2013–2017, while the percentage of "narrowly defined software" has risen to 16.4% in 2013–2017. This demonstrates the growing importance of software in IT.

Of the patents referencing "narrowly defined software," a high number mentioned "algorithm," at 409,373 cases, and "programming" at 163,760 cases, in 2013–2017. This is partly due to the fact that these terms are used to include broadly meaningful concepts and consequently appear more frequently. However, the growing importance of these technologies is undeniable, as shown in the fact that rate also was considerably higher, 13.0 times and 11.8 times higher, respectively in 1993–1997 and 2013–1917. Similarly, mentions of "operating system" and

Table 5.5 Patents by IT Field (1993–2017)

Year		(1) 1993–1997			(2) 1998–2002			(3) 2003–2007		
		Japan	Others	Ratio %	Japan	Others	Ratio %	Japan	Others	Ratio %
	Software	63,226	4,703	93.1	1,21,231	42,766	73.9	1,80,272	2,23,136	44.7
	Hardware	49,465	3,275	93.8	81,642	27,487	74.8	1,37,206	1,49,938	47.8
	Software hard™	1. 3			1.5			1.4		
Narrowly defined software	Application software	3,145	124	96.2	10,851	1,154	90.4	15,393	4,708	76.6
	Neural network	4,409	272	94.2	3,322	828	80.0	4,367	3,222	57.5
	Middleware	62	10	86.1	1,317	524	71.5	2,988	5,191	36.5
	Operating system	10,876	751	93.5	27,891	4,989	84.8	41,108	21,936	65.2
	Artificial intelligence	951	64	93.7	1,025	328	75.8	1,682	2,099	44.5
	Neuro computer	436	12	97.3	88	10	89.8	60	18	76.9
	Deep learning	0	0		0	0		0	1	0.0
	Algorithm	34,396	3,136	91.6	54,628	23,992	69.5	75,253	119, 960	38.5
	Programming	13,584	2,952	82.1	20,452	14,187	59.0	26,246	63,090	29.4
	Compiler	2,082	331	86.3	3,380	1,607	67.8	3,528	6,905	33.8
	Assembler	941	92	91.1	1,116	453	71.1	1,043	2,043	33.8
	High-level language	695	42	94.3	712	144	83.2	795	1,700	31.9
	C language	2,068	90	95.8	2,399	529	81.9	2,821	2,289	55.2
	Sub total	73,645	7,876	90.3	1,27,181	48,745	72.3	1,75,284	2,33,162	42.9
		7.50%			10.70%			12.80%		
Technologies in fusion areas of software and hardware	Firmware	2,922	217	93.1	7,071	3,023	70.1	18,101	26,434	40.6
	Machine language	66	9	88.0	172	36	82.7	226	225	50.1
	Microprocessor	35,403	7,382	82.7	38,639	24,798	60.9	45,212	80,541	36.0
	Central processing unit	6,904	137	98.1	9,566	602	94.1	13,825	2,185	86.4
	Logic circuit	18,362	1,940	90.4	16,900	6,671	71.7	17,882	16,529	52.0

(continued)

Table 5.5 (continued)

Year		(1) 1993–1997			(2) 1998–2002			(3) 2003–2007		
		Japan	Others	Ratio %	Japan	Others	Ratio %	Japan	Others	Ratio %
	Orciit design	5,774	380	93.8	8,259	2,724	75.2	9, 552	14,627	39.5
	Memory	2,42,172	20,226	92.3	3,00,208	1,02,257	74.6	3,89,190	3,83,394	50.4
	Dedicated circuit	1.087	49	95.7	1,473	531	73.5	3,090	3,279	48.5
	Mask pattern	7,829	270	96.7	10,302	2,634	79.6	13,723	13,045	51.3
	Non linear circuit	416	20	95.4	264	113	70.0	182	239	43.2
	Sub toal	3,20,935	30,630	91.3	3,92,854	1,43,389	73.3	5,10,983	5,40,498	48.6
		32.2%			32.5%			33%		
Narrowly defined hardware	Semiconductor	2,04,730	13,870	93.7	2,55,773	84,956	75.1	3,14,137	3,10,909	50.3
	Exposure apparatus	15,593	300	98.1	2 3,602	2,659	89.9	32,483	12,020	73.0
	Semiconductor thin film	3,069	103	96.8	3,565	957	78.8	3,904	3,952	49.7
	Transistor	1,08,413	10,321	91.3	1,06,322	46,595	69.5	1,23,387	1,71,776	41.8
	Quantum device	144	3	98.0	76	27	73.8	185	155	54.4
	Superconductivity device	748	31	96.0	299	73	80.4	372	236	61.2
	Neuro device	74	1	98.7	17	2	89.5	11	1	91.7
	Sensor	2,74,374	23,684	92.1	3,21,814	83,050	79.5	4,10,817	3,24,959	55.8
	Optical sensor	4,043	406	90.9	5,056	1,956	72.1	8,290	8,338	49.9
	Sub total	6,11,188	48,719	92.6	7,16,524	2,20,275	76.5	8,93,586	8,32,346	51.8
		60.4%			56.8%			54.2%		
Total		10,05,768	87,225	92.0	12,36,559	4,12,409	75.0	15,79,853	16,06,006	49.6

(continued)

Table 5.5 (continued)

Year		(4) 2008–2012			(5) 2011–2017			(5) (1)	(5) (3)
		Japan	Others	Ratio%	Japan	Others	Ratio%		
	Software	1,89,483	3,85,487	33	2,19,536	727.599	23.2	13.9	2.3
	Hardware	1,69,839	2,87,946	37.1	214.531	591.403	26.6	15.3	2.8
	Software hard™	1.3			1.2				
Narrowly defined software	Application software	11.670	6,606	63.9	12.454	10.837	53.5	7.1	1.2
	Neural network	3,983	6,797	36.9	4,194	18.407	18.6	4.8	3.0
	Middleware	4,290	12,315	25.8	6,873	22,695	23.2	410.7	3.6
	Operating system	38,082	37,263	50.5	39,770	62.210	39.0	8.8	1.6
	Artificial intelligence	1,772	3,230	35.4	2,188	8.219	21.0	10.3	2.8
	Neuro computer	58	22	72.5	16	20	44.4	0.1	0.5
	Deep learning	0	5	0.0	523	363	59.0	#DIV?0!	886
	Algorithm	74,812	2,10,682	26.2	79,142	4,09,373	16.2	13.0	2.5
	Programming	25,488	94,267	21.3	30,715	1,63,760	15.8	11.8	2.2
	Compiler	3,732	13,021	22.3	4,319	27.357	13.6	13.1	3.0
	Assembler	1,533	3,083	33.2	2,049	5,521	27.1	7.3	2.5
	High-level language	1,007	6,576	13.3	1,235	16,545	6.9	24.1	7.1
	C language	2,047	3,935	34.2	1,395	6,585	17.5	3.7	1.6
	Sub total	1,68,474	3,97,802	29.8	1,84,873	7,51,892	19.7	11.5	2.3
		14.3%			16.4%				
Technologies in fusion area of software and hardware	Firmware	28,302	54,442	34.2	40,327	1,13,776	26.2	49.1	3.5
	Machine language	316	220	59.0	620	659	48.5	17.1	2.8
	Microprocessor	47,874	1,15,856	29.2	57,467	2,11,455	21.4	6.3	2.1
	Central processing unit	12,002	2,967	80.2	11,663	2.994	79.6	2.1	0.9
	Logic circuit	17,936	19,028	48.5	21,758	36,694	37.2	2.9	1.7
	Orciit design	7,133	24,614	22.5	4,866	39,059	11.1	7.1	1.8
	Memory	3,72,333	5,25,141	41.5	3,61,619	8,60,159	29.6	4.7	1.6

(continued)

Table 5.5 (continued)

Year		(4) 2008–2012			(5) 2011–2017			(5) (1)	(5) (3)
		Japan	Others	Ratio%	Japan	Others	Ratio%		
	Dedicated circuit	5,186	5,584	48.2	6,948	7,596	47.8	12.8	2.3
	Mask pattern	12,216	15,264	44.5	7,910	11,797	40.1	2.4	0.7
	Non linear circuit	130	310	29.5	155	598	20.6	1.7	1.8
	Sub toal	5,03,428	7,63,426	39.7	5,13,333	12,84,784	28.5	5.1	1.7
		32%			31.5%				
Narrowly defined hardware	Semiconductor	2,90,248	3,91,895	42.5	2,53,031	4,93,841	33.9	3.4	1.2
	Exposure apparatus	30,079	12,813	70.1	22,823	12,434	64.7	2.2	0.8
	Semiconductor thin film	4,072	5,851	41.0	3,291	5,819	36.1	2.9	1.2
	Transistor	1,15,847	2,14,048	35.1	94,940	3,03,589	23.8	3.4	1.4
	Quantum device	112	218	33.9	91	263	25.7	2.4	1.0
	Superconductivity device	211	233	47.5	119	265	31.0	0.5	0.6
	Neuro device	4	0	100.0	3	2	60	0.1	0.4
	Sensor	4,15,197	6,17,042	40.2	4,27,958	13,28,205	24.4	5.9	2.4
	Optical sensor	9,293	12,696	42.3	11,138	20,413	35.3	7.1	1.9
	Sub total	8,65,063	12,54,796	40.8	8,13,394	21,64,831	27.3	4.5	1.7
		53.6%			52.1%				
Total		15,36,965	24,16,024	38.9	15,11,600	42,01,510	26.5	5.2	1.8

Source EU Patent Office

"middleware" are high, which indicates the importance of building software platforms and acquiring global standards.

"Memory" is the dominant term mentioned in the fusion area of software and hardware (860,159 cases in 2013–2017). This reflects the necessity for larger capacities in computer systems. However, from 2003–2007 to 2013–2017, the growth rate of "memory" reduced 1.6-fold. This seems to reflect both the fact that the importance of memory has decreased and microfabrication technology has reached physical limits.

"Sensor" and "semiconductor" are the most common in narrowly defined hardware. Mentions of sensors also grew 2.4 times from 2003–2007 to 2013–2017. Similarly, the growth rate of "optical sensors" is also high. As the Internet of Things (IOT) era dawns, the importance of sensors is increasing.

Mentions of emerging technologies supporting future microfabrication technologies, such as "neuro device," "quantum device," and "supplier conductivity device," are extremely infrequent, and their growth rate also is low. This can be interpreted as reflecting the current situation that research aiming at breakthroughs in microfabrication technologies are unlikely for the time being.

As described above, the increasing importance of software in IT is demonstrated by the trends in the number of patents. Of course, in absolute terms, this does not mean that the importance of narrowly defined hardware has shrank, only that it has become smaller relative to software. For example, in the IOT age, it is still necessary for hardware and software to evolve together, such that big data can be collected by sensors for analysis using intelligent software.

5.4.2 Japan's Adaptation to the Software-Centered TEP

Next, the author examines whether Japanese companies have properly responded to the growing importance of software.

Table 5.6 shows the top 9 country codes for priority claim numbers for "application software" and "computer architecture" as well as for two technologies in

Table 5.6 Ranking country code of priority claim numbers of patent by technology

Application software		Computer architecture		Firmware		Embedded software	
CN	3,539	US	562	US	4,987	CN	560
JP	2,066	WO	85	CN	3,604	US	143
US	1,740	CN	72	JP	2,208	KR	100
wo	450	JP	36	WO	1,229	WO	29
KR	330	EP	26	KR	906	JP	23
TW	252	DE	22	TW	665	TW	8
DE	103	AU	21	EP	189	EP	7
EP	100	FR	16	GB	175	GB	6
GB	85	CA	10	DE	149	FR	6

Source EU Patent Office

the fusion area of software and hardware (e.g., "firmware" and "embedded software") in the patents issued by the EU Patent Office. The country code is used as an approximation of the applicant's nationality.

First of all, "computer architecture" is predominantly mentioned in the US (562 cases), compared with 85 mentions in Japan, only 15.1% of the mentions in the US. "Application software" is most frequently mentioned in China, followed by Japan and the US. Application software differs from computer architecture in that the latter is a more basic and fundamental technology, and the former is an applied technology that links the latter's output to business. Firms in the US are conducting radical researches that determine the basic nature of the software itself. In other words, US-based research has primarily focused on advances that lead the evolution of the nature of TEP. China and Japan, on the other hand, have accepted this architecture as a basic technology and have used it to develop software tied to business. In other words, Japanese companies are conducting research in accordance with the technological trajectory established by the framework of TEP. However, concentrating on the latter without duly considering the former, which determines the nature of the TEP and the direction of progress with application software, weakens a firm's ability to adapt to a software-driven economy. Thus, the inadequate study of basic principles such as computer architecture in Japan is one of reasons for the country's lack of adaptability to a software-driven economy.

Regarding "firmware" and "embedded software" as technologies that belong to the fusion area of hardware and software, "firmware" is mentioned 4,987 times in US patents versus 2,208 instances in Japan. China ranked first in mentions of "embedded software" whereas the 23 mentions in Japan placed it in 5th place.

Table 5.5 contains patents published by the Japan Patent Office as well as patents published by overseas patent offices. However, as mentioned above, because the Japan Patent Office's Digital Library recently began including overseas patents, but not all overseas patents are available, it is impossible to derive the correct meaning from time-series data for overseas patents using this database. Nonetheless, by analyzing the ratio of patents obtained in Japan and overseas [i.e., number of domestic patents)/(number of domestic patents + the number of overseas patents)] in each technology at a specific point in time, the R&D areas of focus among Japanese companies can be determined.

The average of (number of domestic patents)/(number of domestic patents + number of overseas patents) was 92.0% in 1993–1997, but declined significantly from 75.0% in 1998–002, 49.6% in 2003–2008, 38.9% in 2009–2012, and 26.5% in 2013–2017. As mentioned earlier, this is attributable to the facts that the percentage of missing overseas patents has decreased over the years, and that the rate of increase in patents acquired by overseas companies exceeds the rate of increase in patents acquired by Japanese companies.

Technologies where the ratio is larger than the average ratio for each year can be interpreted as those to which Japan attaches importance, whereas and technologies with a smaller ratio are considered less important. In other words, the former are competitive technologies, and the latter are uncompetitive technologies. In the case of 2003–2007, 49.6% of all patents were issued by Japan Patent Office. In addition, of the patents issued by the Japan Patent Office, 51.8% of technologies belonged to "narrowly defined hardware," 48.6% were "technologies in the fusion area of hardware and software," and only 42.9% belonged to "narrowly defined software." The low ratio of "programming" and "algorithm" among the "narrowly defined software" patents suggests that Japanese companies have not been making sufficient investment in software and its fruits. On the other hand, among "narrowly defined hardware," eight of the nine technologies exceeded the overall average, including the extremely high mention rate of "exposure apparatus" at 70.3%.

This situation had not changed significantly by 2013–2017, during which 26.5% of all patents were issued by the Japan Patent Office. In addition, 27.3% of "narrowly defined hardware" 28.5% of "technologies in the fusion area of hardware and software" were patents issued by Japan Patent Office, whereas only 19.7% of "narrowly defined software" were patents published by the Japan Patent Office. The fact that the ratio of "operating system" and "middleware" are larger than the average may reflect Japanese companies' positive attempts to adapt to the age of computer networking. However, the low ratio of patents published by Japan Patent Office for "algorithm," "programming," and "high-level language" reveals the vulnerability of software development in Japan.

Next, the author extracts patents for automatic driving from those published by EU Patent Office and examines the emphasis on software by country. To this end, 2,107 patents were extracted that included the term "automatic driving" or "autonomous driving" for 2007–2017. Looking at these data in chronological order, as shown in Fig. 5.2, a rapidly since the beginning of the 2010s. This reflects the current situation where automatic driving research is in full swing and practical application is approaching.

Priority claim number ranked in 814 in CN (China), 502 in JP (Japan), 316 in KO (South Korea), 183 in US (US), and 113 in DE (Germany). It can be seen that China's ability to develop automated driving technology is rapidly increasing, and that Japan and South Korea are also seriously developing it. Patent numbers undeniably reflect technological development capabilities to a certain extent. However, it is well known that the US, which is ranks 4th in terms of patent acquisitions in the world, is the most advanced in automatic driving technology. This indicates that the correlation between the number of patents and technology development ability should not be overestimated.

Next, in order to understand the extent to which software is emphasized in Japan's autonomous driving development compared with other countries, terms appearing in the title/abstract of patents granted in three countries (Japan, the US, and China) were extracted up to the top 1,000, and those terms included in the upper rank are regarded as the technologies emphasized in each country. This ranking was used because an evaluation based on the number of occurrences per se tends to underestimate US investment in automated driving technology. Therefore,

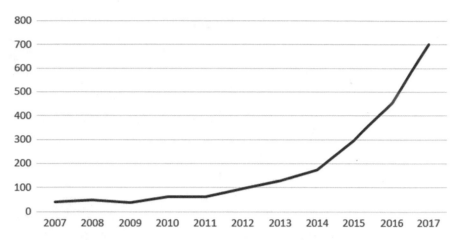

Fig. 5.2 Number of patent related to automatic driving (EU Patent Office) *Source* EU Patent Office

the top terms for each country are regarded as technologies emphasized in that country. The ranking by mentions of technologies such as prediction, recognition, position identification, planning, monitoring, etc., that are decisive for achieving automated driving are shown in Table 5.7, along with ranking and number of occurrences of software necessary for autonomous driving.

First, "algorithm," "program," "software," "architecture," and "learning" were chosen as terms embodying narrowly defined software. Then, within the top 1,000 rankings, 10 terms (including program, algorithm, software, and computer program product) are found in the US. In contrast, both Japan and China had only six terms in the top 1,000. Of terms including "data" or "database," which were selected to embody the meaning of data analyzed by software, 11 terms are found within 1,000 rankings in the US, compared with only 6 in Japan and 5 in China. In addition, if map information as a key technology related to autonomous driving is included in the data, 16 terms are found in the US, compared with 12 in Japan and 6 in China. This reveals that in the development of automatic driving technology, the emphasis is placed on narrowly defined software and data in the US.

Second, the results from including the term "device" as a technology of the fusion area of hardware and software are as follows. The US had only 11 terms compared with China's14 but Japan had as many as 30 terms. It can thus be concluded that Japan is developing more hardware-oriented technologies rather than narrowly defined software. In addition, in terms of the future basic industries underlying autonomous driving, Japan's software development capability is lower than that of the US, while in this paper, author insisted on delay in the softwareization of Japan.

Table 5.7 Appearance ranking of technologies achieving automatic driving

	1–50	51–100	101–500	501–1000
US	Device(41), sensor(40), data(38), location(35), map(18)		Program(7), sensor data(7), database(6), algorithm(6), storage device(5), compute device(5), plan(4), radar(4), location information(4), perception(3), software(3), localization(3). map database(3), object localization(3), position data(3), process device(3), mapping(2), prediction(2), map information(2), reference data(2), user interface device(2), sender device(2), closure device(2), path plan(2), image data (2), measurement data(2), communication device(2), object map(2), object data(2), perception system(2), traffic control device (2), computer program product(2),	Monitor(1), predictor(1). learning(1). architecture(1), recognition(1), program instruction(1), prediction result(1), assistance device(1), point data(1), user device(1), map building(1), management compute device(1), radar detection equipment(1), vehicle radar perception(1), curvature estimation algorithm(1), driver assistance device(1), path predictor(1). mapping information(1), control console device(1), prediction component(1), radar system(1), user device display(1), present user location(1), output device(1), data fitting algorithm(1). software program(1), perception information(1), prediction subsystem(1), plan view occupancy map(l), radar perception information(1), navigation plan(1), control device(1)
Japan	Device(264), program(60), sensor(53).	Data(38), control device(38), support device(38), map(37), plan(29), computer program (28), recognition(22).	Location(19), drive device(18), vehicle control device(18), drive control device (16), monitor(14), assistance device(14), prediction(ll), navigation device(11), database(10), map information(10). travel control device(10), display device(8), communication device(6), vehicle device (6), map data(6), illumination device(6), learning(5), vehicle control program(5), radar(4), image data(4), map database(4), assist device(4)	Safety device(3), support control device(3), vehicle sensor(3), notification device(3), road map data(3), operation device(3), perception(2), control parameter setting device(2), drive support device(2), learning recording device(2), attitude adjustment device(2), plan generation unit(2), recording device(2), surround recognition sensor(2), measurement device(2), flow passage device(2), map update determination system(2). navigation ECU

(continued)

Table 5.7 (continued)

	1–50	51–100	101–500	501–1000
				(2), object course prediction method(2), power storage device(2), precision map unit(2), information recognition unit(29), control program(2), support program(2), recognition information(2), storage device (2), radar wave(2), prediction unit(2), surround recognition sensor(2), output comparison device approach(2), change control device(2), map update determination unit(2), map unit(2), brake control device(2), vehicle operation device (2), sensor data(2)
China	Device(372), data(158), sensor(139)	Drive device(51), moiiitor(47)	Map(40), radar(37), control device(34), algorithm(33), plan(28), program(20), location(20), sofhvare(20). recognition(19), path plan(ll), database(10)	Power fraction device(9). learning(8), perception(8), programming(7), power drive device(7), computer program(7), state control device(6), monitor device(6), communication device(6), monitor system (5), wing device(5), data information(5), detection device(5), storage device(5), prediction(4), alarm device(4), transmission device(4), safety device(4)

Note () means appearance ranking
Source EU Patent Office

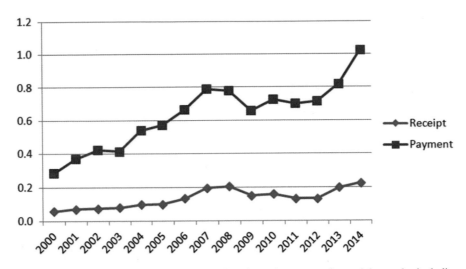

Fig. 5.3 Trade balance of software etc. *Note* Receipt and payment of copyright royalty including software. *Source* Ministry of Economic, Trade and Industry (2015). p. 11

5.5 Software Competitiveness in Japan

5.5.1 Actual Situation of Low Competitiveness in Japanese Software Industry

Data on the number of patents and academic business articles examined in the previous section has revealed the weakness of Japan's software development capabilities. This examines the current state of Japan's software competitiveness and the reasons behind this situation.

The most obvious indication of the low competitiveness of Japanese software is the country's software trade balance. As Fig. 5.3 shows, Japan's software trade balance deficit is large. In 2001, software exports and imports were 0.1 trillion yen and 0.3 trillion yen, respectively; by 2014; they were 0.2 trillion yen and 1.0 trillion yen, respectively, indicating that the deficit is growing.

The data also show Japan's inferior level of R&D investment. For example, of the 2,880 leading IT companies involved in the software business, only 1,006 were conducting R&D in 2016, and their ratio of R&D expenses to sales were a mere 3.2% of sales.[1]

The inferiority of Japan's software technology level has a negative impact on the country's macroeconomic growth. Figure 5.4 shows the impact of investment in software on Japan's GDP growth. According to this figure, software contributed less than 0.1% to Japan's GDP growth in 1985–2000, less than in the US and many

---

[1]Global ICT Strategy Bureau, Ministry of Internal Affairs and Communications (2018), p. 24.

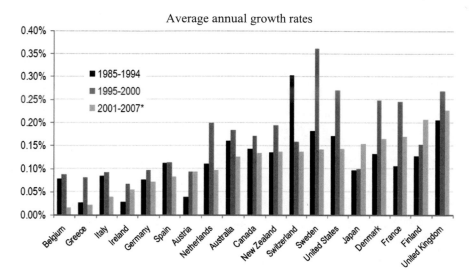

Fig. 5.4   Contribution of software investment to GDP growth. *Source* Lippoldt and Stryszowski (2009), p. 9

other countries. By 2001–2007, its contribution exceeded that in the US, indicating that the Japanese economy had begun to become software-led. In the 21st century, however, software is becoming more intelligent. As described in the previous section, software intelligence began rapidly advancing around 2012 as a result of advances of emerging technology called deep learning.

In the future, it will be necessary to analyze the effect of AI's introduction on GDP growth. It is well known that Japan's research on AIs has lagged far behind that of other countries, especially the US. Table 5.8 compares the degree of penetration of AI among Japanese and US companies. For this comparison, a questionnaire survey was conducted among 2,211 people in Japan and the United States engaged in four job types with high possibility of mechanization: ((1) clerk, (2) driver, (3) production/construction worker(4) ④ restaurant staff), and four job types with low mechanization possibilities ((1) doctor/pharmacist, (2) faculty staff, (3) system engineer, (4) nursing/long-term care staff). According to the survey results, the degree of penetration of AI in each group of four occupations was much greater in US companies than in Japanese companies (Nomura Research Institute 2016, p. 26).

These findings offer further evidence that TEP's core technology has been shifting from hardware to software, as described repeatedly. Furthermore, intelligent software is particularly becoming central. Today, the Japanese economy and Japanese companies must undertake further adaptation to this new paradigm.

Average annual growth rates

Table 5.8 Current status and planning situation of introduction of artificial intelligence into the workplace

	Japan		US	
	(1)	(2)	(1)	(2)
Number of respondents	553	553	551	554
Introduced and used already	1.8	2.0	4.5	6.1
Introduced already, never used so far	2.9	3.3	7.8	8.8
Not introduced yet, but there are plans to be introduced in the future	4.2	7.1	14.0	19.0
Not introduced yet, and there are no plans to be introduced in the future	62.0	64.2	54.4	49.5
Unknown	29.1	23.5	19.2	16.6

Note (1) Occupation with high possibility of mechanization, (2) occupation with low possibility of mechanization
Source Nomura Research Institute (2016), p. 26

5.5.2 Reasons for Weakness of Japanese Software

The Japanese economy was able to adapt well to the TEP in 20th century, a time when hardware technologies such as microfabrication technologies in IT were of great importance; the country was also able to enjoy the high international competitiveness in the memory and home appliances markets. In the 21st century, however, the networking of computer systems has continued to progress, and the importance of software is increasing. Nevertheless, Japan's software development capacity and the widespread penetration of software into society have lagged, and its competitiveness has been rapidly eroding. This paper will conclude with a clarification of future challenges by considering the causes behind the delay in the Japanese companies' software responses.

First, the end of the Japanese bubble economy in 1991 and subsequent long-term deterioration in business performance weakened Japanese companies' investment capacity. As Table 5.9 shows, the growth rate of IT investment in the 1990s was not significantly different from that in the US, but in the 2000s, the growth rate of IT investment by Japanese companies stagnated compared with that of US companies, a divergence that reflects the deterioration of Japanese firms' financial condition in the 21st century due to performance deterioration, particularly among electrical equipment manufacturers. This deterioration is symbolized by Panasonic's large operating loss in 2001–2002 and Sony's stock price crash in 2003.

Moreover, software accounts for a much smaller proportion of IT investment than in the US. As shown in Table 5.9, in 1994, software received only 31.5% of IT investment in Japan, compared with 42.9% in the United States.

One of the reasons why Japanese companies disregarded software was a failure to recognize the shift in IT from hardware to software that occurred at the turn of the century. One example is the television manufacturing industry. At the end of the

Table 5.9 IT Investment Trends in Japan and the US

Year		1994	1995	1996	1997	1998	1999	2000	2001	2002	2003	2004
US	Software	0.09	0.09	0.1	0.13	0.15	0.18	0.21	0.21	0.21	0.22	0.23
	Hardware	0.12	0.14	0.15	0.17	0.18	0.21	0.24	0.2	0.17	0.17	0.18
	Software + hardware	0.21	0.23	0.25	0.3	0.33	0.39	0.45	0.41	0.38	0.39	0.41
	Ratio of software	42.9%	39.1%	40.0%	43.3%	45.5%	46.2%	46.7%	51.2%	55.3%	56.4%	56.1%
Japan	Software	4.6	4.7	5.4	6.1	7.1	7.6	8.2	9.2	9.7	9.7	9.9
	Hardware	10	12	14.1	13.9	11.8	11.8	11.8	10.3	8.7	8.5	8.2
	Software + hardware	4.6	4.7	5.4	6.1	7.1	7.6	8.2	9.2	9.7	9.7	9.9
	Ratio of software	31.5%	28.1%	27.7%	30.5%	37.6%	39.2%	41.0%	47.2%	52.7%	53.3%	54.7%

Year		2005	2006	2007	2008	2009	2010	2011	2012	2013	2014	2015
US	Software	0.25	0.26	0.28	0.29	0.29	0.29	0.31	0.32	0.33	0.36	0.37
	Hardware	0.18	0.19	0.21	0.19	0.17	0.18	0.18	0.18	0.18	0.18	0.18
	Software + hardware	0.43	0.45	0.49	0.48	0.46	0.47	0.49	0.5	0.51	0.54	0.55
	Ratio of software	58.1%	57.8%	57.1%	60.4%	63.0%	61.7%	63.3%	64.0%	64.7%	66.7%	67.3%
Japan	Software	10.1	10.2	10.2	10.4	9.6	9.5	9.2	9.7	10	10.1	10.5
	Hardware	8.3	7.9	8.2	7.9	6.3	5.8	6.5	5.7	5.8	6.1	5.7
	Software + hardware	10.1	10.2	10.2	10.4	9.6	9.5	9.2	9.7	0	10.1	10.5
	Ratio of software	54.9%	56.4%	55.4%	56.8%	60.4%	62.1%	58.6%	63.0%	63.3%	62.3%	64.8%

Note US trillion dollar, Japan trillion yen
Source Ministry of Internal Affairs and Communications (2018), p. 30

twentieth century, the quality of television pictures was more determined by software than by the quality of the CRTs or LCD panels. This shift was accelerated by the digitization of television programming. However, Japanese companies continued to focus on improving the quality of their hardware. This focus was not limited to televisions. Improvements in the functioning of many home appliances have been determined by software. However, Japanese companies were less aware of the importance of developing such software, which in turn contributed to the decline of the Japanese consumer electronics industry.

In this way, the contraction of the Japanese market and deterioration of corporate performance, as well as the misunderstanding of trends in technological progress, deprived companies of incentives to invest in R&D or make capital investments in software. As a result, although by the 2000s, the ratio of software investment to total IT investment in Japan reached a level comparable to that in the US, the total amount of IT investment did not increase. For that reason, software investment in the US grew 1.8-fold between 2000 and 2015, compared with a 1.3-fold increase in Japan. This hindered Japanese companies' ability to adapt a software-driven economy.

The second reason for the delay in Japanese companies' adaptation to the new TEP, i.e., the delay in software development, is their inability to understand that global trends in software development were moving toward open architecture and packaged software, which meant they failed to abandon the old system of custom-made software development tailored to each customer's needs. In 2016, Japan's custom software sales were 7,966.5 billion yen versus packaged software sales of 1,093.0 billion yen; it is estimated that the market size of packaged software was only 13.7% that of custom software.[2] In 2016, custom and embedded software accounted for 88.3% of total software sales in Japan, compared with the 11.7% share of packaged software sales. In comparison, in 2016, the percentage of custom software and packaged software in the US was 53.8% and 46.2%, respectively. It can thus be seen that Japan is overly biased toward custom-made software.[3]

Certainly, compared to software development based on packaged software, an integrated development system that develops new custom software for each customer and controls such software from upstream to downstream enables high-quality software. On the other hand, because the majority of software costs are personnel costs, developing new software separately for each customer by a single company inevitably greatly increases the total cost. This strategy thus weakened the international competitiveness of the Japanese software industry. Even if packaged software has inferior quality, it can significantly reduce costs. The tendency to emphasize packaged software is consistent with the fact that the global computer industry has shifted from mainframes to personal computers as well as shifted from stand-alone computers to an open network called the Internet.

---

[2]Global ICT Strategy Bureau, Ministry of Internal Affairs and Communications (2018), p. 12.
[3]Ministry of Internal Affairs and Communications (2018), p. 30.

In addition to under-investing in software and failing to recognize the trend toward packaged software, Japanese companies face new challenges today. This is because software intelligence has made advancements in the 21st century, with AI spreading rapidly to industrial activities and civic life.

Deep learning as evolved from machine learning is the driving force behind the current advances in AI technology. Deep learning, which first spread rapidly around 2012, allows feature quantities to be detected without first defining those features. This is possible because the neural network has a multilayered intermediate level that increases the possible amount of information processing without increasing the number of calculations required. This makes it possible to obtain a much richer and more accurate knowledge than conventional machine learning.

In Sect. 5.2, AI was described as a general-purpose technique, which also holds true for deep learning, which is widely used for automated driving, medical imaging, speech recognition, and natural language processing, among others. The accuracy of deep learning improves as more data are available. As shown in Fig. 5.1, analysis of big data leads to a solution, with further data collected and analyzed based on that solution. This iterative process makes it possible to increase the accuracy of the found solutions.

5.6 Conclusion

In the era of a knowledge-based economy, software has become more important than hardware, which means that R&D and other investments in software play a greater role than ever in maintaining corporate international competitiveness and producing macroeconomic growth. This shift to hardware was spurred by several factors. First, there is a physical limit to the ability to further miniaturize the linewidth of IC chips, which makes it impossible to improve the degree of integration, as done in the past. Second, limited customer needs for mass-produced, mass-consumption products, and the uncertainty and variability of today's newly emerging customer needs give software an advantage in terms of ability to respond both flexibly and at low cost, compared with the high cost and time frame required to change hardware specifications. Third, to identify customer needs that are difficult to understand and predict, the intellectual software (AI) that accumulates and analyzes large volumes of data has begun to evolve at a remarkable pace.

These factors have led to an increasing importance of software worldwide in R&D and other investments. The analysis in this paper has demonstrated that the inability of Japanese companies to adequately respond to this change was one factor behind the prolonged stagnation of Japanese companies from the end of twentieth century through to the 21st century. Today, AI is rapidly evolving and increasingly penetrating a wide range of fields, including autonomous driving, robotics, social infrastructure, and medical care. Within this context, there is no doubt that software is becoming increasingly important, meaning that Japanese companies must take new measures in the future to adapt to these ongoing changes and increasing prominence of software.

References

Cantwell, J., & Santangelo, G. D. (2000). Capitalism, profits and innovation in the new techno-economic paradigm. *Journal of Evolutionary Economic, 10*(1). 131–157.

Cavanillas, J. M., Curry, W., & Wahlster, W. (2016). Big data value opportunity. In J. M. Cavanillas, W. Curry, & W. Wahlster (Eds.), *New horizons for a data-driven economy*. Berlin: Springer, 3–11.

Center for Global Communication, International University of Japan. (2018). *Jinkou Chinou to Nippon 2018 (Artificial intelligence and Japan 2018)*. Tokyo: Glocom, 1–157 (in Japanese).

Cockburn, I. M., Henderson, R., & Stern, S. (2017). The impact of artificial intelligence on innovation. In *Paper Prepared for the NBER Conference on Research Issues in Artificial Intelligence*. 1–38.

Drechsler, W., Kattel, R., & Reinert, E. S. (Eds.), (2009) *Techno-economic paradigms: Essays in honour of Carlota Perez*. London: Anthem Press, 1–420.

Freeman, C. (1987). *Technology policy and economic performance: Lessons from Japan*. London: Pinter Pub Ltd, 1–150.

Freeman, C. (2009). Shumpeter's business cycles and techno-economic paradigm. In W. Drechsler, R. Kattel, & E. S. Reinert (Eds.), *Techno-economic paradigms: Essays in honour of Carlota Perez*. London: Anthem Press, 125–144.

Global ICT Strategy Bureau. (2018). Ministry of internal affairs and communications. *Results of the basic survey on the information and communications industry, 2017*. Tokyo: Ministry of Internal Affairs and Communications, 1–74.

Hayashi, T. (1995). Amerika gijutsu taikei no henka to gijutsu kanren (Changing mechanism of the American technologies and the inter-relationships of the technologies). *Rikkyo Economic Review, 49*, 57–74 (in Japanese).

Hayashi, T. (2007). Shinseihin kaihatsu purosesu niokeru chishiki sozou to ibunka manejimento (Knowledge creation and multi cultural management in the new product development process). *Rikkyo Business Review, 1*, 16–32 (in Japanese).

Kodama, F. (1992). Technology fusion and the new R&D, *Harvard Business Review*, July-August 1992. 70–78.

Komoda, F., Nishiyama, K., & Hayashi, T. (1996). *Jouhou tsuushin to gijutsu renkan bunseli (Information telecommunication and technology linkage analysis)*. Tokyo: Chuo keizai, 1–203 (in Japanese).

Komoda, F., Nishiyama, K., & Hayashi, T. (1997). *Gijutsu paradaimu no keizaigaku (Economics of technology paradigm)*. Tokyo: Taga publishing, 1–260 (in Japanese).

Komoda, F. (2000). *Kagaku gijutsu to kachi (Science/technology and value)*. Tokyo: Taga publishing, 1–269 (in Japanese).

Kuhn, T. (1962). The structure of scientific revolution. Chicago: Chicago University Press, 1–210.

Lyko, K., Nitzschke, M., & Ngomo, A-C. N. (2016). Big data and acquisition. In J. M. Cavanillas, W. Curry, & W. Wahlster (Eds.). *New horizons for a data-driven economy*. Berlin: Springer, 39–61.

Lippoldt, D., & Stryszowski, P. (2009). *Innovation in the Software Sector*. OECD, 1–88.

Ministry of Economic, Trade and Industry (2015). *Tsusho hakusho: 2015 (White paper on international trade: 2015)*, Tokyo: Ministry of Economic, Trade and Industry, 1–371 (in Japanese).

Ministry of Internal Affairs and Communications (2018). *Information and communication in Japan, White paper*. Tokyo: Ministry of Internal Affairs and Communications, 1–380 (in Japanese).

Nomura Research Institute (2016). *ICT no shinka ga koyou to hatarakikata ni oyobosu eikyo ni kansuru chousa kenkyu (Survey research on the impact of evolution of ICT on employment and workplace)*. Tokyo: NRI, 1–72 (in Japanese).

OECD (1996). *The knowledge-based economy*. OECD, 1–26.

OECD (2014). *Data-driven innovation for growth and well-being*. OECD, 1–86.

Perez, C. (1985). Microelectronics, long waves and world structural change: New perspectives for developing countries. *World Development, 13*(3), 441–463.
Perez, C. (2004). Technological revolutions, paradigm shifts and socioinstitutional change. In E. S. Reinert (Ed.), *Globalization. Economic Development and Inequality*. Cheltenham: Edward Elga, 217–242.
Schumpeter, J. A. (1912). *Theorie der Wirtschaftlichen Entwicklung*. Leipzig: Verlag von Dunker & Humblot, 1–548.

Dr. Fumio Komoda (Ph.D., Kyushu University) is Professor Emeritus of International Business at the Faculty of Economics, Saitama University, Japan, and visiting professor of International Business at the Faculty of Economics and Business Management, Saitama Gakuen University. His main research interests is developing text mining methods that can contribute to building a company's technology strategy. He received his PhD in Economics from Kyushu University. He has published widely works and academic papers in the domain of management of technology and text mining.

Chapter 6
Paradigm Shifts in the TFT-LCD Industry and Japan's Competitive Position in East Asia

Kazuhiro Asakawa

Abstract This chapter analyzes Japan's changing competitive position in the TFT-LCD industry over the past several decades. Based on the analysis of multiple factors that affected Japan's decline, I identified paradigm shifts in the TFT-LCD industry that have directly or indirectly led to a change in Japan's competitive position. The chapter shows that declining competitiveness of a country and a firm cannot be fully understood without closely relating it to the shifts in the paradigm of the industry. In addition, this chapter shows how the paradigm shifts of the industry, the changing competitive positions of a country and a firm, as well as the changing locus of innovation for a company, i.e. domestic vs. international and in-house versus collaborative, are aligned with one another.

Keywords Paradigm shifts · TFT-LCD industry · Japan · Competitive position · Locus of innovation · East Asia

6.1 Introduction

In today's transnational economy, national competitive advantage is becoming less stable and sometimes ephemeral in many industries (Doz et al. 2001). Knowledge and technology migrate from advanced to less advanced countries, meaning that countries with less technological progress can catch up with those with more advanced technologies through learning and imitation (Kim 1997; Lee 2005; Mathews 2006). A nation's competitive advantage may not be sustainable under these circumstances. The latecomers outperform the incumbents by leveraging cost advantage of less-developed countries.

We know that firms based in a declining home country have difficulty accepting the reality and shifting the locus of innovation from being home country-driven to

K. Asakawa (✉)
Graduate School of Business Administration, Keio University, Yokohama, Japan
e-mail: asakawa@kbs.keio.ac.jp

© Springer Nature Singapore Pte Ltd. 2019
J. Cantwell and T. Hayashi (eds.), *Paradigm Shift in Technologies and Innovation Systems*, https://doi.org/10.1007/978-981-32-9350-2_6

overseas-driven (Doz et al. 2001). This chapter is focused on such a change, with a particular attention to firms struggling to cope with their home country's decline in industry competitiveness, by trying to change the locus of innovation from domestic, in-house to international and open (Doz et al. 2001; Cantwell and Mudambi 2005; Asakawa et al. 2018).

To elucidate such a shift in the competitiveness, I shed light on Japan's thin-film transistor liquid crystal display (TFT-LCD) industry which has gone through a series of major transformation in the past few decades. US initially enjoyed its competitive advantage in the TFT-LCD industry in the 1970s since RCA developed the first flat panel display in 1968 (Hu 2012). However, in spite of their supremacy in R&D capability, US firms stopped pursuing innovation in this industry due to the time-consuming nature of commercialization from the state-of-the-art technologies at the time. In response, Japanese firms, i.e., Sharp, NEC and Seiko, absorbed technologies from the US and nurtured the competencies throughout the 70 and 80s (Murtha et al. 2001; Hu 2012; Johnstone 1999; Polgar 2003). Widely known as "spiral management," Sharp engaged in technology absorption and application over several decades to develop a wide range of products using TFT-LCD technology. Japan's production volume of TFT-LCD panels reached over 90% of the worldwide share in the 90s, when Japan became the world's center for the TFT-LCD industry. By then, US firms had lost interest in the industry with the exception of IBM which formed a Joint Venure (JV) with Toshiba (DTI) for further joint development of technologies and products. Applied Materials also formed a JV with Komatsu (AKT), among a few US survivors in the industry. By then, Japan became the leader in TFT-LCD industry (Murtha et al. 2001).

However, Japan's dominance in TFT-LCD production share has weakened over time, with Korea and Taiwan taking over the leading position. The success behind Korean and Taiwanese firms lies in the fact that they have learned knowledge from Japan and adopted the global best-supplier policy for equipment and materials, regardless of nationality (e.g., Samsung). Korea and Taiwan succeeded in rapid technological catch-up through intensive learning from Japan, combined with the ease of access to financial capital in each country (Linden et al. 1998; Lee and Lim 2001; Lee and Mathews 2012). The 2000s was a decade of Korean and Taiwanese dominance in the TFT-LCD industry, with each achieving approximately 40% of worldwide share of production as Japan's share plummeted rapidly (Asakawa 2007). Subsequently, the 2010s was a decade of further reshuffling in the competitive arena in the industry, manifested by a rapid surge of China as a major threat to Korean and Taiwanese rivals (IHS Markit 2018). Based on a powerful backup of the Chinese government, Chinese firms aggressively invested in huge volume of production lines in large size TFT-LCD panels, up to Generation (Gen) 10.5 which is by far the largest as of April 2019 (HIS Markit 2018). Such aggressive large-scale investments in fabrication facilities (fabs) have drastically lowered the price of the panel, thus making the incumbent Korean and Taiwanese manufacturers less price competitive. China also invests in small- to medium-size panel production for mobile devise such as smartphones (HIS Markit 2018).

Fierce competition among the East Asian countries pushed Japan further down in a marginal production share. While the catch-up of Korea and Taiwan has been explained by the flying geese model which explains how Japan contributed to the development of other East Asian countries (Linden et al. 1998; Kojima 2000), such a model lost its validity by the 2010s, as Japan, once the leader in TFT-LCD industry in East Asia, has turned to a subordinate position by the end of the second decade of the twenty-first century. Japan's TFT-LCD panel makers have stopped investing in fabs beyond Gen 3.5 by the end of the 90s, and only Sharp has continued to invest in large size TFT-LCD panels (Murtha et al. 2001). Hitachi, Toshiba and Sony have merged their TFT-LCD divisions and set up Japan Display Inc. (JDI) specializing in small- and medium-sized panel in 2012 funded by Innovation Network Corporation of Japan (INCJ), the public-private fund (currently INCJ Ltd.). Basically Sharp and JDI have been the only two major Japanese players in TFT-LCD panels in the twenty-first century. The 2010s marked a new era of Japan's TFT-LCD industry: these national players ended up falling under the control of foreign actors. Sharp has been acquired by Foxconn in 2016 and has become a subsidiary company under the control of Taiwanese firm. And JDI, a national merger of Hitachi, Toshiba and Sony, is likely to fall under the control of a group of investors from China and Taiwan according to the news as of early April 2019. Such a drastic decline in Japan's presence can be analyzed along the shifting paradigm in the industry over the past several decades.

The objective of this chapter is to interpret various factors behind the decline in Japan's share of TFT-LCD panel production over the past several decades from the lens of paradigm shifts in innovation strategy. Looking back the history of industry evolution in the past several decades, we identify changes at differing dimensions: source of competitive advantage, the nature of technology, country leadership, product, panel size, geographic scope, and the locus of innovation. What kinds of paradigm shifts have taken place in the TFT-LCD industry in the past decades? This chapter illustrates the paradigm shifts in the TFT-LCD industry in the past several decades and discusses how and why Japan's competitive positioning has changed so drastically.

The rest of the chapter is organized as follows: I first analyze various factors that have led to Japan's decline in the TFT-LCD industry in two consecutive periods, the first and the second decade of the twenty-first century (Sect. 6.2). Based on the analysis of Japan's decline in the TFT-LCD industry, I introduce a set of paradigm changes in this industry and relate it to Japan's decline in the industry (Sect. 6.3). Then I summarize how such decline in Japan's TFT-LCD industry has led Japan to shift the locus of innovation from domestic/in-house to international/collaborative (Sect. 6.4).

6.2   Evolution of Japan's Competitive Position in TFT-LCD Industry

6.2.1   Japan's Leadership in TFT-LCD Industry (Until the Mid-1990s)

Japan established its competitiveness in the TFT-LCD industry across the stages of value chain (Murtha et al. 2001). During the golden decades of Japan's electronics and electric industry in the 1980s and the early 1990s, electronics manufacturers enjoyed their dominant position in the global market by leveraging their superior technology, corporate brand, capable and hardworking engineers, and loyal customers of the Japanese products. Japan's domestic economic growth lasted until the bubble crashed in the early 90s. This economic boom encouraged Japanese consumer electronics manufacturers to develop laptop PCs with TFT-LCD display panels. Stable market demand allowed the Japanese firms to invest in R&D to enhance the product quality appreciated by the Japanese consumers. Branding was successfully done based on the high reputation of the Japanese product. TFT-LCD panel makers initially supplied their panels to their own PC and TV product divisions to serve as source of their own competitive advantage. The panel makers collaborated closely with upstream equipment and material manufacturers within Japan based on long-term trust (Murtha et al. 2001; Johnstone 1999; Nugamami 1999).

6.2.2   Decline of Japan's Production Share in the TFT-LCD Industry Since the 2000s

Japan's decline in leadership in terms of production volume of TFT-LCD display panels was influenced by various conditions.

Economic recession:
    First and foremost, Japan's economic recession altered the trajectory of Japan's TFT-LCD firms. Since the existing business model of Japanese' TFT-LCD was largely based on the assumption of a stable economy, a serious economic recession generated concern for continuing investment in large-scale factories. In a way, managers lost their nerve in the Asian financial crisis due to the firms' insufficient capital, and most Japanese companies refrained from taking a risk to invest in future generation fabs beyond Gen 3 (Murtha et al. 2001). No Gen 3 fab investments were made after Sharp and Display Technologies Incorporated (DTI), the IBM-Toshiba manufacturing alliance, and then Hitachi's Gen 3.5. Instead, they relocated their manufacturing function to the lower-cost country, Taiwan. Such a move shows a remarkable departure from the previous model in which Japanese TFT-LCD companies created value within Japan through tight coordination across the value chain (Asakawa 2007). While relocating part of the value chain would have been

avoided at all costs under the previous paradigm, the economic recession triggered by Asian financial crisis has changed the organizing principle of Japan's TFT-LCD industry.

Isomorphic approach to Taiwanese firms:

Japanese TFT-LCD companies' isomorphic approach to Taiwanese firms was intended to contain rapid expansion of Korean firms (Murtha et al. 2001; Shintaku and Yoshimoto 2009; Tabata 2014). In 1999, everyone rushed to follow ADI. In the wake of the Asian financial crisis in 1998, mass-production of TFT-LCD panels by Korean firms lowered the price substantially. Japanese firms' market share in Taiwan plummeted, and Japanese firms, including Mitsubishi, Toshiba, IBM Japan, Sharp, and Matsushita, had to start producing TFT-LCD locally to lower the production cost, and they ended up transferring technologies to Taiwanese partners. Korean firms rapidly followed Sharp, DTI, and Hitachi in investing in large-scale panel fabs when no other Japanese companies were doing so. They rapidly accumulated core technological know-how by hiring Japanese engineers formally or informally over the weekend (Murtha et al. 2001). Japanese TFT-LCD manufacturers saw this as a major threat, and decided to transfer technology to Taiwanese firms to deter the Korean move (Shintaku and Yoshimoto 2009; Asakawa 2007). ADI, a Mitsubishi group member and a joint TFT-LCD panel-production venture between Asahi Glass and Mitsubishi, was among the first, in collaboration with Chung Hwa Picture Tubes (CPT) (Murtha et al. 2001). This shows a remarkable departure from a previous paradigm in Japan's TFT-LCD industry, in which technology was retained within Japan and collaboration with overseas partners was avoided.

Underestimating the value of old-generation knowledge transferred to Taiwan:

Another departure from the previous model in existence until the 90s concerns the mobility of engineers and knowledge across national borders in a rather careless manner. At the time of setting up older generation fabs in Taiwan, Japanese firms underestimated the value of the knowhow which was shared to Taiwan firms that launched the lagged generation factories. Since these Taiwanese companies were starting from scratch, older generation knowledge was more useful and important for them to catch up (Murtha et al. 2001; Suarez and Utterback 1995). Taiwanese firms which introduced technologies from Japanese firms, including CPT from ADI (Mitsubishi), ADT (established by Acer in 1996) from IBM Japan, Unipac from Fujitsu (2000) MVA technology, and Unipac also from Matsushita, chose to adopt the lagged 3 Gen technology while Japan had 3.5 Gen; Taiwanese engineers were sent to Japan to learn Japan's SOP (standard operating procedure), and Japanese engineers also visited Taiwan to supervise the local operations. Japanese firms had little concern for leaking core technologies, primarily because they were transferring old generation technologies. Although Taiwan firms preferred to lag, they did not lag too much, to make sure they can learn and catch up quickly enough, by maintaining the "fast follower strategy" (Mathews et al. 2011; Mathews 2005; Shintaku et al. 2006; Mathews and Lim 2001). This phase of history in Japan's TFT-LCD industry shows a remarkable contrast with the previous stage until the end of 90s, in terms of the mobility of knowledge-holders across borders. Japan was obliged to engage in lower-cost manufacturing in Taiwan, thus technology transfer through human interface was a necessary part of the process. However, Japan did not find technological leakage risky because the technology transferred was older and no longer special in Japan (Asakawa 2007). Japan's propensity for advanced technology led them to underestimate the value of older generation knowledge when the firms with lagged technological capacities could learn a great deal more from the older generation technologies.

Japan's risk-averse decision-making style:

In fact, what differentiated Japan's aversion to continuing investment in larger-size TFT-LCD panel manufacturing from Korea's and Taiwan's aggressive investment relates to the traditional decision-making style of Japan's large, diversified electronics firms: Economic growth in the 80s facilitated business diversification, so that senior managers could not keep them abreast of the most recent situation of all those businesses. Since top management was not necessarily updated on the most recent situation of the TFT-LCD business, they avoided making a hasty decision to invest in future generation fabs during the period of high economic uncertainty (Murtha et al. 2001). The Asian financial crisis forced Japanese firms to delay or stop such long-term investment in future generation fabs. In contrast, Taiwan and Korean firms had better access to capital thanks to the abundant financial support of Korean Chaebolsand less strict Taiwanese equity markets (Mathews 2005).

Risk-taking investment by Korean and Taiwanese firms:

The bold risk-taking strategy implemented by Korean and Taiwanese firms contributed to Japan's decline: Taiwanese and Korean firms took a bold risk by entering during industry downturns within the crystal cycle as resources become available for challengers in the downturns (Mathews 2005; Lee and Mathews 2012). Such a strategy is only effective for the new entrants as downturns provided them with a small "window of opportunity" for raising investment while the incumbents could not do so (Mathews 2005). In contrast, Japanese firms, the dominant incumbent players at the time, were risk-averse given the uncertainty in their surrounding environment.

Mass production in China by new entrants:

The Japanese TFT-LCD industry had relied on stable growth in the downstream market via TV set segment. However, new entrants in this business, both from Japan and outside Japan, have expanded the market size significantly and have lowered the price of TV sets. These new entrants chose China as a location for manufacturing TV sets. The massive entry of Chinese makers of TV sets contributed to the declining price of TV set price by mid-2000s. And these latecomers could accelerate the design of TFT-LCD TV sets by sourcing technology more easily and quickly without accumulating their own proprietary technologies. Digitalization and modularization of technologies in TFT-LCD TV sets allowed them to source external technologies that are modularized, such as LSI. As Chinese and other East Asian firms could draw on external modularized technologies relatively easily, they could save time required to launch the large-scale production and sales of their TV sets (Shintaku and Yoshimoto 2009). Consequently, Japanese TV set makers lost their price competitiveness vis-à-vis other East Asian competitors and fell out of competition. Such a decline in cost leadership of Japan's TFT-LCD TV products had a significantly negative impact on Japan's TFT-LCD panel manufacturers in terms of the shrinking market share.

Japan's equipment and material makers:

Japan's decline in the share of TFT-LCD panel production was further precipitated by the independent moves by Japan's equipment and materials makers which were anxious to sell their products to wider overseas market when Japanese firms stopped investing. Although equipment and materials embody substantially all of the critical knowledge necessary to start up new fabs, the Japanese equipment and materials makers tried to avoid knowledge leakage as much as possible by entering the new East Asian markets through exporting or FDI (Murtha et al. 2001). Nevertheless, export and FDI were never perfect for avoiding the erosion of the proprietary knowledge of specialized advanced technologies (Asakawa 2007).

Equipment and materials by Korean firms:

Korean companies such as Samsung could not afford to rely on inefficient strategies for supplying equipment and materials from Japan. Samsung rapidly internalized the manufacturing of materials and equipment for TFT-LCD and substituted certain technologies so as not to rely exclusively on Japanese suppliers. For example, Samsung already internalized most procurements of core materials production within its group companies or its suppliers networks in the 2000s (Song 2006). Up to the Gen 5 line, Samsung Electronics (SEC) internalized glass substrates, color filters, and driver IC. For the Gen 7 line, SEC internalized backlight production. SEC also constructed glass substrate plant next to its own LCD plant. For LG-Philips and CMO, polarizers can be procured from their own group companies. The Korean and Taiwanese governments also tried to raise the percentage of internal procurement of material and equipment (Korea: 50% by 2005; Taiwan: 70% by 2008) (Song 2006; Shintaku et al. 2006).

The role of the government:

The surge of Korean and Taiwanese firms is not exclusively based on the company-level efforts. The Korean and Taiwan governments played major roles, including Taiwan's government-owned Industrial Technology Research Institute (ITRI), which developed multiyear plans to foster TFT-LCD industry and led to the rapid growth of the industry. (Murtha et al. 2001). At the same time, spin-off of ITRI staff to TFT-LCD industry also contributed to the rapid growth of the industry (Akabane 2004; Shintaku et al. 2006). Spin-off projects from ITRI were significant and included CMO, UMC and TSMC (Shintaku et al. 2006; Tabata 2014); Topply was also a spin-off of a national research institute.

Japan's attachment to technological upgrading:

Japan's tendency to prefer technological upgrading to large-scale investment in manufacturing lost its advantage in the advent of intensive price competition in East Asia. Nevertheless, Japanese firms continue to adhere to the technological upgrading by putting priority on moving up to concentrate on new technologies with higher-value added, such as LTPS. Japanese firms outsourced less sophisticated manufacturing of TFT-LCD panels to Taiwanese suppliers (Akabane 2004). Such a propensity to stick to advanced technology was consistent with the logic of value chain disaggregation on the smile curve of the value chain (Mudambi 2007), in which firms that control value-adding upstream R&D activities have strong incentives to increase the efficiency and effectiveness of such activities by relocating the standardized activities in the middle to emerging market economies (Mudambi 2008). However, in the case of TFT-LCD industry, the middle of the value chain, i.e., manufacturing of TFT-LCD panels, was a major source of competitiveness for East Asian panel producers which concentrated on the large-volume production of price-competitive panels.

6.2.3   Further Changes in Japan's TFT-LCD Industry Since 2010—Present

The TFT-LCD industry manifested a series of further changes in the 2010s. Major changes in the industry include the rapid growth of the small size panel market triggered by the proliferation of the smartphone; a surge of organic light-emitting diode (OLED) as a major competitor of TFT-LCD technology; the major growth of China in the TFT-LCD industry, and the changing source of competitive advantage from technological differentiation to cost and price reduction. Sharp was acquired by the Taiwanese Foxconn in 2016, and JDI, a merger of Hitachi, Toshiba and Sony since 2012, is on the verge of falling under the control of a group of investors from China and Taiwan as of April 2019. These two incidents indicate a symbolic decline of Japan's TFT-LCD industry, in both large and small- to medium-size panel categories.

The rise of the smartphone panel display market:
    Due to the rapid growth of the smartphone market, small-size panels saw rising shares as well. Japan Display Inc. (JDI), merged by display divisions of Sony, Toshiba and Hitachi and sponsored by Innovation Network Corporation of Japan (INCJ), the public-private fund (currently INCJ Ltd.) in 2012, focuses on small-size TFT-LCD display panel for smart phone. JDI continues to emphasize on technological competencies in TFT-LCD (Nikkei Business 2016, 05.30), and it has remained as a Japanese national team, despite fierce price competition in Asia. However, here, the strong presence of China is clear. Based on the shipping volume, Century is ranked number one (11.2%), followed by BOE (10.8%), Hannstar, Samsung and Tianm (IHS Markit 2018). China's aggressive investment in mass production has lowered the price substantially, and made JDI far less price competitive in the face of Chinese rivals. JDI continued to believe in the power of technological competencies while overlooking the efficiency aspect of competitiveness. Furthermore, its primary customer Apple's decline in its iPhone sales has

severely damaged JDI's shipment of the panels. Apple's decision to adopt OLED for iPhone's screen has further damaged JDI which was not agile in shifting its attention to OLED in spite of its increasing market share vis-à-vis the TFT-LCD (IHS Markit 2018; Nikkei 2019a, 4.4., p. 15). Due to its administrative heritage of being funded by the public sector, the company refrained from drastic restructuring by closing down underperforming domestic fabs (Nikkei 2019b, 4.4., p. 15). As of early April 2019, JDI is most likely to fall under the control of a group of investors from China and Taiwan (Nikkei 2019b, 4.4., p. 1). This indicates the decline of Japan's TFT-LCD industry in the small- and medium-sized panels, along with the decline in the large TFT-LCD display panel.

TFT-LCD rivalry:

PDP's shipment share has declined since 2006 at its peak, and OLED continues to grow. According to IHS Markit (2018), TFT LCD revenues are declining whereas AMOLED revenues are growing. Apple's move to adopt OLED for iPhone's display panel should affect such a trend much further. The OLED market is virtually monopolized by Korean firms: Samsung Display for smartphone and LG Display for TV. OLED TV manufactures source panels from LG Display, including Sony and Panasonic. Since Gen 3 in TFT-LCD, Japanese firms have refrained from investing aggressively in OLED-related R&D in the wake of the Lehman shock in 2008. The OLED divisions of Japanese firms Sony and Panasonic merged in 2015 as JOLED which developed its unique and innovative printed OLED displays for the first time in Japan in 2017 with a potential to realize a simple, low cost manufacturing process. JOLED targets both large and small OLED panels (JB Press 2018). Nevertheless, despite their superior technology, Japanese firms lag behind Korea in the product development and manufacturing of the OLED panels.

Surge of China's presence:

The emergence of China in the TFT-LCD industry in the 2010s shows a remarkable evolution of the industry since 2000s when Korea and Taiwan were the two dominant giants. Today, China is a major threat to Korea and Taiwan which enjoyed their dominant position in the 2000s. Japan cannot compete against China in terms of cost advantage, and China's technological standard has further threatened other East Asian countries. According to HIS Markit (2018), by 2022, China will have more than 19 TFT-LCD fabs in Gen 8.6 through 10.5 and 20 OLED fabs. Gen 10.5 line is already in the ramp-up phase in BOE, or under construction by China Star. Foxconn-Sharp is building its Gen 10.5 line in China as well, and China has undoubtedly become the world center of large size panel production. China has become a center of Gen 8 through Gen 10.5 investment. While most 65" are made in Korea (Gen 8) and Taiwan (Gen 6), new entrants come from China (BOE and China Star) and elsewhere in Gen 10.5 (IHS Markit 2018).

China also intends to internalize FPD's equipment based on the initiative "Made in China 2025," in which China intends to create the entire value chain in FPD locally. Such a move should present a major threat to Japan and Korea. (IHS Markit 2018).

6.3   Paradigm Shifts in the TFT-LCD Industry

So far I have discussed how the TFT-LCD industry has evolved in the past several decades, and have shed light on various factors that have led to Japan's declining position in the industry. Based on the discussion above, we can identify a number of paradigm shifts in the industry that have directly or indirectly led to a change in Japan's competitive positioning over the past decades.

From technological advantage to cost advantage:
    The source of competitive advantage has changed over time, from technological differentiation capability to cost reduction capability. As discussed above, TFT-LCD industry has been led by firms with superior technological capabilities. Innovation in TFT-LCD panel manifested itself in the form of upgrading across generation (Gen), which basically meant the war of enlarging substrate size and further enhancing the high resolution of the screen. However, since Korea and Taiwan began to invest in massive scale of panel production in the fabs since the 2000s, followed by China since the 2010s, cost advantage has taken over as an alternative source of competitive advantage for the manufacturing of flat panel display. Japan has lost its industry leadership because of its continuing attachment to high-spec technological sophistication. Interestingly, today's Korean firms, as represented by Samsung, are facing the same challenge when pursuing their technological sophistication at the expense of price competitiveness (Nikkei 2019c, 4.6., p. 10). The industry seems to have reached a limit in size and the quality of the panels which consumers would expect. Commoditization of TFT-LCD panel, which has taken place so rapidly (Nikkei 2019d, 3.31), has made the latecomers easier to learn from Japan and apply the acquired knowledge much easier. While technological capability continues to be necessary, it alone no longer serves as an absolute source of competitive advantage given the high standard of technological sophistication of the existing products. The Chinese latecomers are catching up in technological innovation and are outperforming the Japanese and other East Asian incumbents in cost advantage.

Modularization of TFT-LCD technology:
    Traditionally, source of competitive advantage in this industry has adhered to the tacit, human-embodied knowledge contained within a production site, for rapid size war required the firms to keep investing in higher generation fabs, and there was no codified knowledge available for rapidly and frequently upgrading higher Gen fabs (Murtha et al. 2001). However, the size war seems to have reached its limit at Gen 10.5, and the required technology for operating panel production has been widely shared among Korean, Taiwanese and Chinese firms. Furthermore, digitalization facilitated the modularization of the TFT-LCD industry, and such a trend makes Japan's traditional integral approach in manufacturing less competitive. Japan's strength based on integral production architecture has begun to lose its competitive edge in the TFT-LCD. (Nakagawa et al. 2016). While Japan's technological supremacy remains, the digital revolution has shifted the source of competitive advantage away from Japan's traditional manufacturing method pursuing high

quality improvement through long-term efforts to the cost advantages of leveraging AI and robotics. Modularization of TFT-LCD technology opened a window of opportunity for new entrants into the game without a long history of their own internal technological development.

From exploiting the advanced TFT-LCD technologies to exploring the opportunities in OLED:

Evolution of the TFT-LCD industry can be summarized from the aspect of technology and product category. TFT-LCD replaced the cathode-ray tube (CRT), a dominant technology until the end of the twentieth century. The TFT-LCD industry has grown in Japan since the early 90s with the use of LCD displays for laptop computers and desktop monitors. Sharp engaged in the first production of laptop size TFTs (Linden et al. 1998). LCD revenues passed those of CRTs in 2002 (Semenza 2012), and the arrival of the twenty-first century marked the beginning of the large-size screen panel of LCD TVs. Plasma display (PDP) surged as an alternative display for large flat-panel TVs, with the symbolic introduction of a 42-inch PDP in 1997. However, after a period of rivalry with TFT-LCD, PDP lost its momentum since hitting the peak of its revenues in 2006. The TFT-LCD panel market continued to grow, with aggressive investment in higher Gen lines throughout the 2000s, with Korea and Taiwan engaging in fierce competition of large size TFT-LCD panels. After a decade of the fierce race of upgrading the fabs among Korean, Taiwanese and Japanese (Sharp) firms, until the Gen 10.5 line, built by Sharp in 2009, the 2010s marks a new era of even fiercer competition involving China as a major competitor. In addition, another decade of manufacturing high-performance displays for mobile applications has arrived across the 2010s (Semenza 2012). In the meantime, OLED surged as a new competing technology, although Kodak initially developed it in the 80s. Forecasting suggests that the shares of TFT-LCD to those of OLED will change from 8:2 in 2018 to 6:4 in 2025 (IHS Markit 2018). In sum, a paradigm change in technology manifested itself as a transition from small- and medium-sized TFT-LCD displays for laptop PCs in the 90s to a greater size of TFT-LCD panel displays for large TVs during the 2000s, then to a surge of market for smaller screens for mobile handheld devices such as smartphones and tablet PCs along with the continuing scale of large TVs since the 2010s (Semenza 2012). The surge of small screen display for mobile devices requires different technological innovation distinct to smartphone use, and this opens the windows of opportunity for further technological innovation and competitions.

While TFT-LCD remains the dominant category, its relative share is declining while the share of OLED is increasing since 2010s. According to a survey by IHS Markit 2018, the proportion of TFT-LCD to OLED is likely to change from (HIS 8:2 in 2018 to 6:4 in year 2025 (HIS Markit 2018). Although Japan has done R&D on OLED since early stage, its priority has been put on TFT-LCD for Sharp and JDI, the only two main survivors of TFT-LCD manufacturers based in Japan. As the share of TFT-LCD plummets, however, the panel makers are directly affected negatively. Facing the fierce rivalry against Chinese companies in the TFT-LCD panel, Samsung has shifted its direction toward OLED by adopting OLED panel to its own smartphone. In contrast, Sharp and JD were slow to react, and Sharp has been acquired by Foxconn in 2012. JDI is likely to be acquired by a group of investors from China and Taiwan so that mass production of OLED panel shall be realized in China (Nikkei 2019d, 4.4., p. 15).

Evolving country leadership:

The evolution of the TFT-LCD industry can be summarized from the standpoint of country leadership. As summarized above, the dominance of US and Europe shifted to Japan throughout the 90s when the major electronics firms invested in up to Gen 3 TFT-LCD fabs; then all but Sharp stopped after Gen 3.5 lines. Korean dominance was a threat to Japanese firms, which tried to contain Korean expansion by transferring technologies to Taiwan firms manufacturing panels in Taiwan. The 2000s was a decade of Korean and Taiwan dominance in the TFT-LCD industry in terms of the production volume. Japanese firms specialized in advanced technology and partnered with Korean and Taiwanese firms for low-cost manufacturing, but most Japanese firms eventually lost their market share in TFT-LCD display panels and withdrew from the game. The Japanese firms specialized in new, alternative LCD advanced technologies such as LTPS for small- to medium-sized product displays, while Korean and Taiwanese latecomers focused on the more conventional technologies used in flexible LCD panels and LCD TVs (Hu 2012). As the latecomers complemented Japan technologies by complementary knowledge to technology leaders (Nekar 2003), East Asia's technological range in TFT-LCD industry has further expanded. Sharp continued to invest in the race of higher generation TFT-LCD fabs to compete with Korean and Taiwanese firms throughout the 2000s, up to the Gen 10.5 line. Although Sharp was acquired by Foxconn in 2016, it is the only surviving player from Japan in the fierce race of generational upgrading of fabs. In the meantime, the rapid advancement of generations in TFT-LCD fabs has reached a limit, currently at Gen 10.5 line.

In the meantime, during the 2010s, Chinese latecomers aggressively invested in multiple fabs in TFT-LCD and OLED, thus China ended up having fab Gen 10.5 lines and a forecast says China will have 19 Gen 8- Gen 10.5 fabs in TFT-LCD and OLED by 2022, and 19 Gen 5.5- Gen 6 fabs in a-Si, TPS, and Oxide by 2022 (HIS Markit 2018). Since 2017, China has become the largest FPD manufacturer in the world (HIS Markit 2018). China's aggressive investment in fabs has expanded the production lines and generated overcapacity in production. Taiwan and Korea slowed down investments, and some of the firms (especially Korean) were obliged to close down the older generation fabs while concentrating on newer generation lines. Obviously Foxconn Sharp was not big enough to deter Chinese influence. With TFT-LCD technology becoming more modular, technological learning has become less challenging for the latecomers like China (Shintaku and Yoshimoto 2009).

China has also become a center of production of TFT-LCD panels for other Asian companies: Japanese, Korean and Taiwanese companies operate factories in

China. However, China has not reached a stage of advanced technologies in equipment and materials which are still dominated by Japanese firms.

The competition in both large and small panel display markets:
    While the main competition in the past has been exclusively around the enlarging size of the substrate, today's competition occurs both in the large size screen for the TV sets and the small size screen for the smartphone. The war of substrate size in the 2010s has reached the Gen 10.5, and the 2010s is characterized by the competition in both the large and the small- to medium-size panels. For example, even among the major Korean firms, LG display focuses on large panel for TV sets, while Samsung shifted its focus onto the small panel for the smartphone. Japan's JDI focuses on small size panel for mobile applications (Nikkei 2019d, 4.4). Indeed, in the 2010s the smartphone market has rapidly grown, but so are the growing number of competitors in this sector. And the demand of the small screen panel depends on Apple's sales as well as its decision to adopt OLED screen for the iPhone. (HIS Markit 2018).

From national to regional innovation:
    The TFT-LCD industry has initially been tied to a national innovation system, in which Japan, Korea and Taiwan has manifested differing patterns of industry development. Japanese electronics companies have kept the core knowledge within Japan throughout the 90s. Sharp's black box strategy represents this approach, by shutting out non-Japanese players from its production facility in Japan. However, since Japan transferred technology to Taiwan to start local production, the industry has become more regional. Japanese companies, as once represented by Sony-Samsung, started collaborating with Korean and Taiwanese companies to lower production cost, secure stable manufacturing, and supply panel for their products. Furthermore, the 2010s marked a new stage of Japanese Sharp being acquired by Taiwanese Foxconn. Current negotiation indicates a likelihood of JDI being acquired by a group of investors from China and Taiwan, as of April 2019. This is a symbolic transformation of the industry from national to regional. Such evolution of industry indicates a clear shift in paradigm from nation-based innovation to region-based innovation, in which no single country can fulfill the entire value chain activities satisfactorily (Cantwell and Iammarino 2003). Japan put priority on the left end of the smiling curve (Mudambi 2007; Everatt et al. 1999) focusing on R&D, design, etc. while leaving the middle of the value chain, i.e. low-cost manufacturing etc., to other East Asian countries (Mudambi 2008). The smile dynamic explanation (Mudambi 2018) fits the situation of the regional nature of TFT-LCD innovation, in which Korea and Taiwan in the 2000s and China in the 2010s have been on a catch-up mode, targeting at the left end of the smiling curve from the middle of the value chain; Japan situated in the left end of the smiling curve has incentives to relocate the less value-creating tasks that can be modularized in the middle of the value chain, i.e. low-cost manufacturing, to other East Asian locations (Mudambi 2018). Japan, despite its declining position in the mass-manufacturing of TFT-LCD display, remains dominant in the upstream equipment and materials sector. Thus East Asia covers a full range of tasks on the

Table 6.1 Paradigm shifts in the TFT-LCD industry

		Old paradigm	New paradigm
	Period	Until the 1990s	Since the 2000s
Paradigm shifts in the TFT-LCD industry	Source of competitive advantage	Technological advantage; firm-specific technological capability and tacit knowledge	Cost advantage as for panel manufacturing; Technological advantage as for equipment and materials
	Production architecture: Modularization of TFT-LCD technology	Closed Integral (trans-generational upgrading of fabs required tacit human-embodied knowledge)	From closed integral to open modular (as for mass production of panel)
	Exploitation and exploration of technology	From CRT to TFT-LCD and PDP; Exploitation of existing TFT-LCD technology	TFT-LCD and OLED; PDP declines; Exploration of emerging opportunity in OLED
	Evolving country leadership	Japan as the leading country for technology, production and branding	East Asian region (Aggressive production expansion of Korea and Taiwan during the 2000s; rise of China since the 2010s; Japan for upstream advanced technologies)
	Final product and panel size	First for laptop PC; then for large LCD TV sets	For large LCD TV sets competing with large PDP TV sets until the mid-2000s; Racing over generations (larger substrate size) until the end 2000s; For small mobile (smartphone) display panel since the 2010s
	Geographic scope of innovation	National innovation	Regional innovation
Paradigm shifts in Japan	Japan's position	Dominant	Declining
	Locus of innovation for Japanese firms	"Domestic & in-house"; Vertical integration or collaboration across value chain stages within a business group in Japan	"International & collaborative"; Collaboration across a business group across borders

smiling curve of the value chain (Mudambi 2017), with each country playing a differentiated role on its strength.

Table 6.1 summarizes the paradigm shifts in the TFT-LCD industry and how Japanese firms have changed their competitive positions.

6.4   Transition in the Locus of Innovation for Japanese TFT-LCD Firms: From Domestic/In-House to International/Collaborative

So far this chapter shows that Japan's declining competitiveness is associated with the shifts in paradigm in the TFT-LCD industry. Paradigm shifts have made certain sources of competitive advantage for Japanese firms obsolete while opening a window of opportunity for new entrants like China to establish their competitive position. In this section, I show how the locus of innovation, i.e. domestic versus international and in-house versus collaborative, also shifts along the changing competitiveness of a country and a firm and a changing paradigm in the industry. In other words, I show how the paradigm shifts of the industry, the changing competitive positions of a country and a firm, as well as the changing locus of innovation are aligned with one other. For example, a shift from the domestic/in-house approach to the international/collaborative one is strongly related to the declining competitive position of a country and a firm, which is tightly linked to the shifts of paradigms in the industry.

This section focuses on the changing locus of innovation in the TFT-LCD industry for Japanese TFT-LCD firms, with a particular attention to the following dimensions: *geographic scope* and *organizational boundary*. The former refers to the locus of value-added activities to be conducted in a single country vis-à-vis across national borders. The latter refers to the extent of collaboration with external parties in conducting value-added activities. Geographic scope depends on the home-country's industry competitiveness, whereas organizational boundary depends on firms' competitiveness. Four categories can be composed based on these dimensions: *domestic/in-house, domestic/collaborative, international/in-house,* and *international/collaborative*, respectively.[1]

The following logic can be drawn, as presented in Fig. 6.1: First, if both home country advantage and company-specific advantage exist, the company is likely to adopt the *domestic/in-house* approach. Second, if home country advantage resides in the home country's industry but no company-specific advantage exists, the company is likely to adopt the *domestic/collaborative* approach. Third, if company-specific advantage exists but no home country advantage resides in the home country's industry, the company is likely to adopt the *international/in-house* approach. Forth, if neither home country advantage nor company-specific advantage exists, the company is likely to adopt the *international/collaborative* approach.

In Fig. 6.2, I show that Japanese companies depart from *domestic/in-house* approach to either *domestic/collaborative* or *international/collaborative* approach. As the country's presence declines facing the low-cost mass production of

---

[1]This two-dimensional framework is consistent with Mudambi (2018) which classified strategic choice by geographic location (i.e. concentrated or dispersed) and control (i.e. vertical integration or specialization).

Fig. 6.1 The locus of innovation in the TFT-LCD industry

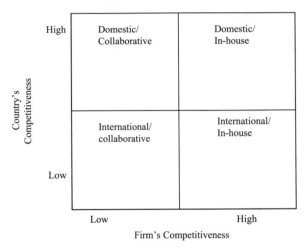

Fig. 6.2 Change in the locus of innovation of Japanese firms

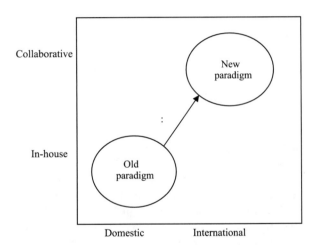

TFT-LCD panels in China, Taiwan and Korea, Japanese companies shifted their locus of innovation toward *international/collaborative approach.*

Domestic/In-house:

Japanese TFT-LCD companies enjoyed dominance in the industry, with the market share as high as over 90% in the early 1990s. Japanese firms such as Sharp can be plotted in the lower-left quadrant, where both the home country advantage and company-specific advantage were strong. However, due to the erosion of relative technological strengths vis-a-vis Korea and the lack of cost advantage vis-à-vis Taiwan, the validity and sustainability of the black box approach has been questioned.

When both home country advantage and company-specific advantage were strong and sustainable, as represented by the upper-right quadrant, the firms could internalize advanced learning through vertical integration, known as the"black box" approach. Alternatively, they could engage in advanced learning through arms-length collaboration with domestic suppliers. In reality, pure vertical integration was rare, even for Sharp which tried to extend their "black box" approach to the supplier relations in its Kameyama plant (Nakata 2007). In fact, Japanese firms in the TFT-LCD industry manifested less vertical integration than Korea (Lehmberg 2017). As the companies' competitiveness eroded, however, the black box approach through vertical integration became less effective.

Domestic/Collaborative:

Domestic/collaborative is more appropriate when the competitive advantage of the home country and industry remain strong but that of the company becomes weak. Japanese TFT-LCD panel makers other than Sharp stopped investing beyond 4 Gen in the 1990s due to the economic recession and financial crisis in Japan. As runners-up to Sharp, they chose to collaborate themselves, given the remaining competitive advantage of Japan's TFT-LCD industry, at least in the high end of the spectrum. For example, when Matsushita and Toshiba announced that they would integrate their TFT-LCD business in October, 2001, they were both suffering from low-price competition by Korean firms, and found it reasonable to integrate their TFT-LCD business to cope with the Korean rivalry (Asakawa and Kim 2007).[2] Another case was the foundation of IPS Alpha Technology, Ltd.,[3] a joint venture among Hitachi, Matsushita and Toshiba, along with Hitachi Displays Ltd., founded in January 1, 2005. This new company sought to develop, manufacture, and sell amorphous TFT-LCD panels larger than a 23" TV, with increased production capacity and reduced cost (Asakawa and Kim 2007). These companies fell behind major players in TFT-LCD industry such as Sharp and Samsung Electronics in terms of the generation-upgrading race. Other examples include Sharp-Toshiba alliance in 2017, with Sharp's technology in TFT-LCD matched with Toshiba' system LSI for TV in a complementary way (Nakata 2009). Furthermore, Sharp-Sony alliance in 2018, with Sharp supplying its TFT-LCD panel to Sony's TV with stability and continuity. Sony can benefit from the alliance by also stably procuring panel from Samsung and Sharp for its TV sets (Nakata 2009).

Although they found sophisticated technological niches, they were threatened by low-cost pressure by Korean rivals. Thus, they chose to form domestic alliances to compete against Korean catch-up. For example, to deter market expansions of Korean and Taiwanese firms, Sharp acquired the LCD business unit from Fujitsu, which represented Japan's strong technology in flexible panel displays technologies (Hu 2012). However, as the gap in national industrial competencies between Japan

---

[2]TMD (Toshiba Matsushita Display Technology Ltd.) was founded on April 1, 2002, with 60% Toshiba and 40% Matsushita, in charge of development, manufacturing, and sales of TFT-LCD, STN-LCD, and OLED for small-medium size panels, aiming at cost-breakthrough by economies of scale (Asakawa and Kim 2007).

[3]"IPS" stands for "In-Plane-Switching," the technology for widening the view angle originally developed by Hitachi (Nikkei Newspaper, November 1, 2004; Asakawa and Kim 2007).

and Korea/Taiwan became narrower, pure domestic collaboration did not make sense. In reality, pure domestic collaboration among material and equipment makers, panel makers, and the end producers has been uncommon in the twenty-first century. Along the decline in national and company competitiveness, firms even use cooperative arrangement as a way to reduce commitment and to facilitate exit from the industry (Lehmberg 2017).

International/In-house:

In theory, this approach is most effective when a company tries to buffer the core technology internationally for the company from competitors. However, in reality, companies chose to partner with foreign firms outside of their technological core, as seen in Sharp's business alliance with Quanta in Taiwan to cope with low-cost competition (Asakawa and Kim 2007; Asakawa 2007). This mode is consistent with the logic of the smile curve of value creation (Mudambi 2007) in which firms that control the upstream end of value chain choose to relocate the standardized activities at the middle of the smile curve to less developed countries to enhance efficiency and effectiveness of high value-added activities (Mudambi 2008). While core technological knowledge is contained within the firm and within Japan, peripheral or older-generation production technologies that can be standardized are shared with its Asian operations[4] mainly to cope with low-cost production pressure. Therefore, the international vertical integration is has not been commonly chosen and implemented by Japanese TFT-LCD firms.

International/Collaborative:

International/collaborative also became common for Japanese electric and electronics firms which were hesitant to remain in a fierce competition over the low-cost production of TFT-LCD panel. Our conceptual framework predicts that firms are likely to adopt this mode when the competitive advantages of a country's industry and company-specific advantage are gradually eroding. Sony, despite its competitiveness as a company, had been focused on CRT and thus lagged in TFT-LCD technologies. It was relatively easy for Sony to look for international collaboration in this case, because of the obvious technological shift from CRT to TFT-LCD, and because of the obvious recognition that the company lagged behind Sharp in this technology. Sony formed S-LCD, the joint venture company established with Samsung Electronics in 2003, to secure a stable supply of large TFT-LCD panels beyond Generation 7 which was believed to be much more cost-efficient than in the earlier generations (Asakawa 2007; Asakawa and Kim 2007: 14). Sony took advantage of its collaboration with Samsung to overcome its weakness in the stable supply of large panels and in cost efficiency, until S-LCS dissolved in 2011. In fact, many Japanese panel makers engaged in much looser inter firm collaboration than vertical integration or joint venture arrangements. Typical examples include Japan's TFT-LCD makers that transferred technology to Taiwanese firms for OEM

---

[4]This includes OEM from Taiwan, LCD module and LCD TV assembly in China, Malaysia, Mexico, and Poland, as for Sharp (Nakata 2007).

manufacturing with cost advantage to deter fast-growing Korean makers. The case of technology transfer includes the following: from ADI (Mitsubishi Electrics) to CPT; from IBM Japan to ADT (founded by Acer); from Toshiba to Hannstar; from Sharp to Quanta Display (QDI); from Matsushita to Unipac (Akabane 2004). For China, Sharp agreed on the first Chinese-Japanese joint TFT-LCD fab in Nanjing in Gen 8 by licensing its advanced technologies to China Electronics Corp. (CEC), which is the parent firm of CEC-Panda (Hsieh 2013; Negishi 2013).

More recently, the Foxconn/Sharp takeover deal came when the TFT-LCD industry was under fierce competition from Korean, Taiwanese and Chinese firms, when the industry was facing the challenge of overcapacity and the need for product diversification. The Foxconn-Sharp deal was intended to make both companies become a powerhouse in TFT-LCD industry (Hsieh 2016).

On the upstream front, materials and equipment companies, no matter how competent they may be, cannot survive without collaborating with the panel makers as buyers. Therefore, pure in-house approach does not make sense. At the same time, equipment and material makers cannot afford to ignore their overseas customers in Korea and Taiwan, although the government wishes to protect the core technologies from leaking to Korea and Taiwan (Murtha et al. 2001). Many firms engage in collaboration with Korean makers. The *international-collaborative* strategy is the high-risk, high-return approach, because it may run into intellectual property right (IPR) risks. Nevertheless, Japan's equipment makers are inclined to collaborate with Korean and Taiwanese firms due to their large volume of production. For example, 70% of CMO's R&D was conducted jointly with Japan's materials and equipment makers, whereas 30% with Taiwan's universities, government, and/or within CMO in Taiwan (Shintaku et al. 2006).

6.5   Conclusions

Japan's dominance in the TFT-LCD industry in the 1990s has been replaced by Korea and Taiwan in the 2000s, and then by China in the 2010s. This chapter was intended to present various factors behind the decline in Japan's leading position in the TFT-LCD industry from the lens of paradigm shifts in innovation strategy. Looking back the history of industry evolution in the past several decades, I summarized multiple changes in Japan's TFT-LCD industry from the 1990s to the 2000s and the 2010s in light of Japan's declining position in the industry. Based on this analysis, I then identified paradigm shifts in the TFT-LCD industry that have directly or indirectly led to a decline in Japan's competitive position: changing source of competitive advantage, modularization of TFT-LCD technology, evolving country leadership, product and panel size, and geographic scope of innovation. This chapter illustrates the paradigm shifts in the TFT-LCD industry in the past several decades and discusses how and why Japan's competitive positioning has changed so drastically.

The contributions of this chapter include the following. First, this chapter shows that Japan's declining competitiveness is associated with the shifts in paradigm in the TFT-LCD industry. In other words, the chapter shows that declining competitiveness of a country and a firm cannot be fully understood without closely relating it to the shifts in the paradigm of the industry. Paradigm shifts have made certain sources of competitive advantage for Japanese firms obsolete while opening a window of opportunity for new entrants like China to establish their competitive position.

Second, this chapter show the locus of innovation for a firm, i.e. domestic vs. international and in-house versus collaborative, also shifts along the changing competitiveness of a country and a firm and a changing paradigm in the industry. In other words, I showed how the paradigm shifts of the industry, the changing competitive positions of a country and a firm, as well as the changing locus of innovation are aligned with one other. For example, a shift from the domestic/in-house approach to the international/collaborative one for a Japanese firm is strongly related to the declining competitive position of Japan and the firm, which is tightly associated with the paradigm shifts in the industry.

In this chapter, the competitive position of Japan and the Japanese firms has been examined. Future study can broaden the scope by covering the competitive position and the locus of innovation of firms from other East Asian countries.

Acknowledgements I would like to thank Stefanie Lenway and Thomas Murtha for their insights into the early stage of this research through our collaboration at Alfred P. Sloan Foundation Industry Studies Project. I thank RIETI for its generous support to conduct the early stage of this research. I also thank Keio Business School's Mitsubishi Chairship Fund and the JSPS Grant-in-Aid (B) No. 15H03384 for the financial support.

References

Akabane, J. (2004). The Taiwanese TFT-LCD industry: The role of Japanese companies and the Taiwanese government in the development process. *Asian Studies, 50*(4), 1–19 (in Japanese).

Asakawa, K. (2007). Metanational learning in TFT-LCD industry: An organizing framework, *RIETI Discussion Paper* (07-E-029). Tokyo, Japan: Research Institute of Economy, Trade and Industry.

Asakawa, K., & K. Kim (2007). Japan's TFT-LCD Display Industry. Keio BusinessSchool Case #90-06-5225 (in Japanese).

Asakawa, K., Park, Y., Song, J., & Kim, S. (2018). Internal embeddedness, geographic distance, and global knowledge sourcing by overseas subsidiaries. *Journal of International Business Studies, 49*(6), 743–752.

Cantwell, J., & Immarino, S. (2003). *Multinational corporations and european regional systems of innovation.* Oxfordshire: Routledge.

Cantwell, J., & Mudambi, R. (2005). MNE competence-creating subsidiary mandates. *Strategic Management Journal, 26*(12), 1109–1128.

Doz, Y., Santos, J., & Williamson, P. (2001). *From global to metanational.* Boston: Harvard Business School Press.

Everatt, D., Tsi, T., & Cheng, B. (1999). The Acer Group's China manufacturing decision. Version A. Ivey Case Series #9 A99M009. Richard Ivey School of Business, University of Western Ontario.

Hsieh, D. (2013, July 1). Sharp and CEC-Panda: The first Chinese-Japanese joint TFT LCD fab, rAVe Publications, online.

Hsieh, D. (2016, March 31). HIS: Foxconn-sharp deal creates TFt-LCD production powerhouse, News Freedom, news online.

Hu, M.-C. (2012). Technological innovation capabilities in the thin film transistor-liquid crystal display industries of Japan, Korea, and Taiwan. *Research Policy, 41,* 541–555.

IHS Markit. (2018). Display market outlook for 2019, presentation by D. Hsieh at the Display Innovation China 2018, Beijing Summit.

JB Press. (2018). Japanese firms are fighting a last-ditch battle in OLED falling behind the Korean (Kankokuzei ga senkousuru yuki-EL de nihonkigyou ga toru haisuinojin), in Japanese. Nikoniko News, 20180911, 06:00. JB press.

Johnstone, B. (1999). *We are buring: Japanese entrepreneurs and the forging of the electronic age.* New York, NY: Basic Books.

Kim, L. (1997). *Immitation to innovation: The dynamics of korea's technologial learning.* Boston, MA: Harvard Business School Press.

Kojima, K. (2000). The "flying geese" model of Asian economic development: Origin, theoretical extensions, and regional policy implications. *Journal of Asian Economics, 11*(4), 375–401.

Lee, K. (2005). Making a technological catch-up: Opportunities and barriers. *Asian Journal of Technology Innovation, 13*(2), 97–131.

Lee, K., & Lim, C. (2001). Technological regimes, catching-up and leapfrogging: Findings from the Korean industries. *Research Policy, 30*(3), 459–483.

Lee, K., & Mathews, J. A. (2012). South Korea and Taiwan. In E. Amann & J. Cantwell (Eds.), *Innovative firms in emerging market countries* (pp. 223–245). Oxford: Oxford University Press.

Lehmberg, D. (2017). The process of industry exit in the Japanese context: Evidence from the flat panel display industry. *Journal of Management & Organization, 23*(1), 92–115.

Linden, G., Hart, J., Lenway, S., & Murtha, T. (1998). Flying geese as moving targets: Are Korea and Taiwan catching up with Japan in advanced displays? *Industry and Innovation, 5*(1), 11–34.

Mathews, J. A. (2005). Strategy and the crystal cycle. *California Management Review, 47*(2), 6–32.

Mathews, J. A. (2006). Catch-up strategies and the latecomer effect in industrial development. *New Political Economy, 11*(3), 313–336.

Mathews, J. A., & Lim, C. (2001). Technological regimes, catching-up and leapfrogging: Findings from the Korean industries. *Research Policy, 30*(3), 459–483.

Mathews, J. A., Hu, M. -C., & Wu, C. Y. (2011). Fast-follower industrial dynamics: The case of Taiwan's emergent solar photovoltaic industry. *Industry and Innovation, 18*(2), 177–202.

Mudambi, R. (2017). Offshoring: Economic geography and the multinational firm. *Journal of International Business Studies, 38*(1), 206.

Mudambi, R. (2018). Location, control and innovation in knowledge-intensive industries. *Journal of Economic Geography, 8,* 699–725.

Murtha, T., Lenway, S., & Hart, J. (2001). *Managing new industry creation.* Stanford University Press.

Nakagawa, T., Kutsukake, T., & Nakashima, T. (2016). Materials industry under the radically changing competitive environment, (Kyosokankyou ga gekihensuru sozaisangyo), Chitekishisansouzou (Intellectual Asset Creation) February 2016), in Jpanese: 38–53.

Nakata, Y. (2007). Japan's competitiveness in TFT-LCD industry: Analysis on the decline and a proposal for core national management. RIETI DP (in Japanese).

Nakata, Y. (2009). Inter-organizational knowledge creation from the aspect of industrial architecture: The analysis of competitive strategies from the cases of TFT-LCD, semiconductor, solar cells, and automobile industries. Unpublished doctoral dissertation, Graduate School of Technology Management, Ritsumeikan University.

Negishi, M. (2013). Sharp licenses advanced LD technology to its Chinese partner: The Japanese electronics maker will take a stake in a Nanjing factory. *Wall Street Journal*, June 27, 2013, 7:12 a.m. E.T.

Nekar, A. (2003). Old is gold? The value of temporal exploration in the creation of new knowledge. *Management Science, 49*(2), 211–229.

Nikkei Shimbun. (2019a). The JDI negotiation reaching the finale (JDI Shien kosho odume), in Japanese. 3.31.

Nikkei Shimbun. (2019b). JDI falling under the umbrella of the Chinese-Taiwanese group: The national TFT-LCD initiative came to a deadlock (JDI taichuzei sanka ni: hinomaruekishou ga tonza), in Japanese .4.4, p. 1.

Nikkei Shimbun. (2019c). JDI falling under the umbrella of the Chinese-Taiwanese group: Complacency in TFT-LCD caused delay in its engagement in OLED (JDI taichuzei sanka ni: ekisho kashin yuki el deokure: ekishokashi yuki el dokure), in Japanese, April 4, p. 15.

Nikkei Shimbun. (2019d, April 6). Samsung losing profit: Dark clouds looming over its counterattack (Samsung geneki: haakou ni kage), in Japanese, p. 10.

Nikkei Business. (2016). Beyond the smile curve (sumairu kabu wo koero), in Japanese, Nikkei Business. 5.30: 38–39.

Numagami, T. (1999). *History of technological innovation in TFT-LCD display. In Japanese.* Tokyo: Hakuto.

Polgar, L. G. (2003). Flat panel displays. *Business Economics, 38*(4), 62–68.

Shintaku, J., Kyo, K., & So, S. (2006). Development of Taiwanese TFT-LCD industry and corporate strategy. MMRC-J-84, Discussion Paper, The University of Tokyo.

Shintaku, J., & Yoshimoto, T. (2009). TFT-LCD TV and panel industry: International division of labor. In J. Shintaku & H. Amano (Eds.), *Global strategy of manufacturing management: Industrial geography in Asia* (pp. 83–110). Tokyo: Yuhikaku.

Semenza, P. (2012). The display industry: Fast to grow, slow to change. *Information Display, 5&6* (12), 18–21.

Song, J. (2006). Samsung and Korea's Flat Panel Display industry, presentation at RIETI seminar, July 2006.

Suarez, F., & Utterback, J. (1995). Dominant designs and the survival of firms. *Strategic Management Journal, 16*(6), 415–430.

Tabata, M. (2014). The rose of Taiwan in the TFT-LCD industry. *Journal of Technology Management in China, 9*(2), 190–205.

Kazuhiro Asakawa (Ph.D., INSEAD) is Professor of Global Innovation Management at Graduate School of Business Administration, Keio University, Japan. He received his Ph.D. and M.Sc. from INSEAD and his MBA from Harvard Business School. His research interest lies in the areas of global innovation and R&D management, global knowledge sourcing, global open innovation and frugal innovation. He is a Fellow of the Academy of International Business (AIB). His research appeared in major international journals such as *Journal of International Business Studies, Global Strategy Journal, Journal of World Business, Research Policy,* and *Journal of Product Innovation Management.* He has served as an Associate Editor for *Global Strategy Journal* and *Asia Pacific Journal of Management*; is an Advisory Editor for *Research Policy,* and sits on the editorial board of *Journal of International Business Studies, Journal of World Business, Journal of International Management,* and *Asia Pacific Journal of Management,* among others. He was president of Japan Academy of Multinational Enterprises, has served as Japan Chapter Chair of the AIB, and was a representative-at-large in Global Strategy at Strategic Management Society.

Chapter 7
Business-University Collaboration in a Developing Country in the Industry 4.0 Era—The Case of Hungary

Annamaria Inzelt

Abstract Emerging economies, as well as Hungary in Europe, are frequently host countries to multinational companies. For both parties it is a great challenge to build relevant knowledge- generating capacities which are attractive in respect of collaboration. In such collaborations those countries on the receiving end of foreign direct investment have the ambition to become more than mere pools of knowledge assets for multinational companies. An insight into Hungarian attempts to achieve this aim under difficult circumstances might be useful for other countries in similar situations. In recent years the majority of Hungarian business research and development expenditure has come from companies wholly-, or majority-owned by foreign interests. This high proportion indicates the significant role of foreign companies in the Hungarian research agenda and in business-university collaboration. This chapter focuses on how foreign companies are shaping business-university collaboration in research and experimental development and touches upon the role of government as facilitator. The subjects of research and development contracts and collaboration depend on the environment and on both potential partners—that is to say, by the types of demand generated by companies, and by how relevant are the competences and capabilities of universities in meeting these demands. Are they moving towards the cutting edge agendas inherent in Industry 4.0 and globalisation? Method of research: analysis of available data; information from websites and interviews with key actors who are partners in collaboration. The chapter also summarizes a few lessons which may be relevant for other economies too.

Keywords Business—university collaboration · Research and development episodes · Host-country of multinational companies

A. Inzelt (✉)
IKU Innovation Research Centre, Financial Research Co., Budapest, Hungary
e-mail: inzelt.annamaria@penzugykutato.hu

© Springer Nature Singapore Pte Ltd. 2019
J. Cantwell and T. Hayashi (eds.), *Paradigm Shift in Technologies and Innovation Systems*, https://doi.org/10.1007/978-981-32-9350-2_7

List of Abbreviations

BME	*Budapest* University of Technology and Economics
BUC	Business-university collaboration
BERD	Business R&D expenditure
CWUR	Centre for World University Rankings
EIT	European Institute of Innovation and Technology
ELTE	Eötvös Loránd University
EPO	European Patent Office
EU	European Union
FDI	Foreign direct investment
FIEK	Higher Education and Industry Cooperation Centres
GDP	Gross domestic product
HCSO	Hungarian Central Statistical Office
HE	Higher Education
HEIs	Higher Education Institutes
HERD	Higher education R&D
HQ	Head quarter
HUF	Hungarian Forint (currency)
Industry 4.0	Fourth Industrial Revolution
IPRs	Intellectual Property Rights
MSTI	Main Science and Technology Indicators
MNC	Multinational Company
NIS	National innovation system
NKFIH	National Research, Development and Innovation Office
OECD	The Organisation for Economic Co-operation and Development
R&D	Research and development
RDI	Research, development and innovation
S&T	Science and technology
SME	Small and Medium Enterprise
TTO	Technology Transfer Office
TH	Triple Helix

7.1 Introduction

The globalization of innovation brings both new opportunities and challenges to all countries. Europe is one of the leaders in this trend, with its European research and innovation programmes (such as Horizon 2020). The other major research, development and innovation (RDI) spenders, multinational companies from many different countries, are also searching for ambitious ideas and partners.

Companies are increasingly opening up their organizational boundaries to tap into external sources of knowledge and to strive for better innovative performance (Berchicci 2013). Multinational companies (MNCs) have substantially expanded their global innovation networks, and their aim to collaborate with universities located abroad has been identified as one of the main drivers of the internationalization of their research and development (R&D) centres.

Although the internationalisation of R&D through FDI is not a new phenomenon, its rapid growth and scope have changed dramatically (Raymond and Taggart 1998; Cantwell and Molero 2003; Narula 2014; Foray 2006). The purchase or outsourcing of R&D (whether domestically or internationally) is now a serious complement to in-house R&D as a part of corporate innovation strategy (OECD 2008a; EU 2005).

The globalisation of R&D affects more the R&D intensive sectors which are key players in disruptive technologies such as artificial intelligence, big data, block-chains, genome editing, robotics, drones, 3D printing, the Internet of Things, and autonomous vehicles (OECD 2017). The new era of disruptive technologies often named as the Fourth Industrial Revolution (Industry 4.0) is still in its infancy, but its effects are already having an impact on the nature of competition and corporate strategies in many industries. Industry 4.0, as well as upgrading internationalisation, is pushing the actors of the national innovation system (NIS) towards self-transformation and collaboration to capture new opportunities.[1]

Collaborative research involving businesses and universities has a long history, although its intensity, character and the types of business partner are changing in the twenty-first century. There is no single model of interaction between industry and science. The formal and informal channels of collaboration differ by country and time (Bonaccorsi and Piccaluga 1994; Inzelt 2004). Some countries pay more attention to one particular channel than do others.

In collaboration the role of MNCs differs in line with the home- or host-country's perspective. The main drivers of the internationalization of R&D from a home-country's perspective are the opportunities for global sourcing and pooling, which may increase the capacity of scientific excellence, optimize the innovation process and find effective solutions to global changes.[2] Traditional cross-border R&D sought to *adapt* products and services to the needs of the host countries and to support the foreign investors' local operations. In this case, host-countries can benefit from the presence of foreign-owned firms in their research, development and innovation (RDI) activities. The MNCs which are burdening their markets may contribute to increase R&D expenditure and knowledge spill-overs in host countries. However they hardly broaden R&D opportunities for domestic universities for collaboration with industry.

---

[1]From the actors in innovation system we focus on two kinds of organisation: universities and businesses. In Hungary the term 'university' covers so-called science universities, applied science universities and colleges, each of which may be public or private, although the latter may follow a different financial regime. In Hungary business organisations belong to different size categories, similar to other countries, and they may be owned by foreigners, private and public domestic owners and a mixture of these.

[2]In Europe a crucial principle is non-discrimination between domestic and foreign-owned firms.

Contemporary multinational companies seek the opportunities to generate new knowledge in other countries. They need more and more access to highly skilled scientific personnel and to tap into worldwide centres of knowledge (Edler 2008; Inzelt 2008; Taggart 1998).

MNCs have increasingly moved R&D activity across borders within their global value chain and rely on outside inventions for new products and processes. Large companies (mostly MNCs) increasingly adopt innovation networks which link networks of people, institutions (universities, government agencies and other companies) in different countries to solve problems and produce ideas (Cook 2005),[3] and so collaboration with local universities is becoming more and more important for MNCs. In host countries this process is generating FDI-led R&D.

In our epoch it is increasingly important for all universities to attract more business investment in R&D from all over the world. The rising level of internationalisation in business R&D benefits and strengthens the host countries capabilities in science and technology (S&T).

Hungary is mainly a host-country to MNCs, and so the internationalisation of R&D, the business-university collaboration (BUC) is very important. The international BUCs are great challenges to developing countries—as we know from many sources (such as Dachs et al. 2013). Business-university linkages are stronger among more advanced EU countries than between them and moderately or modestly innovative countries (e.g. Hungary). Beyond many positive effects foreign presence may also have negative impacts on the host countries if the foreign companies are simply pooling knowledge (such as relocating top researchers or regularly commercialize inventions in other countries). It is a delicate issue for policy-making to back up potential positive effects and minimize any threatening negative effects.

This chapter investigates business and universities R&D collaboration, devoting special attention to multinational companies as partners with universities in the Industry 4.0 era from a host-country perspective. The next section sets the topic in the context of its literature, and we then clarify the methodology. Section 7.4 deals with external and internal influencing factors, whilst. Section 7.5 looks at Hungarian BUC in the European mirror. Section 7.6 examines R&D collaboration with universities, following which we offer some conclusions and a few lessons for other host countries of MNCs.

7.2 The Literature Context

This topic is linked to various branches of literature. Due to size limitations, we do not give a detailed overview of the literature, but merely systems highlight the interdisciplinary nature of the topic.

---

[3]These two paragraphs were published in Inzelt (2010).

The main literature framework is the innovation systems approach (Freeman 1988; Lundvall and Johnson 1994; Nelson 1993; Edquist 1997).

Investigating the development of BUC over time, however, may support approaches from different standpoints:

- The literature on the Third Mission of Universities extends the traditional Missions of Teaching (1st) and Research (2nd), exhibiting a vice versa impact on both (Mollas-Gallart et al. 2002; Pinheiro et al. 2015). The Third Mission's approach focuses on collaboration from a university perspective not only on economic utilisation but on other dimensions of university collaboration with society. It devotes attention to economic and social impacts and economic and social relevance and investigates the development of universities' economic and social relationships. The Third Mission is multifaceted, as it may have both economic and social dimensions, such as technology transfer, the transfer of competences through new graduates to industry, contracts with industry and public bodies which contribute to regional development and involvement in social and cultural life (Govind and Kütim 2016; Gulbrandsen and Slipersaeter 2007; Inzelt et al. 2006; Laredo 2007; Nedeva 2008; Perkmann et al. 2013; Shimoda 2008).
- The Triple Helix (TH) model concentrates very specifically on university-industry-government relationships, a few of the factors which influence innovation processes. Among these it devotes special attention to universities and their restructuring as organisations (Technology Transfer Offices/ TTOs/, spin-offs, entrepreneurial universities, regulations, governance) and governmental rules (such as regulating Intellectual Property Rights/IPRs/ originated from publicly financed research), policy instruments (tax incentives, innovative public procurement, public risk capital, and programs) (Etzkowitz and Leydesdorff 1997, 1998; Etzkowitz 2008; Inzelt 2004; Ranga and Etzkowitz 2013). Development of the TH idea led to a concept of *entrepreneurial universities* (Clark 2015; Etzkowitz 2004, 2008) which are particularly active in the USA where the environment was especially supportive in strengthening the commercialization of knowledge and of moving towards the business-like operation of Higher Education Institutes (HEIs) such as by patenting, licensing and spin-off entrepreneurship.

Farinha and Ferreira (2013) give a good summary of different generations of TH models from the state-centric to the model in which different institutional spheres take the role of others and hybrid organisations emerge.

- Innovation literature takes into account universities as one of the sources of innovation and partners in innovation processes (Koschatzky and Stahlecker 2010; Wen and Kobayashi 2001). In the business innovation-related literature there is an emphasis on linkages within the R&D and innovation processes. There are three different types of industrial R&D: exploratory, exploitative, and imitative. Each type has the potential to create or drive competitive advantage, albeit with varying degrees of novelty, depending on the company's capabilities and business strategy. The characteristics of collaboration according to the type of industrial R&D have a strong impact on possible industry-university linkages. Collaborations are more important for exploratory and exploitative research than for others.

The central idea behind *open innovation* (Chesbrough 2003; Chesbrough et al. 2006) is that, in a world of widely disseminated knowledge, companies cannot afford to rely entirely on their own research, but should instead buy or license processes or inventions from other players, either from other firms or from universities, research organisations. Although the firms have many different partners beyond knowledge producing organisations in the innovation process, we are dealing only one of them: the universities. The Open Innovation idea looks on universities as one of the knowledgeable partners, for companies or for innovation alliances (Chesbrough 2003), although it should be kept in mind that R&D collaboration is a means and not an end (Mowery 1998). The end is innovation, that "takes place when the invention or new idea is exploited", (Swann 2014, p. 213) either in commercial or in non-commercial spheres.

Common in these approaches are how they broaden the relationships among the actors (Literature review in Perkmann et al. 2013; Ankrah and Al-Tabbaa 2015). The internationalization of collaboration has some specificity to national collaboration (Caloghirou et al. 2001). The literature is revealed the linkages when approaching collaboration from the perspective of internationalisation and multinational companies.

The role of cross-border interdependence in the innovation process, as well as interdependence between home and host countries firms and scientific organisations are discussed in the literature that examines the interdependence of globalization and technological innovation exploring knowledge elsewhere (Archibugi and Lundvall 2002; Archibugi and Filippetti 2015; Cantwell 2017; Guimón and Narula 2017; Inzelt 1999, 2000, 2010; Leydesdorff and Sun 2009; Narula 2003; Pinherio et al. 2015; Singh 2007).

Globalisation is bringing the dramatic new challenges in a world where competition is becoming increasingly knowledge-based and global. Innovation and internationalisation have become interconnected—that is, leading more and more open international innovation systems. The internationalisation of business R&D is a relatively new phenomenon compared to the internationalisation of manufacturing. The geographic dispersion of R&D gives access the companies to diverse pools of knowledge (Cantwell 1989; Cantwell and Molero 2003; Cantwell and Piscitello 2007; Dunning 1988, 1992).

International business networks and subsidiaries of MNCs are important actors to engage in interaction with the local universities, so accessing local knowledge capabilities in host countries. Knowledge demand created by foreign actors encourages the potential industrial R&D partners of universities collaborate. This additional business R&D demand is important for developing countries with good scientific capabilities in some fields but with a shortage of innovative local business.

However, this raises the issue of how can these collaborations contribute to upgrading local innovativeness beyond their contribution to global value chains.

Govind and Küttim (2016) provide a systematic literature review of international knowledge transfer from university to industry. How business-university cooperation emerges and develops into international relations was described by Sorensen and Hu (2014), Mégnigbéto (2015).

7.3 Methodological Notes

This chapter uses interview research conducted at two research universities—the Eötvös Loránd University (ELTE) and the Budapest University of Technology and Economics (BME)—and with two MNCs, which are cited as MNC-1 and MNC-2 because of their requests. Both MNCs are collaborating with two interviewed universities and with other Hungarian universities in education and R&D. It also relies on university websites and reports of companies which contain information on the type, length, inputs and outputs of collaboration.

It employs also several one-off surveys on Hungarian BUCs at country level which are mainly available from international sources.

Statistics on inputs and outcomes on BUCs scarcely exist, either in official Hungarian statistics at national level, or in university archives. Data on collaborative projects and contracts exist in universities' databank but such information is not summarized either at faculty or university level. There is no access to basic data for research purpose, and so we have limited hard information for analysis.

A combination of all available information, in which interviews (anecdotal evidence) played a major role) helped to improve our understanding of business-university collaboration.

7.4 Influencing Factors of Business-University Collaboration

There are many external and internal factors influencing university-industry interaction, their type, frequency and level of intensity.

7.4.1 External Factors

Important external factors are governmental policy, programs and initiatives to stimulate BUC in R&D. Governments can facilitate the process of BUC: to provide direct and indirect incentives for innovation and to remove obstacles to collaboration. In addition is the provision of consultancy services relevant to innovation processes, e.g. technology transfer, commercial information and legal advice (Edquist 2018).

Here we highlight only a few elements of legislation, governmental programs and supporting schemes for BUC in Hungary. Some of these directly forced changes at university level to prepare them for mutually fruitful collaboration with industry. Let us take a brief look at them.

7.4.1.1 Stop-Go Legislation

The *legislation process* which significantly widened the gateway for BUCs appeared in 2004/05 (Law on Innovation, 2004 and Law on Higher Education 2005) (Overviews on STI legislation see in: OECD 2008b; EU 2016a). These laws stated that HEIs have to set up technology transfer offices, introducing them into university management. The function of TTOs, as in other countries, is to support the exploitation of university research results through patenting, licensing, other kinds of transfer to industrial firms, or launching start-ups.

Early modern laws (2004/05) introduced Bayh-Dole type regulation[4] (put into practice in 2007). By this law, the HEIs should reregulate the intellectual property of research findings (following Bayh-Dole principles) and introduce norms for sharing income within a university derived from patents and inventions. The 2004/05 laws were modified several times in a Stop-Start process. There was a step backwards to the old regulation of intellectual property in 2014 (Amendment 2014). Intellectual property (patents) belonged again to the state if a publicly financed university or R&D organisation was involved, and these entities can only exercise rights in the name of the state.[5]

This modification became a serious obstacle to taking universities by companies as co-owners of patents. Companies became very reluctant in respect of co-patenting if they financed only a part of the research.

At the end of 2018 the law has changed once more and follows again the Bayh-Dole type regulation which improved the opportunities for co-patenting with business organisations. However the impact may be seen only in future years as it takes time for actors to trust the durability of legislation.

---

[4]The Bayh-Dole Act originates from the US (1980, amended in 1986). It rationalized and simplified federal policy toward the patenting and licensing by non-profit institutions of the results of publicly funded research. Since 1980 many countries have followed this form of regulation.

[5]This is not unrelated to how universities' autonomy fluctuates: non-autonomy through semi-autonomy to full autonomy over time.

7.4.1.2 Government Programs

Various *governmental programs* have been launched to support, set up and continue collaboration. The programs had many advantages to support the nurturing of BUCs. The common aims of these governmental programs are: to upgrade universities research capabilities, to make them more attractive for cutting edge business partners and to accelerate the commercialization of new technologies transferring research results from universities to industry and supporting the creation of a common technological "vision" that can guide R&D and related investments by public and private entities. The programs and incentives are also encouraging business actors to collaborate with universities taking into account the international race among universities to attract cutting-edge industrial partners whether domestic or foreign-owned companies. The companies must also compete for the best researchers world-wide (These programs are detailed by Inzelt and Csonka 2016).

In contrast to the advantages of the programs, there were a few shortcomings. The length of the program period was too short for each side: they could start something together but could not reach the self-financing stage. The frequent changes in programs (without evaluation) and the structure of program management —whether university-led or business-led—were far from working well (EU 2016a; Inzelt and Csonka 2016).

Here we mention two still running programs that have an important role in BUC with MNCs (Some elements of these programs are also important for collaborating small and medium enterprises /SMEs/).

The government launched a new, better institutionalized program in 2015, building on the accumulated experience of previous programs and reflecting new challenges (Industry 4.0, digitalization, sustainable development), known as 'Research infrastructure development of Higher Education and Industry Cooperation Centres '(Hungarian abbreviation FIEK). This program has targeted upgrading the infrastructure of universities for BUC as well as for the First and Second Missions of Universities, and making the consortium durable. Members of the consortium must be 1 university and a maximum of 4 business partners. Among the consortium members there are domestic firms and foreign, mainly multinational companies. The Consortium is closed by the contract but the FIEK organisational units that were created by universities are open to other organisations also. The aims of the program are to optimise links between university and industrial R&D within a new institutional approach, and to speed up the transfer of scientific results into practice. The FIEK centres could continue working after state support ends (It also enables education to come closer to industrial needs and contribute to the vocational training of industrial staff). FIEK organisations are single window solutions long desired by collaborating companies.

An additional program was launched in 2017 (Program for Excellence HEIs) with the aims: to improve research conditions in HEIs for increasing their research capacities (such as strengthening research and innovation consciousness within HEIs; to improve the conditions for young talents; to increase scientific outputs).

Both universities investigated are among the winners and focusing on basic research in the fields of science relating to disruptive technologies. This program supports the capabilities for BUC.

Beyond national programs the European *transnational initiatives* launched by European Union have been playing an important role in networking in several industries and universities. Since the beginning of the participation scheme, Hungarian universities have been inducted into the BUC concept. Without going into detail, we have to mention that in previous years multilateral projects offered good opportunities to universities for studying workable BUC, even if Hungarian universities were not in direct contact with participating businesses.

7.4.1.3 Tax Incentives

Tax incentives are important factors to encourage generally business R&D expenditure and especially collaboration with HE and other research organisations.[6] There are several tax incentives which indirectly support BUC. These incentives have been modified over time, with effects on BUC.

In the case of BUC contracts for outsourcing or collaborating on R&D, there are three types of tax and their related allowances: (1) corporate tax, (2) local operation tax, (3) innovation contribution.

To be legally entitled to any of them, the R&D has to belong to the economic activity of a tax-paying company (except for so-called supporting contracts).

Merely to avoid any penalties for misrepresenting activities as R&D, the companies can apply to the National Office of Intellectual Property to ask expert opinion as to whether the activity may be classified as R&D and be entitled to tax reduction.

(1) *Corporate tax*

- The **tax base** may be reduced by 3 times the direct cost of R&D minus state support for the jointly performed R&D contract with academic organisations, and this reduction may be up to a total of HUF 50 m.
- **A Development Tax** reduction is possible if a company invests HUF 100 m into R&D equipment. In this case the maximum value of the reduction is 80% of the annual corporate tax. However the following years may also reduce the tax.
- The company can also enjoy a 50% tax base advantage if it has **supporting contract** with HEIs. This support may have many different purposes from scholarships to students, purchasing books, laboratory equipment or university R&D activity. The R&D does not have to relate to the economic activity of the tax-paying company.

---

[6]http://www.nav.gov.hu/magyar_oldalak/nav/ado/tarsasagi/kf_20120319.html (03/19 2012).

(2) *Local operation tax base* may be reduced by the direct cost of R&D in the given year. The companies also could reduce their local tax base if collaborating with universities or public research organisations.

(3) *Innovation contribution* is a form of levy which has existed since 2003 and feeds the Innovation Fund. The basis for calculating the innovation contribution is the local operation tax.

By the original law (2003) only medium and large companies had to pay the innovation contribution. Companies could reduce the innovation contribution by the cost of relevant R&D expenditure if the company has collaborated with universities or public research organisations or is merely contracting out R&D without participation. These opportunities for reduction were cancelled by the new law (2014) which took effect in 2015. The reason was the frequent misuse of this part of the contribution. The rule for reducing the innovation contribution by the cost of in-house R&D expenditure and of R&D contracted with universities was also withdrawn. The promise was to replace this lost financial resource for R&D with higher allocated public sources for R&D bids.

Another change occurred at the beginning of 2019 when a group of small companies also had to pay this levy who were small in themselves but, when counting them together with their partners or linking companies, reach the medium-size category (such as subsidiaries of MNCs, members of a domestic group of companies).

The modifications of the rule for Innovation Contribution have had a visible impact on contracting and collaborative R&D as companies must pay the R&D costs from their own income. They are more reluctant to invest in high-risk basic research—and even in applied research. MNCs as potential partners for universities were affected in two ways: they cannot cover the cost of R&D for using a part of their Innovation Contribution for their own research purposes. Several groups of MNCs (small firm category) were obliged to pay the contribution.

7.4.1.4 Business R&D Investment

From the point of view of a university, an important external factor is the presence of innovation-hungry and R&D-involved business organisations. International data tracked by OECD shows 42 per cent of global R&D is performed by just 200 multinational companies in the 2010s (2018).

In Hungary the rate of business R&D expenditure (BERD) to GDP has increased dynamically, since 2004 (when Hungary joined the EU) from 0.32% to 0.71% by 2017, but the rate is still low.[7] The majority of Hungarian BERD comes from multinational companies, similar to the worldwide phenomenon. According to available Hungarian statistics, the foreign and majorly foreign-owned business organisation provide more than two-thirds of business R&D expenditure and so

---

[7]Source: HCSO, www.ksh.hu/htm/1/indi1_3_1.html.

they might play an important role in BUC.[8] Foreign-owned businesses mostly fund their research activities from their own funds. It should also be noted that business funds are very important for universities as public support for higher education R&D (HERD) has declined in relative and absolute term in 2010s.

7.4.1.5 Attention to Innovation

If we investigate how hungry are Hungarian enterprises for innovation, we can see that the country is far behind the European innovation leaders and strong innovators (EU 2016b). The highest proportions of innovative enterprises during the period 2012–2014 exceeded 60% of all enterprises in Germany, Luxembourg, Belgium, Ireland and the United Kingdom, whilst in Hungary the figure was little more than 25% of all enterprises—slightly ahead of Poland, Latvia, Bulgaria, and Romania. Although the Hungarian innovation performance has improved a little (0.4%) between 2008 and 2015 the country's performance relative to the EU average has declined.

In Hungary, as in other EU countries, a higher proportion of large enterprises have introduced innovations compared to SMEs (EU average: 77%, Hungary 56%). The large companies are the main R&D investors and their higher level of participation in must also mean more potential innovative partners for R&D collaboration with universities than for SMEs.

There are four sectors in Hungary where the proportion of innovative enterprises is at least 50%: pharmaceutical-, information and communication-, oil manufacturing- and computer, electronic and optical products. Most of them belong to the Industry 4.0 sector (NKFIH 2019). However, one or two Industry4.0 related sectors are missing (Fig. 7.1).

Not all forms of innovation are based on R&D results. 'New to the market" product innovations usually employ R&D results whilst 'new to company' product innovations need only limited R&D services—if any. Novelty is a matter for R&D contracts. According to Eurostat data (Science, technology and digital society, Eurostat, online data code: inn_cis9_prod) Hungarian innovative performance is far below the EU average by the proportion of enterprises which have either new to the market products or only new to the enterprise products. 'New to market' product innovations mean a higher level of novelty than others. In this field Hungary reaches little more than 50% of the EU average, which is a sign of a limited number of innovative companies which are hungry for new R&D results. This limited innovative performance of business does not create a high demand for any kind of research: it means that Hungarian business can create little impulsive demand for R&D collaboration with universities.

---

[8]The proportion of BERD's foreign funding was more than one fifth at fully or majority foreign-owned companies, whilst that proportion was only 5% in all other companies in 2017. (HCSO 2017).

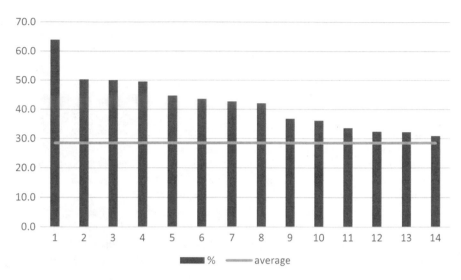

Fig. 7.1 Ranking of Hungarian innovative sectors, 2017 (The proportion of innovative businesses by economic activity) *Note* Figure 7.1 contains only those economic activities where the share of innovative business at least 30%. Most of them belong to manufacturing sectors.
Key to signs 1. Pharmaceutical products; 2. Information and communication; 3. Coke and refined petroleum products; 4. Computer, electronic and optical porducts; 5. Chemicals and chemical products; 6. Financial and insurance activities; 7. Electrical equipment; 8. Motor vehicle; 9. Beverages; 10. Paper products; 11. Electricity, gas, steam and air conditioning supply; 12. Rubber and plastic products; 13. Basic metals; 14. Machinery and equipment.
Source NKFIH, Hungary 2019 based on HCSO

7.4.2   Internal Factors

From the standpoint of collaboration, two topics are briefly highlighted here: the institutionalisation of technology transfer within universities and intellectual property issues.

7.4.2.1   Changing University Management: Institutionalize Technology Transfer Function

Key mission of TTOs is moving their universities towards an entrepreneurial culture. A fully-fledged TTO assists researchers in the patent application process, licensing agreements, obtaining licence fees and royalties, in their search for partners and funding resources, and for training and supporting faculty members and students in the creation of university-based spin-offs (Correa and Zuñiga 2013; Schaeffer and Matt 2016). Management of university with TTOs has a dual role: supporting the exploitation of scientific results and making the universities attractive for future research investment.

Launching effective TTOs is a time-consuming process. First of all it requires changes in university management to find the proper place for a new organisation in the university structure and set up the rules and procedures for collaborations with inside (faculties, departments, teams, researchers) and outside (business) actors. Also the new organisation needs financial resources.

In the early phase, TTOs were burdened by several factors: hardly any changes in university management (TTOs existed at a low level of the university management hierarchy), a lack of managerial capabilities, slow acceptance by university staff, and old, surviving routines even at universities with accumulated experience in BUC. During this early period the officers of TTOs have to overcome on cultural and linguistic barriers of both faculty members and business actors. For example it is not easy to have accepted with researchers some restrictions in the dissemination of foreground results (publication, conference presentation, discussion with colleagues from other organizations) if the sensitivity of the application, or interest of financing/buying business need secrecy for the shorter or longer term.

Not surprisingly, it took almost two decades for TTOs to become a widespread institutional mechanism.

The position of Hungarian TTOs started to change in the university hierarchy following a much debated legal change in 2014 which introduced the chancellorship into university management.[9] The chancellor is responsible for property management in universities including managing and utilising intellectual products. This changed the position of TTOs who now belong to the highest management level (the Rectorate) in the majority of universities, and the organisations are linked to the chancellor.[10] Universities are now approaching the single gateway model which simplifies the procedure for companies. The changes in the hierarchical position of TTOs give much better opportunities to negotiate with companies if the company needs a solution which requires collaboration by different disciplines or departments within a university. Each HEI deals with technology transfer in its own way, but following government strategy.

TTO staff is not very large at HEIs, and those fully devoted to technology transfer activities number from 2–10 in different universities (Bene et al. 2018; p. 44).

TTOs still have several problems, such as faculty members being reluctant to accept TTOs as supporters in commercialization; a shortage of competent experts for different technology transfer tasks; uncertain financial conditions for organisation.[11] TTO staff have to have good knowledge of research assets and forthcoming

---

[9]The Amendment (2014) to Higher education law 2011 aimed to facilitate the professionalization of institution management sharing the previous tasks of rectors between rectors and chancellors.

[10]At several universities chancellors are political appointees rather than professionals—which may lead to the somewhat fragmented functioning of TTOs.

[11]The early staff members of TTOs were naturally inexperienced. There were many changes in their ranks due to poor conditions for TTOs, and the reluctance of faculty members to collaborate caused chaos at several TTOs. Financial conditions for TTOs were linked to projects and this damaged their stability, further eroding the staff.

inventions at the organisations and have to be familiar with intellectual property, licensing, business development, and other TT related legal matters. Marketing capabilities are also crucial, but the cyclically available resources make it difficult to keep trained staff, to build good relationships within a university and with collaborating business.

Nowadays financial conditions are improving for TTOs, and this has generated some stability in relation to increasingly knowledgeable staff.

7.4.2.2   Intellectual Property Issues

The Hungarian law deals with intellectual property within universities in a variety of ways and, as mentioned earlier, it has a turbulent history from first introduction of Bayh-Dole-type regulation in 2004/05. Since the law has changed twice regarding the ownership of inventions originating in universities, the regulations on treating intellectual property is also changing—including approved forms of contract with businesses. One more or less stable element from 2004/05 is worth mentioning here: if a company wishes to sign a tight contract with a university which excludes the names of the inventors from the patent application, this can be blocked by the university. This is an important change from previous practice.

If companies contract for R&D activities, they prefer to become the owners of all R&D outputs. In the case of government-/EU program- based BUC, a university has some possibility to be a co-owner. The involvement of the Innovation Centre of ELTE from the stage of the draft contract can upgrade the chances for university to become the co-owner of the patent. However the majority share still remains at the company. If the consortium agreement excludes the university from patenting the Centre does not have further role in that collaboration (Interview at ELTE Innovation Centre).

An MNC collaborating with two universities has expressed its intention to be 100% owner of all research findings. They have not shared any fully or partly university-based inventions in patent applications, but this company is co-patenting from time to time with universities in its home country (Interview at MNC-2).

Now HEIs are revising their own regulations following the changes at the end of 2018, and we do not record the details of their internal rules since these are again under review.

Income from patented BUC-related inventions is usually treated in a standard way at MNCs. They offer the University the same fee for inventions as to an employee of the company for his/her (official working time) invention. The university may then share this sum according to its own rules.

A common aim of HEIs is to encourage faculty members and students to be interested in the valorisation of their inventions and increase their patenting consciousness. The sharing of income from the commercialization of research results either through licensing or assigning is a delicate issue within universities. One thing, however, has been established: the inventor(s) is entitled to a pre-defined share of the income rather than to some post-negotiated, ad hoc bonus. This income sharing varies among Hungarian universities and the proportions differ by amount of income, type of intellectual property and type of utilising organisation (matured companies, university supported spin-offs, or other start-ups). Two main types of sharing model can be identified: (1) Sharing the income simply between inventor and university; (2) Sharing the income among various actors who may have an indirect impact on inventions. In this latter model the largest proportion goes to the inventors, and then some proportion may go to the faculty, another fraction to the organisation to which the inventors belong (department, institute or group) and to the university as a whole. A clear system of sharing income is crucial to make university staff interested in commercialization and limits neglected commercialization or the leaking of intellectual assets from a university.

The rules are concern not only to faculty member but project-based researchers, foreign guest researchers and students. Students don not have to share their income from inventions if their patentable results were created outside subject duties and they did not use the infrastructure of university.

Annual patent applications and patents granted usually show together the science producing capabilities, the interest in intellectual property protection and the awareness of HEIs. As universities are competing for business's demand for R&D contracts and acquiring intellectual property, this indicator is important. However this application-based patent indicator is usually lower than inventor-based ones in mostly the host-countries of MNCs (Inzelt 2014). In Hungary there was an additional factor between 2014 and 2018: a legal step-back reduced the chances of universities co-patenting with business, if the latter was co-financing the research or had bought the research findings.

7.4.2.3   Capacities of Hungarian Universities

By the number of students and number of faculty members, Hungarian universities belong to the small-to-medium HEI size group. Size is an issue for collaboration in several ways. As smaller universities have a smaller quantity of brains than the larger ones, they may not be able to mobilize research capacity at other than below the critical mass.

Over the last few years expenditure on higher education R&D (HERD) has declined. From 2010 to 2016 the public expenditure fell by 29% and business expenditure for HE by 45%. The increases may be seen from foreign sources (41%) and in the small category of sources: non-profit (30%). Because of these changes the HERD to GDP ratio, which was around 0.23% between 2007 and 2012 has fallen to 0.13% by 2016, and slightly increasing in 2017 (0.18%) but it is still far below the 2012 level. Not surprisingly, in this period R&D personnel decreased by 7% in HE.

Comparing the HERD to GDP ratio in several OECD countries and in some Asian countries which feature in OECD statistics, we can observe opposing tendencies: most of the selected countries have increased their relative (and absolute)

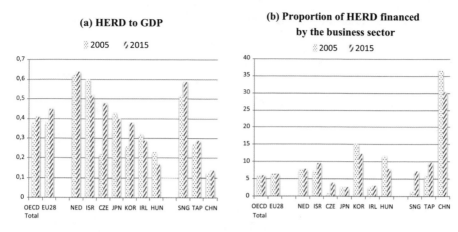

Fig. 7.2 Changes in financing HERD, 2005 and 2015 (or nearest year) in some countries (%) *Source* ECD (2018), Main Science and Technology Indicators: Volume 2018/1

expenditures except Israel, Japan, Ireland and Hungary. The drop was largest in Hungary from a lower level than two of these countries (Fig. 7.2).

From the standpoint of a BUC, another performance indicator is the HERD financed by the business sector. The business proportion has grown in several countries because they are more innovative than others or because business investment changed more slowly than other sources which cut their R&D expenditure in HE (Fig. 7.2).

The funding gap between the world's or Europe's top universities and Hungary's has a great influence on the gap in performance between front and rear runners. In Hungary there are some good research universities, but they are not world-class universities. They have good accumulated capabilities and produce excellent research in some fields, but not in all fields.

According to various world rankings, 6 or 7 Hungarian universities out of 27 are among the second half of 'The 1000 World's Best HEIs' since the quality of Hungarian universities is mixed.

The Centre for World University Rankings (CWUR) selected the first thousand for the 2018/19 ranking from 18 thousand HEIs. CWUR employs seven ranking criteria. These are: (1) Quality of Education, (2) Alumni Employment, (3) Quality of Faculty, (4) Research Output, (5) Quality Publications, (6) Influence, and (7) Citations.

By the overall rankings, 6 Hungarian universities are on the list, but not by all criteria. Our two universities interviewed have good positions: ELTE are among the Top 1000 by 6 criteria from 7 and BME by 5 criteria.

Looking at the position of well performing Hungarian universities, all 6 are among the best 1000 in respect of "influence" and "citation" (namely: ELTE, the Semmelweis University, Debrecen University, Szeged University, BME and the University of Pécs). Except for Pécs, the others are also in the Top 1000 for "research outputs" and "quality of publications". Two universities are in the Top

Table 7.1 Number of patent applications and primary patent applications

Name of Universities	2014		2015		2016		2017	
	All	Primary	All	Primary	All	Primary	All	Primary
	Applications							
Univ. of Szeged	16	9	11	5	13	1	4	4
Univ. of Pécs	12	5	6	5	3	1	12	1
ELTE	3	1	1	1	2	0	6	1
Semmelweis Univ.	3	2	3	2	6	3	3	1
Univ. of Debrecen	14	4	16	15	4	3	8	10

Note 'All' means invention disclosure, IP disclosure that was provided to TTO by researchers. *Source* Bene et al. (2018), p. 44

1000 for "quality of education" (ELTE and BME) and two by "quality of faculty" (ELTE, University of Szeged). None of them is among the Top 1000 for "alumni employment" (CWUR 2018).

From the perspective of BUC, an important performance criterion is not included in the CWUR, namely *invention activities of universities* measured by patents. Limited data are available directly on patentable performance but not for all R&D related Hungarian universities. Table 7.1 shows a university's patent applications (The lack of a university's data does not mean that performance is lacking).

As can be seen, the number of applications is very limited at data- providing universities. The number of patents does not accurately reflect the invention performance of a university. Discrepancies exist between performance in patentable invention of universities and their owned patents, as may be seen clearly if we compare applicant-based and inventor-based data. There is a remarkable difference between two data-sets. In 2006 the Hungarian performance was 44% higher by inventor indicator than by applicant by EPO data. At the same time the EU average was 2% less by inventor address than applicant address (Inzelt 2014). More than a decade later the difference between the location of a patent's inventors and applicants (ownership) was larger: –63% in Hungary whilst it was only –40% for the Czech Republics, –36% in Poland, –15% in Portugal, –1% in Denmark, and they were remarkably higher by applicant than inventor-based data—for example, in Ireland +13%, in Finland +14%, and in Sweden +19% (Hernández et al. 2018).

Among the reasons, the headquarters factor is significant, but the external and internal regulation of intellectual property's ownership is also an important reason —at least in Hungary.

The utilisation of intellectual products is broader than patenting. Table 7.2 shows the income of some universities from utilization.

The income from utilization (commercialization) of intellectual products fluctuated between 2014 and 2017. In 2017 the sum was larger than in 2014.

Table 7.2 Incomes of universities from utilization intellectual products (HUF)

Name of universities	2014	2015	2016	2017
Univ. of Szeged	2 882 352	4 599 436	3 963 111	3 794 733
ELTE	749 326	2 197 815	9 206 845	4 717 821
Semmelweis Univ.	1 869 708	256 194	1 390 968	20 160 000
Univ. of Debrecen	n.a.	9 000 000	27 734 200	n.a.

Source Bene et al. 2018, p. 44

7.5 Hungarian BUC in the European Mirror

Scientific capabilities and performance by Hungarian universities put the country on the map of foreign firms for R&D, even if the quantity of relevant human capital is limited. So Hungarian universities can benefit from these opportunities, but many burdening factors still exist.

Comparing the main types of Hungarian BUC to the EU average by a recent European Commission-financed study, we can see that the types are far less developed in Hungary than the EU average.[12] Looking at the importance of different types of collaborative activity, Hungarian businesses mostly engage in collaboration in R&D, student mobility and several forms of education. This is similar to our European counterparts, but the valorisation of R&D results in any form is below the European average. Not surprisingly, Hungarian BUCs are much less developed than their European counterparts (Orazbayeva et al. 2018).

Even if some crucial legal and financial barriers are eliminated and the government facilitate collaboration, there are still some de-motivating factors on the university side—which hardly supports collaboration.

Table 7.3 puts together the main barriers according to the responses from collaborating and non-collaborating academics and from business actors.

Table 7.3 shows that there are some commonly agreed barriers by three different types of actor, even if the ranking orders are slightly different. A group of frequently mentioned barriers relates to funding. It is an important motivator for universities to obtain funding from business for conducting research at their institution. A shortage of funding is a burdening factor for SMEs, but financial issues are not a great barrier

---

[12]A recent EU-financed study focused on BUC in 33 European countries. From Hungary 620 academics, 120 representatives of HEI and 42 business representatives (among whom 21% were MNCs) responded to the on-line surveys. (Orazbayeva et al. 2018, www.uni-engagement.com) These surveys focused on European trends and devoted less attention to national pictures. Most Hungarian organisations responded to fewer questions than organisations from advanced countries, and so the comparison between Hungary and the European average may be rough. That study dealt with the second most important engagement—'student mobility'—whilst, according to interviews, collaboration in education is most important and student mobility is only a part of BUC in education.

Table 7.3 Top 5 barriers for I-U cooperation as viewed by various actors

Factors	Academics		Businesses
	Collaborating with business	Not collaborating with business	
Financial factors	① Lack of university funding for BUC		
	② Lack of government funding for BUC		① Lack of government funding for BUC
	④ Limited resources of SMEs		
Motivational factors		② Differing motivation/values between university and business	
		③ The focus on practical results by business	③ The focus on producing scientific outcomes (e.g. papers) by universities
			④ Differing time horizons between universities and business
	⑤ Insufficient work time allocated by the university for academics' BUC activities		
Managerial factors	③ Bureaucracy related to BUC (in university)		⑤ Bureaucracy related to BUC in universities
Others		④ Difficulty in finding the appropriate collaboration partner	

Note Numbers show the rank of barrier by type of respondents
Source Author compilation on the base of Orazbayeva et al. (2018)

for MNCs. However, if funding barriers are removed, they do not necessarily create BUC; rather it makes the collaboration possible (Orazbayeva et al. 2018).

Another group of barriers relates to the divergent focus (and time-horizon) of university and business and to differing motivation/values between them. If the motivations for collaborations are not sufficient, cooperation is unlikely to occur. Even if some important managerial changes (TTOs, regulation of intellectual property and sharing the income from licensing) have happened at universities, the academics' job description and evaluation system have not changed yet. BUC activities do not belong to an academics' job description, although an important declared aim for universities and faculty members is to work in close collaboration with business. There is no reduction in academic duties in order to explore entrepreneurial opportunities, the insufficient working time allocated for academics' BUC activities. Some other problems are well-known: business's lack of awareness of university research activities and a strong business focus on producing practical results. The opposing needs of business (confidentiality) and the academics (publication, open access) are not easy to harmonize. Unchanged evaluation systems used by academics and universities which do not include BUC-related performances among output and outcome indicators cannot help the development of BUC. It is not easy task to revise the existing evaluation system based on academically acceptable, but the task is no longer to be overlooked.[13]

Apart from their European counterparts, Hungarian businesses are much less motivated to obtain new discoveries at an early stage from universities. It is no accident but another sign that the Hungarian inventions produced by the universities are still less accessible and less appealing from the business perspective. There have been some initiatives to make them more visible (such as universities' science days, innovation days, exhibitions, publishing on websites and TTOs efforts for dissemination).

We may add to the survey results the main motivators for MNCs in Hungary to collaborate in various forms of education are related to access to qualified graduates. There is a world-wide competition among high-tech industries for the best, fresh brains—which is why education, including student mobility is so important. Collaborative and business supported education give better access to bright students. Some kinds of involvement in education leads to company related Ph.D. programs and education of best talents through research exercises, mentoring for nurturing talents can provide an excellent workforce to companies and these fresh graduates' knowledge better fits company needs. Much BUC means only collaboration in education and its related activities (The estimated proportion of companies collaborating only in education is around 30%. At some (non-research) universities this proportion may reach 80% of total collaborators).

Beyond qualified students and graduates many companies are interested in new technologies and knowledge. These R&D collaborating MNCs are usually partners in education too. Their multi-faceted collaboration might strengthen and modernize the First Mission of universities and upgrade the quality and employability of new graduates. The R&D collaborations are contributing to the Second Mission of universities bringing new insights into research topics and better conditions for R&D. Those academics who are cooperating with business are motivated to obtain financial resources to continue their research and, by the benefits for their own research, such as in advancing their scientific knowledge, improving their teaching activities with real life examples and also in graduate employability (Interview at MNC-1 and 2 and BME).

According to the survey and interviews, Hungarian academics as well as businesses which have already relationships show very strong commitment to continue the collaboration, although the circle of R&D collaborating MNCs is not very wide and very slowly broadening. Some BUC active MNCs are usually collaborating with more than one university, whilst some locally R&D conducting companies have no university-related collaboration in Hungary. It is an open question if the circle of businesses collaborating with universities will broaden. However it is a

---

[13]European arguments also favour the inclusion of BUC activities in evaluation, but Hungary has introduced none of these notions.

positive sign that partnerships are usually durable. Next we focus on R&D collaboration with MNCs.

7.6 Collaboration in R&D

Business-university R&D collaboration is still limited if we look at the level of business R&D expenditure in HE. According to the OECD data (MSTI 2018/4) business share was 15.5% in 2009 and from this pick it has dropped to 8% to 2015.[14] The drop may somehow explain by a slow recovery from the 2008 economic crisis. However the real drop might be lower in business financed R&D since mass is missing in measuring BERD to HE.

Statistics cannot yet cover the contracts which are not subject to a university budget. The reason for budget avoidance is simple: both business and university actors would be very unhappy with processes such as hugely time consuming public procurement. Some contracts with business are handled by university-linked non-profit organisations to circumvent the burdening factors of cooperation caused by the legal status (publicly financed institutions) of universities. The participation of an intermediary organisation allowed them to avoid the strict public procurement system and to speed up the purchase of new research equipment, professional software, and research material. It is also important to pay regularly (monthly) the hired experts, PhDs and Master graduates on the projects. Both businesses and universities are interested in much faster processes to make procurement and R&D activities less time-consuming—and so to improve the competitive positions for BUC actors.

This avoiding technique may be used if state grants are not involved to the BUC project. Leading researchers at universities consider this technique workable (Consultation with several faculty members at various universities).[15] We do not know the total size of the missing mass, but we recognise its presence.

Generally, MNCs are involving local universities within host-countries in their R&D agenda through their affiliations.[16] Traditionally, subsidiaries near to their manufacturing site have used local HEIs or other available organisations (such as Technology Support Institutes) to adapt existing products and processes to local circumstances.[17] The collaboration might be a simple extension of the volume of R&D capacities.

---

[14]It is difficult to compare internationally as the Hungarian definition differs from that of most OECD countries.

[15]Of course this avoiding technique has not only advantages but disadvantages also.

[16]BUC is important in education and training also, as mentioned earlier, but this short chapter focuses only on R&D.

[17]Most R&D collaboration targets the adapting of products new to the local company and to different types of customer. Adaptation-related activities are also important but are less relevant for research universities. Their investigation goes beyond the scope of this chapter.

In a globally open world MNCs more frequently use local universities to solve R&D tasks to their economic activities. The competence-creating sub-units with their own R&D facilities are candidates for local inter-organisational collaboration. They become engaged in greater knowledge-based interactions with the local universities, since the development of new or improved products and processes require more active collaboration. The local subsidiaries may be outsourcing R&D or involving universities in joint R&D activities. Collaboration for inventions and innovation has evolved over time. The demand may be focused much more on experimental development-related issues (testing, prototyping) than on research.

Another level of MNC collaboration across countries relates to acquiring competences from any location. If the local scientific capabilities are strong, the MNCs may involve universities in their key research agenda either through subsidiary or contracting directly from head quarter (HQ) or one of the node-companies. Universities and departments with good intellectual capacity (and infrastructure) may attract foreign business partners in R&D in the fields of Industry 4.0 from all over the world.

In Hungary the extension of the volume of R&D capacities of MNCs goes parallel with acquiring additional competences from this location. The advantage of these parallel purposes is 70–80% of running R&D collaborations belong to disruptive technologies (Estimation of the interviewee from MNC-1). According to ELTE's experience, MNCs are approaching universities in such fields of disruptive technologies (e.g. big data, artificial intelligence) where the company has only limited capacities and they are very reluctant to invest in in-house R&D laboratory because of high risks.

One of the European initiatives we mention here as match-makers among Hungarian scientific and business organisations and foreign universities as well as HQs or nodes of MNCs is the European Institute of Innovation and Technology (EIT) that located in Budapest. EIT has launched BUC type projects in three important fields: EIT-Digital, EIT-Health and EIT-Climate.[18] Both universities investigated are active in EIT initiated projects. Table 7.4 gives some examples on EIT-Digital networks.[19]

As examples show, MNCs from their home countries participate in such collaboration under the umbrella of a European Joint Research Centre, EIT.

The significant presence of BUC collaborations in the field of disruptive technologies may be considered a positive evaluation of universities' capabilities and the openness of companies to embed Hungary's leading universities into cutting-edge technology-related R&D activities.

---

[18]EIT is one of the European Joint Research Centre. www.eitdigital.eu.

[19]The EIT Digital Budapest is a Pan-European Organisation which is located in the Central and Eastern European region. The consortium includes 2 leading universities (ELTE and BME) 1 research centre of HAS and 8 leading IT companies (4 corporates: Ericsson, T-Com, OTP bank, Nokia, 3 SMEs and 1 start-up). EIT Digital is networking with leading European ICT institutions: 42 universities, 26 research institutes and 74 industrial partners, involving the CEE region into EIT Digital's efforts to leverage European digital innovation.

Table 7.4 Examples of BUC in the frame of EIT-digital

Program	Collaborating partners
Smart Content delivery and Storage • Open source library for deterministic source coding for efficient content distribution in mobile environments. Traffic management for ISPs and network coding for improved handover of media-streams	Trento University, KTH, TU Berlin, Orange, Institute Mines, ELTE and BME
FITTIING • IoT facility—resource browsing, cross testbed measurement and data repository and extended federation	Institute Télécom, UPMC Paris, Fraunhofer, INRIA, TU Berlin, Trento ELTE and BME
eBIZ • eBIZ (Business Information Zone) introduces a new, disruptive, non-traditional banking service by OTP Bank Plc. first in Hungary, later in the CEE region for SMEs & start-ups to safely and efficiently perform financial and administrative tasks anywhere, anytime to focus on business opportunities	OTP Bank Plc. ELTE, SMEs
Medical CPS Environments • Data Mining Application	Siemens, TUM, DFKI, Trento Rise and ELTE

Source Várhalmi, Zs presentation, EIT Digital @ELTE, Budapest, 16/10/2018

7.7 Are MNC-University Collaborations Redeploying the Structure of R&D Episodes?

Although there is no clear separation between different episodes of R&D, the fact is that collaboration in "R" raises different issues and poses different challenges than collaboration does in "D" or in R&D. One of the delicate issues is how the MNCs influence the share of R&D activities by episode.

Description of three types of R&D base in the Frascati Manual (OECD 2015):

Basic research is the most risky episode. It is undertaken primarily to **acquire new knowledge** without any particular application or use in view. The outcome is uncertain and not very frequently financed by business (Public funding is important here). The time horizon of basic research is usually longer than companies' interests in collaboration. Basic research is not the main issue for BUC—even in most innovative countries with world class universities.

Applied research is also original investigation undertaken to **acquire new knowledge**, but it is directed primarily towards a specific, practical aim or objective. It is usually less time consuming than basic research and it has a medium or short-term perspective and is more closely oriented to application. Naturally, companies are much more interested to collaborate with universities for applied research than basic.

Both kinds of research may lead to inventions.

Experimental development is systematic work **using knowledge** gained from research and practical experience and producing additional knowledge, which is directed to producing new products or processes or to improving them. So 'D' leads to application (innovation) or decisions on further research or aborting any further related activities. Experimental development is frequently fed by testing pre-pre-production development.

Statistical information on the proportion by R&D episode of business-financed activities is not available either for the total HE sector or for the disaggregated level,[20] although this would be an important indicator of how and why university's knowledge capabilities are used.

As mentioned earlier, some companies—including MNCs—sponsor universities and a part of this sum goes to R&D. This sum is usually not very large but sponsorship provides support for research and for the development of university teams and faculties whilst focusing on non-targeted (basic) research. From the standpoint of carrying out basic research, sponsorship is a more frequent (but still limited) source than R&D contracts.

Fields of science differ in terms of the strength, weakness or total absence of a clear-cut border between basic and applied research and experimental development. Connections between knowledge generation and the solution of problems may be faster or slower among the three, influenced by the particular features of the scientific fields. In some emerging fields such as artificial intelligence or biotechnology the borderline between basic and applied research is almost eliminated. According to some experts, not only are basic and applied research but experimental development is also interwoven in the field of artificial intelligence. However, this does not lead to BUC contracts to cover all episodes at university. The company may decide to outsource applied research and experimental development, only the latter or a few elements of the experimental development. The episodes of R&D partnership and its importance differ by sector (high or low tech) and by age of companies (start-ups, matured) and fields of science at universities. Some types of testing might be a challenging task that can lead to important feedback for further research. MNCs may contract out to universities only just make BUC to test for further developments of the novel pre-product or pre-process based on company's in-house research findings.[21]

At BME-FIEK the collaboration topics are closely linked to industrial projects, such as high-speed wireless communication for Nokia; increased energy efficiency for hybrid drives for Siemens; balanced integration of green energy into energy

---

[20]The administrative data on contracts would be a good basis for information on BUC and its details. The universities are reluctant to process them for research purposes.

[21]Without going into more detail on definition problems, we would like to emphasise here that not all types of testing belong to experimental development. Some are derived from R&D activities, although universities usually count them as R&D activities.

distribution networks for MVM; and improving the stability and curative effect of medicines for Richter Gedeon Pharmaceuticals.[22] These are hot R&D topics today.

However according to head of FIEK there is no basic research in these collaborations and 50–50% goes for applied research (research documentation, trials of proof concepts) and experimental development (prototyping, testing). Testing might be 30% of total projects (Here we used how university classified the types). Several companies—either members of consortium or others—are using the university labs only for testing or validation. Most of the testing process are knowledge demanding, sometimes 3–4 departments have to collaborate to perform the tasks.

BME as well as ELTE has one or two collaborations with MNCs outside FIEK in disruptive technologies and other fields. New technological solutions and effective methods, proven by the university's laboratory experiments, can be utilized by the industry to accelerate development.

In the estimation of a senior figurer at BME, a breakdown of all BUC related R&D by research type shows roughly 20% basic research, 40% applied research and 40% experimental development (BME collaborates with more than 200 companies—both domestic and, MNCs).

From the fragmented information from ELTE we may assume that the proportion of basic research is higher in BUC related R&D—not unrelated to the different structure of main scientific fields of two universities (Natural science is one of the key fields in ELTE, while engineering science is the main field in BME).

The interviews at the leading Hungarian universities may help us to obtain a 'fuzzy' picture of collaboration by R&D episode. According to oral information from faculty members, TTOs and from project websites, there are only a few examples of business-financed basic research. There are many more examples when Hungarian universities carry out applied research with or for business in the frame of collaboration. Experimental development (such as prototyping, testing) is very frequent. Some MNCs contract out to universities only for testing. There are numerous reasons why companies collaborate with universities only in "D" and why universities are ready to be involved in "R" marginally in a host country. Testing can contribute to the better exploitation of up-to-date valuable capacities and bring in income for the further development of capabilities. Knowledge-demanding testing could be a good entry point for deeper R&D collaboration. However it is important to know why there are some changes in business decisions by episode in BUC.

- In several fields the significantly upgraded research infrastructure made the universities more attractive for different kinds of experimental development

---

[22]BME-FIEK program includes 5 university-laboratories that are working on collaborative projects. The laboratories belong to the faculties for their long-term sustainability. They participate in education (First Mission), academic research (Second Mission) and collaborative R&D with companies (Third Mission).

BME-FIEK function is joining forces in R&D, serving the demand of consortium member businesses (Richter, Siemens, Nokia, MVM Group) and manages the demand of new business clients, connecting them with researchers. FIEK encourages synergies between different areas.

(Same instruments, equipment may serve different purposes: education, research, testing).

- Structural changes in the Hungarian economy—growth of the automotive industry—create more and more demand for experimental development for BUC.
- BUC-related government programs are also encouraging to move toward those R&D activities which are closer to application.
- Changing regulations for Innovation Contribution made companies less willing to pay for risky basic or applied research. New regulations do not allow the companies to allocate 'soft money' to university since 2015. They have to pay the full compulsory contribution into the Innovation Fund. Previously they could share the compulsory sum between the Innovation Fund and University R&D.

"The company had large R&D contracts with BME that used to include basic research too. Since this reduction is not available anymore, the company is contracting only for R&D projects that are leading to results in the short term, less risky and are closer to introducing as innovation." (Interview at MNC-1)

- Time matters in companies' R&D investment. Several managers clearly expressed their need for such kinds of R&D tasks which generate knowledge for utilization within 1–3 years.

The level of R&D by episodes is a crucial matter for universities. They have to keep or develop a good balance among the episodes of R&D, using BUC for mutual learning. The supply of HE R&D has to meet not only the current but future demand—especially in the field of disruptive technology and Industry 4.0. The companies always wish to solve their actual shortage of R&D personnel for experimental development under the pressure of competition.[23] However much technology can be totally different within 5–10 years and a university has to be capable of doing relevant R&D in the long term.

7.7.1 Linkages Between BUC and Domestic Innovation Activities

One of the purposes of governments when they are encouraging BUC in R&D is to improve the innovative performance of business actors and the country as a whole. The various collaborations may have different effects on R&D within universities and on the innovation performance of the country, and it is worth devoting attention to the relationship between BUC in R&D which targets innovation and drives innovation at the businesses involved.

---

[23]Just to illustrate the impact of the shortage of R&D personnel as a strong motivation for BUC, we may mention that MNC-1 is recruiting Ph.D. degree holders from abroad.

In Hungary the majority of R&D contracts are feeding innovation in the subsidiaries, improving innovative performance in the country on both a larger and a smaller scale. Subsidiaries are the main contractors and they are interested in R&D topics relating directly to their own business interests (Interviews at MNC-1 and MNC-2). Beyond these are BUC in R&D which may not feed the innovation activities in Hungary. Three different types can be identified:

(a) Even if the subsidiary is active but has outsourced research to a university, the R&D based innovation will be introduced in other locations of the MNC where the knowledge absorption or production capabilities are better/cheaper for innovation. This behaviour means that the MNC is searching for the optimal solution for itself.

(b) The HQ or a node of a company may contract with HEIs in the host country for R&D work directly—even if a local subsidiary exists in the country. These types of contract usually do not relate to the activities of the local subsidiary. The MNC is merely using locally available R&D capacity as an additional resource to broaden its knowledge base. This is a positive evaluation of university's capabilities and might be a sign a less advanced presence in manufacturing or in service than in other countries.

(c) Some foreign companies without local activities are making R&D contracts with universities where the knowledge organisation has special, excellent capabilities. These departments of universities can contribute to developing cutting-edge technologies. It gives a positive feedback on a university's R&D. From the standpoint of innovation, however, BUCs may only have their effect in the distant future.

In Hungary those types of BUC in R&D (a–c) are marginal compared to total collaborations. The majority of MNC and university R&D collaboration lead to innovation in the country but there are some exceptions. Type (a) might represent a criticism of the new knowledge absorption capabilities of local subsidiaries or may be a sign of cheaper innovation capabilities in competing countries. Types (b) and (c) do not intend to introduce innovation in the country. What can be deduced from this is that their R&D at their presence signifies a very positive evaluation of the R&D capabilities of scientific organisations.

7.8 Conclusions

The R&D-intensive multinational firms have strong effects on domestic and international BUCs. In Hungary the bulk of MNCs (with or without production lines or services) has been involved in R&D collaboration with universities. This means that there is a demand for BUC. However, their demand differs from their European counterparts. Businesses are much less motivated to acquire new discoveries at an early stage from universities in Hungary. It is a sign that the

Hungarian inventions produced by the universities are still less accessible and less appealing from a business perspective. Greater exploitation of universities' intellectual products was hampered by legislation from 2014 until the end of 2018—which made companies very reluctant to co-patent with universities. University-based inventions have become wholly owned by companies, even if the company has covered only part of the research costs. The absence of universities as applicants has reduced the visibility of universities' capabilities and performance. Recent legal changes (December, 2018) have removed this burdening factor but it will take time to recognise its impact.

The significant presence of collaboration in the field of disruptive technologies is a good sign of university capabilities and the openness of companies to embed them into R&D relating to cutting edge technologies.

It is no coincidence that collaboration hardly exists in basic research, more frequently in applied research and most frequently in experimental development. This observation should draw the government's attention to the tasks of RDI policy and incentive system, to keep a healthy balance among R&D episodes serving not only short- and medium-term business interests but the long-term interests of the economy and society also.

To date neither the Hungarian authorities nor HEIs have devoted enough attention to the problems of the evaluation system—which is another burdening factor in BUC. The evaluation system of academics and universities still follows old practices which do not yet include BUC-related performance indicators. It is not easy to revise the existing evaluation system, but the task can be delayed no longer to support the involvement of academics into BUC activities.

To make BUC mutually more fruitful for business and university actors, it is important to improve the information system. Both sides, as well as the government, need a clear up-to-date picture of how BUC involves the actors in disruptive technologies, Industry 4.0 and how collaboration helps universities to keep and upgrade their intellectual and physical capabilities.

The main lessons to be learnt from Hungary's experience for other economies are that Business-University-Collaboration has many advantages for the innovativeness and competitiveness of economies. However all actors—companies, universities and governments—must work hard to create both the appropriate conditions and capabilities, and to build trust for effective collaboration.

The internationalisation of business R&D is a relatively new phenomenon when compared to that of manufacturing. The geographical distribution of R&D enables companies to disperse pools of knowledge assets, and international knowledge-hungry MNCs are broadening their potential field of university partners for collaboration. Upgrading the BUC factor is a challenge for both developed and developing countries. Typically, multinational enterprises, which dominate the manufacturing and assembly industries, restrict their R&D activities to their home countries (plus a possible handful of locations globally). For host countries of MNCs it is a challenge to build relevant capacities in their universities—to become partners in R&D collaboration. As this process develops, every country has to upgrade the level of internationalisation of their HEIs, strengthening international academic collaboration to make themselves more attractive to companies.

Hungarian experience has demonstrated that trust-building is a long-term process between businesses and universities, and both types of actor need to work seriously at it. Mutual understanding between different cultures and the interests of business and HEIs are the first step in this process, but if the academic culture and the technological innovation infrastructure are inadequate for the task, they will hamper all collaboration.

Governments have their part to play in upgrading the home country as well as international BUC: they can facilitate and encourage national as well as international BUCs in many ways. Promoted by investment in universities, intellectual capabilities can contribute to reaching the critical mass of research which is one of the key issues in collaboration, whilst a reward system for best collaborators should encourage the building of relationships. The critical mass of university laboratory facilities can make them competitive in attracting foreign business organizations to collaboration. An evaluation system both within universities and of faculty members should reward collaboration with business. Policy-making is always a delicate issue, and the Hungarian example has shown clearly how government can stimulate or hinder the collaboration process.

In Hungary as well as, in a bulk of other countries it will be worth launching social science research projects to identify nation-specific burdening factors of BUCs to support policy-making; this will lead to broader and stronger industry-science collaboration in an age of disruptive technologies, and Industry 4.0.

References

Ankrah, S., & Al-Tabbaa, O. (2015). Universities-industry collaboration: A literature review. *Scandinavian Journal of Management, 31*, 387–408.

Archibugi, D., & Lundvall, B. -Å. (Eds.) (2002). The globalizing learning economy. *Oxford University Press*, 328.

Archibugi, D., & Filippetti, A. (Eds.). (2015). *The handbook of global science, technology and innovation* (p. 603). Ltd: John Wiley & Sons.

Bene, T., Liber, N., & Németh, G. (2018). Szellemitulajdon kezelés és a kutatási eredmények hasznosítása a közfinanszírozású kutatóhelyeken. Összehasonlító elemzés és gyakorlati javaslatok, (Handling intellectual property and utilization of research findings at publicly financed research organisation, Comparative analysis and practical suggestions). *Hungarian Intellectual Property Office*, Budapest, p. 47.

Berchicci, L. (2013). Towards an open R&D system: Internal R&D investment, external knowledge acquisition and innovative performance. *Research Policy, 42*(1), 117–127.

Bonaccorsi, A., & Piccaluga, A. (1994). A theoretical framework for the evaluation of university—industry relationships. *R&D Management, 24*, 154–169.

Caloghirou, Y., Tsakanikas, A., & Vonortas, N. S. N. (2001). University-industry cooperation in the context of the European framework programmes. *Journal of Technology Transfer, 26*(1), 153–161.

Cantwell, J. A. (1989). *Technological innovation and multinational corporations*. Oxford: Basic Blackwell.

Cantwell, J. (2017). Innovation and international business. *Industry and Innovation, 24*(1), 41–60. https://doi.org/10.1080/13662716.2016.1257422.

Cantwell, J., & Molero, J. (Eds.) (2003). *Multinational enterprises innovative strategies and systems of innovation* (pp. 234–268). Cheltenham, UK: Edward Elgar Publishing.

Cantwell, J., & Piscitello, L. (2007). Attraction and deterrence in the location of foreign-owned r&d activities—The role of positive and negative spillovers. *International Journal of Technological Learning, Innovation and Development, 1*(1), 83–111.

Chesbrough, H. W. (2003). The Era of open innovation. *Sloan Management Review, 44, 3* (Spring), 35–41.

Chesbrough, H., Vanhaverbeke, W., & West, J. (Eds.). (2006). *Open innovation: Researching a new paradigm*. Oxford: Oxford University Press.

Cook, P. (2005). Regionally Asymmetric knowledge capabilities and open innovation. *Research Policy, 34,* 1128–1149.

Correa, P., & Zuñiga, P. (2013). Public policies to foster knowledge transfer from public research organizations. *Innovation, technology, and entrepreneurship global practice, public policy brief,* World Bank. Washington DC.

Clark, B. (2015). The character of the entrepreneurial university. *International Higher Education* (38).

CWUR. (2018). CWUR World University Rankings 2018–2019, https://cwur.org/2018-19.php, downloaded: 4/12/2018.

Dachs, B., Kampik, F., Scherngell, T., Zahradnik, G., Hanzl-Weiss, D., Hunya, G., et al. (2013). Internationalisation of business investments in R&D and analysis of their economic impact. *Innovation Union Competitiveness papers,* issue 2013/1 European Commission.

Dunning, J. H. (1988). The eclectic paradigm of international production: A restatement and some possible extensions. *Journal of International Business Studies, 19*(1), 1–31. https://www.jstor.org/stable/154984.

Dunning, J. H. (1992). *Multinational enterprises and the global economy*. Addison Wokingham: Wesley.

Edler, J. (2008). Creative internationalization: Widening the perspectives on analysis and policy regarding international R&D activities. *The Journal of Technology Transfer, 33*(4).

Edquist, C. (1997). Systems of innovation technologies, institutions and organizations. *Routledge, London and New York* (p. 432).

Edquist, C. (2018). Towards a holistic innovation policy: Can the Swedish national innovation council serve as a role model. *CIRCLE, Lund University, Papers in Innovation Studies,* Paper no. 2018/02.

Etzkowitz, H. (2004). The evolution of the entrepreneurial university. *International Journal of Technology and Globalisation, 1*(1), 64–77.

Etzkowitz, H. (2008). *The Triple Helix: University-industry-government innovation in action* (p. 161). London p: Routledge.

Etzkowitz, H., & Leydesdorff, L. (1997). Introduction to special issue on science policy dimensions of the Triple Helix of university-industry-government relations. *Beech Tree Publishing.*

Etzkowitz, H., & Leydesdorff, L. (1998). The endless transition: A 'Triple Helix' of university industry government relations. *Minerva, 36*(3), 203–208.

EU. (2005). *The handbook on responsible partnering—Joining forces in a world of open innovation. A guide to better practices for collaborative research and knowledge transfer between science and industry.* EUA, ProTon Europe, EARTO and EIRMA. Retrieved from http://www.responsible-partnering.org/library/rp-2005-v1.pdf.

EU. (2016a). Peer Review of the hungarian research and innovation system. *Horizon 2020 Policy Support Facility*, EU, Directorate-General for Research and Innovation, Luxembourg: Publications Office of the European Union. https://rio.jrc.ec.europa.eu/en/file/10007/download?token=82l5lFpl.

EU. (2016b). The European innovation scoreboard 2016, European Union.

Farinha, L., & Ferreira, J. J. (2013). Triangulation of the Triple Helix: A conceptual framework. https://www.triplehelixassociation.org/working-papers/triangulation-of-the-triple-helix-a-conceptual-framework.

Foray, D. (2006). the economics of knowledge. The MIT Press.

Freeman, C. (1988). Japan: A new national innovation system. Technology and economy theory. *Pinter, London* (pp. 331–348).

Govind, M., & Küttim, M. (2016). International knowledge transfer from university to industry: A systematic literature review. *Research economics and business: Central and Eastern Europe, 8* (2), 21.

Guimón, J., & Narula, R. (2017). When developing countries meet transnational universities: Searching for complementarity not substitution. *Discussion Paper,* Number: JHD**2017-01**. www.henley.ac.uk/dunning.

Gulbrandsen, M., & Slipersaeter, S. (2007). The third mission and the entrepreneurial university model, in Universities and strategic knowledge creation. Specialization and Performance in Europe. In A. Bonaccorsi & C. Dario, *PRIME Series, Edward Elgar Publishing Ltd.* (pp. 112–143).

Hernández, H., Grassano, N., Tübke, A., Potters, L., Gkotsis, P., & Vezzani, A. (2018). The 2018 EU Industrial R&D Investment Scoreboard; EUR 29450 EN; Publications Office of the European Union, Luxembourg. https://doi.org/10.2760/131813, JRC113807.

Inzelt, A. (1999). Transformation role of FDI in R&D: analysis based on a databank. In David Dyker & Slavo Radosevic (Eds./szerk), *Innovation and structural change in post-socialist countries: A quantitative approach* (pp. 185–201), The Netherlands: Kluwer Academic Publisher.

Inzelt, A. (2000). Foreign direct investment in R&D: Skin-deep and soul-deep co-operation. *Science and Public Policy,* August, 4, 241–251.

Inzelt, A. (2004). The evolution of university-industry-government relationships during transition. *Research Policy, 33*(6–7), 975–995.

Inzelt, A. (2008). The inflow of highly skilled workers into Hungary: A by-product of FDI. *Journal of Technology Transfer, 33,* 422–438.

Inzelt, A. (2010). Collaborations in the open Innovation Era. In Ndubuisi Ekekwe (Ed.), *Nanotechnology and microelectronic* (pp. 61–86). USA: IGI Global.

Inzelt, A. (2014). Embeddedness level in central and Eastern European countries as revealed by patent-related indicators. *Prometheus, 32*(4), 385–401.

Inzelt, A., Laredo, P., Sanchez, P., Marian, M., Vigano, F., & Carayol, N. (2006). 3rd mission. In Methodological Guide, *PRIME, Network of Excellence, Lugano,* 125–168. http://www.prime-noe.org.

Inzelt, A., & Csonka, L. (2016). Public-private interaction under fluctuating public support program in: Public-private partnerships in research and innovation: Trends and international perspectives. In K. Koschatzky & T. Stachlecker (Eds.), *Fraunhofer Verlag, Karlsruhe,* pp. 129–158.

Koschatzky, K., & Stahlecker, T. (2010). New forms of strategic research collaboration between firms and universities in the German research system. *International Journal of Technology Transfer and Commercialization, 9,* 94–110.

Laredo, P. (2007). Revisiting the third mission of universities: Toward a renewed categorisation of university activities? *Higher Education Policy, Springer, 20*(4), 46–59.

Leydesdorff, L., & Sun, Y. (2009). National and international dimensions of the Triple Helix in Japan: University—industry—government versus international co-authorship relations. *Journal of the Association for Information Science and Technology Association, 60*(4), 778–788.

Lundvall, B. Å., & Johnson, B. (1994). The learning economy. *Journal of industry studies, 1*(2), 23–42.

Mégnigbéto, E. (2015). Effect of international collaboration on knowledge flow within an innovation system: A Triple Helix approach. *Triple Helix Journal, 2*(16), 21. open access https://doi.org/10.1186/s40604-015-0027-0.

Mollas-Gallart, J., Salter, A., Patel, P., Scott, A., & Duran, X. (2002). *Measuring third stream activities*. SPRU: Report to the Russel group Universities, Brighton.

Mowery, D. C. (1998). Collaborative R&D: How Effective Is It? *Issues in Science and Technology, 15*(1).

Narula, R. (2003). Globalization and technology: Interdependence, innovation systems and industrial Policy. *John Wiley & Sons,* 264 pages (reprinted: 2015).

Narula, R. (2014). Exploring the paradox of competence-creating subsidiaries: Balancing bandwidth and dispersion in MNEs. *Long Range Planning, 47*(1–2), 4–15.

Nedeva, M. (2008). New tricks and old dogs? The 'third mission' and the re-production of the university. In *The World Yearbook of education 2008: Geographies of Knowledge/Geometries of Power: Framing the Future of Higher Education* (pp. 85–105). New York: Routledge.

Nelson, R. (1993). *National innovation systems*. New York: Oxford University Press.

NKFIH. (2019). Kutatás-Fejlesztés és Innováció Magyarországon, (Research, Development and Innovation in Hungary) Budapest, p. 30.

OECD. (2008a). *Open innovation in global networks*. Paris, France: OECD Publishing.

OECD. (2008b). *OECD reviews of innovation policy: Hungary* (p. 228). Paris: OECD Publishing.

OECD. (2015). Frascati Manual 2015, *Guidelines for Collecting and Reporting Data on Research and Experimental Development*, The Measurement of Scientific, Technological and Innovation Activities, OECD Publishing, Paris DOI http://dx.doi.org/10.1787/9789264239012-en.

OECD. (2017). The next production revolution: Implication for governments and business. *OECD Publishing, Paris.* https://doi.org/10.1787/9789264271036-en.

Orazbayeva, B., Davey, T., Prónay, S., Meerman, A., Muros, G. V., & Melonari, M. (2018). The state of Hungarian university-business cooperation: The university and the business perspective. *Study on the cooperation between higher education institutions and public and private organisations, European Commission.* www.uni-engagement.com.

Perkmann, M., Tartari, V., McKelvey, M., Autioa, E., Broströmc, A., D'Este, P., et al. (2013). Academic engagement and commercialisation: A review of the literature on university—industry relations. *Research Policy, 42,* 423–442.

Pinheiro, R., Langa, P. V., & Pausits, A. (2015). The institutionalization of universities' third mission: Introduction to the special issue. *European Journal of Higher Education, 5*(3), 227–232.

Ranga, M., & Etzkowitz, H. (2013). Triple Helix systems: An analytical framework for innovation policy and practice in the Knowledge Society. *Industry & Higher Education, 27*(3), 237–262.

Raymond, S., & Taggart, J. H. (1998). Strategy shifts in MNC subsidiaries. *Strategic Management Journal, 19*(7), 663–681.

Schaeffer, V., & Matt, M. (2016). Development of academic entrepreneurship in a non-mature context: The role of the university as a hub-organisation. *Entrepreneurship & Regional Development.* https://doi.org/10.1080/08985626.2016.1247915, http://dx.doi.org/10.1080/08985626.2016.1247915.

Shimoda, R. (2008). Reform of university research system in Japan: Where do they stand? *National academy of Sciences,* 40–56.

Singh, J. (2007). Asymmetry of knowledge spillovers between MNCs and host country firms. *Journal of International Business Studies, 38*(5), 764–786.

Sorensen, O. J., & Hu, Y. (2014). Triple Helix going abroad? A case of Danish experiences in China. *European Journal of Innovation Management, 17*(3), 254–271.

Swann, P. (2014). *Common innovation: How we create the wealth of nations*. Cheltenham: Edward Elgar.

Taggart, J. H. (1998). Determinants of increasing R&D complexity in affiliates of manufacturing multinational corporations in the UK. *R&D Management, 28*(2), 101–110.

Wen, J., & Kobayashi, S. (2001). Exploring collaborative R&D network: Some new evidence in Japan. *Research Policy,* 30(8), October 2001, pp. 1309–1319.

Dr. Annamria Inzelt is Founding Director of IKU Innovation Research Centre. She is honourable professor at University of Szeged. Her main research interests are systems of innovations, the innovative capabilities and performance of the different actors, business organisations and universities, the role of internationalisation in technology upgrading. She has been the first Hungarian representative in the OECD Working group of the National Experts of Science and Technology Indicators (NESTI) for 11 years, and is an expert of UNESCO on science and technology indicators. She was also involved in OECD collaboration with transition economies. She has been member of the National Committee for Technological Development, the Committee for Industrial Economics at Hungarian Academy of Sciences. She was an editor of several Hungarian professional journals and member of advisory editors to Research Policy, Technological Forecasting and Social Change and still active at Triple Helix Journal. She has published in various international journals, author and editor of books published in Hungarian or English.

Chapter 8
Text Mining Method for Building New Business Strategies

Focusing on the Neurosurgical Robot

Fumio Komoda, Yoshihiro Muragaki and Ken Masamune

Abstract In any time, it has been essential to acquire knowledge of customer needs and global trends of technological progress for proper selection and concentration strategy planning, which is decisive for long-term growth of the company. However, with the change in innovation paradigm, the methods used for its acquisition have also changed. With the era of big data, text mining that gains knowledge necessary for this planning from unstructured natural language with weak affinity with relational databases has attracted attention recently. However, in order to obtain highly accurate and reliable knowledge that can contribute to company decision-making, the current natural language processing algorithm is not sufficient. Current text mining method, which is limited to bird's eye viewing type aimed at capturing the entire text data roughly, is unsuitable for finding out important knowledge written only in a very small part of the text data. Therefore, this paper presents the virtual case of a company planning a new neurosurgical robot project and applies pinpoint focus type text mining technique to acquiring technological knowledge from high-impact peer-reviewed academic journals.

Keywords Text mining · Pinpoint focus type text mining · Innovation paradigm · Text data · Business strategy · Neurosurgical robot

8.1 Introduction

Modern management theories, such as positioning theory (Porter 1980), destructive innovation theory (Christensen 1997), open innovation theory (Chesbrough 2003), and so on, are closely related to the fact that investment in core competence is

F. Komoda (✉)
Honorary Professor, Saitama University, Saitama, Japan
e-mail: techtra@ae.auone-net.jp

Y. Muragaki · K. Masamune
Institute of Advanced Bio-Medical Engineering and Science,
Tokyo Women's Medical University, Tokyo, Japan

© Springer Nature Singapore Pte Ltd. 2019
J. Cantwell and T. Hayashi (eds.), *Paradigm Shift in Technologies and Innovation Systems*, https://doi.org/10.1007/978-981-32-9350-2_8

critical to the growth of company earnings. The long-term stagnation in Japanese companies since the end of the twentieth century can be attributed to a considerable extent to failures in selection and concentration strategies. Typical examples include Sharp's excessive investment in LCD panels, where overproduction was expected to cause low cost competition, and Toshiba, which had succeeded in developing flash memories with promising markets, but lagged behind Samsung in its commercialization.

One reason for the failure of Japanese companies with respect to selection and concentration strategies is that their decision-making depends largely on subjective judgments originated in managers' intuition, their successful experiences in the past, and so on. In other words, when it came to knowledge creation, Japanese companies were more dependent on "tacit" than "explicit" knowledge (Nonaka 1991). Today, however, as globalization and introduction of IT transforms production architecture from integral to modular forms, it has become necessary to integrate a variety of technologies indispensable for developing new products owned dispersively by many companies in the world (Fujimoto 2007). Decision-making under such circumstances requires not only accumulated proprietary knowledge but also actively utilizing objective data created by broadly observing the trends of society as a whole (Veugelers et al. 2010). The innovation paradigm is undergoing major change. As the result, big data, including text data, is becoming to play a crucial role in R&D. Japanese companies seem not to be adapting to this paradigm change flexibly and successfully at present.

One source of objective data is a technology roadmap, which is produced for the purpose of identifying the trajectories of important technologies, and for the purpose of discovering promising business with profitability as company R&D strategy, with priority as government technology policy (Japan Patent Office, home page). A technology road map is produced by combining a variety of knowledge from multiple experts in each area, or by reviewing patent data. However, the roadmap produced in this way is often insufficient to forecast technological progress and business trends across the world. The reason for this is that if knowledge with deep and specific thought possessed by many experts is integrated, it ultimately results in an ordinary conclusion without "personality." In addition, the metadata in patent publications contain insufficient knowledge for planning a business strategy.

Another source of objective data useful for building strategies for selection and concentration is various kinds of text data. Text data may include newspaper articles, industry reports, call center logs, patent bulletins, and research papers. These unstructured natural languages contain much more knowledge than structured relational databases. Therefore, by mining these data, it becomes possible to predict the future potential of the project based on understanding customers' needs, business profitability, and trends of technological progress across the world, etc. Thus, text mining studies have attracted attention in recent years with numerous papers published.

Text mining has often been examined in terms of strategy planning, decision-making, and marketing activities, including finding customer needs.

(Netzer et al. 2012). However, text mining can be applied to more than discovering customer needs. Companies need to know what technologies are necessary to develop new products that satisfy customer needs and what kinds of ideas are necessary to overcome bottlenecks to developing that technology. Nevertheless, as Kim and Bae (2016) states, there are only a few studies that apply text mining to forecast technologies, discover promising technologies, or solve technical challenges compared with research that applies data mining to discover promising products/markets. One reason for this is that it is more difficult to apply text mining to acquiring technical knowledge compared with its application to marketing due to differences between the nature of knowledge about marketing and that about technology.

Knowledge that is generally well-known across the world is insufficient for planning corporate strategy and decision-making. The reason is that "known" or "common" knowledge cannot be the source of monopolistic profit. Knowledge with potential that is unknown to anyone, or that is only known to some extent, enables companies to gain monopolistic profits. In this regard, either marketing or technical knowledge remains in the same.

The difference in knowledge about marketing and technology is as follows. Marketing knowledge is often expressed in the language of daily life; therefore, in many cases, it is easy for analysts to understand the results of mining as they already have accumulated personal knowledge. On the other hand, knowledge about technology is expressed by esoteric technical terms, and thus to understand that, analysts are required for background knowledge about current technical trends. For example, exploring concepts such as "weight reduction of camera" or "expansion of refrigerator capacity" is enabled by data analysts' common knowledge. On the other hand, technological knowledge is based on expert knowledge, jargon, and equations, which are impossible to understand solely based on common knowledge; therefore, the text mining results are required to provide a high degree of specificity and accuracy. For example, in studying glycosylation, an important metabolic theme in biology, in order to clarify the mechanism of glucose conversion from the mining results, it is essential that data analysts have prior relevant expertise such as glycosylated molecule, glycosylated position, catalytic enzyme and so on (Rzhetsky et al. 2008). In other words, in the case of the customer needs context, rough and ambiguous text mining results are enough to understand the meanings. In the same way, text mining results in the technical context require the analyst to have a sophisticated level of knowledge in order to deliver meaningful information.

For the above reasons, the methodology used in text mining differs according to the nature of the required knowledge. Categories extracted in mining aimed at understanding widespread phenomena in the context of marketing activities are generally broad, like "shades related preference" in fashion, i.e., relative preference to light colors against dark colors, and to cold against warm color. This requires a "bird's-eye-view" type text mining. On the other hand, a "pinpoint focus" type text method is often required in the case of mining for engineering purposes, where engineers want to find ways to overcome technical obstacles and to discover useful technical ideas.

Most text mining research aims to analyze enormous volumes of text and to present a bird's-eye-view showing the overall trends. In other words, using statistical analysis methods, such as multivariate analysis, the overall image or pattern is grasped in general.

On the contrary, technological knowledge, for example, ideas for overcoming technical bottlenecks, is not widely shared and well-known in society but for a limited number of individuals. In addition, it is usual that such knowledge is described only in a small part of the text. Therefore, detailed technical knowledge is necessary for pinpointing the related parts of all the text data and link them together, i.e., a pinpoint focus type text mining method.

There are various techniques used for text mining, for example: (1) information extraction; (2) topic tracking; (3) summarization; (4) categorization; (5) clustering; (6) concept linkage; (7) information visualization; (8) sequestration-assisted mining; and (9) association rule mining (Gupta and Lehal 2009; Patel and Soni 2012). Among them, topic tracking, clustering, concept linkage, etc. are aimed at finding the overall tendency based on a bird's-eye-view of the whole. Information extraction plays a role in searching necessary parts; however, from our experiences, it is understood that information extraction cannot provide sufficient detail. Therefore, it is required to start with a bird's-eye-view of the whole through combining pinpoint focus types text mining methods and then to gradually extract necessary parts accurately, and finally to connect them. As above, text mining is classified into two types: (1) bird's-eye-view and (2) pinpoint focus. The pinpoint focus type is a mining method aimed at acquiring more precise knowledge by limiting the analysis target narrowly as compared with the bird's-eye-view type.

Therefore, this paper presents the virtual case of a company planning a new neurosurgical robot project and applies text mining to acquiring technological knowledge from high-impact peer-reviewed academic journals. To succeed in the surgical robot business, this company is required to have a pinpoint focus approach to extract text data that can reveal developing trends in the surgical robot industry and the kinds of technology useful for solving technological bottlenecks.

8.2 Methodology

8.2.1 Co-occurrence Relationship Analysis Based on Text Blocks

Taking as an example medical research involving surgical robots, we survey previous research aimed at distinguishing the two text mining methods.

Typical examples of bird's-eye-view type text mining are often found in studies that describe research trends as a whole (Kurata and Takigawa 2010). Pinpoint

Fig. 8.1 Text Block
Consisting of 5 Sentences.
Note S = Sentence
Source Created by author

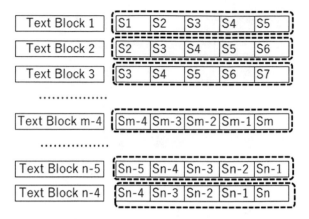

focus text mining is applied to studies, for example, to discover relationships between genotype and disease (Koike 2007), or to determine the toxicity and efficacy of candidate compounds in the early stages of R&D projects in the high-risk drug discovery (Kaneko and Ootake 2010). Studies that find the relationship between treatment and prognosis also apply pinpoint focus type text mining (Abe et al. 2005; Kushima et al. 2012).

One of the difficulties in applying pinpoint focus text mining method is the need to cleanse a huge amount of data in order to extract the data of interest for a detailed analysis. In addition, because these critical parts are scattered throughout the data, it is not easy to combine them together and discover their meaning.[1]

In this paper, in order to overcome these difficulties, the author divides all of the acquired text into small blocks. A "text block" (TB) is created by applying the N-gram approach, proposed by Shannon, to sentences instead of words. This is illustrated in Fig. 8.1.

The reasons for dividing all text into TBs is that most of the data include scientific papers, industry reports, and patent bulletins, which range from a few pages to tens of pages; however, only a few portions contain the required knowledge. Mining such large texts presents difficulties in finding the critical information. In addition, if using each article as a mining unit, co-occurrence relations of words are analyzed, and thus two contextually unrelated words are considered as co-occurring, leading to misunderstanding the meaning of the text. It is considered that "clumps of meaning are discovered within 5–10-sentences (Mima 2006). It is, therefore, appropriate to analyze co-occurrence on the basis of TBs consisting of about five sentences in order to meaningful co-occurrence relationships without omission.

---

[1]A typical example of knowledge obtained in this way is the discovery of specific leukemia therapies using IBM's Watson Genomic Analytics.

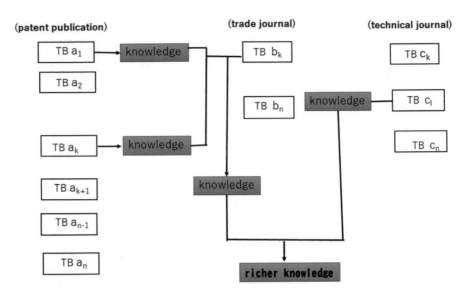

Fig. 8.2 Linked text blocks. *Source* Same to Fig. 8.1

Needless to say, TBs containing the knowledge being sought can be found in a variety of texts including patent publications, trade journals, and technical journals. Thus, by combining these TBs correctly, more in-depth knowledge can be retrieved (Fig. 8.2).

8.2.2 The Dataset

This section describes the data of interest in the pursuit of the technical knowledge necessary to develop and commercialize new neurosurgical robots.

In general, text data used to understand R&D trends are patents, academic research papers, etc. In this study, the data are composed exclusively of academic research papers instead of patents.

In general, text data suitable for acquiring a bird's-eye-view of R&D trends come from patent bulletins, patent applications, academic research papers, etc. The first reason why patent data are insufficient is that most R&D results for surgical robots fail to reach the stage of patent application or commercialization; thus, it is impossible to gain an in-depth understanding of the current state of technological progress and to forecast its future by only looking at patents. The second reason is that in comparison with academic journals, the descriptions written in patent publications are often insufficient to describe the fundamental direction of the R&D or to express customer needs or technical functions achieved.

This study analyzes six academic journals: three journals that cover the neuro-surgery field in general and three that are related to the development of surgical

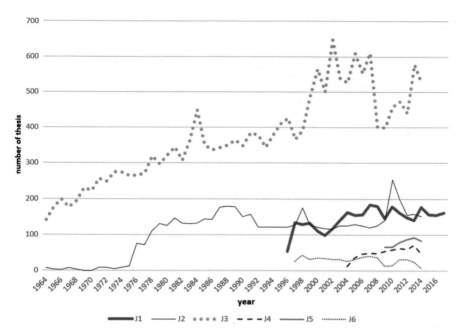

Fig. 8.3 Number of theses by year. *Source* J1 = Neurosurgical Focus, J2–6 = JDream3

robots, including computer and endoscopic techniques, which are establishing the technological foundation of surgical robots.

The two categories are represented as follows:

(A) Neurosurgery in general
   "Neurosurgical Focus" (JI) 3,194 full texts 1996–2015
   "Neurologia Medico-Chirgica" (J2) 5,451 abstracts 1964–2014
   "Journal of Neurosurgery" (J3) 20,333 abstracts 1950–2014
(B) Surgery using a surgical robot or computer
   "The International Journal of Medical Robotics + Computer Assisted Surgery" (J4) abstracts 547 2004–2014
   "Computer Aided Surgery" (J5) 534 abstracts 1997–2014
   "International Journal of Computer Assisted Radiology and Surgery" (J6) 479 abstracts 2009–2014

Neurosurgical Focus is obtained from home page of American Association of Neurosurgical Surgeon. Other five theses are obtained from JDream3, which is scientific paper database provided by G-Search.[2]

The trend of number of theses by year in each journal is shown in Fig. 8.3. Surgery using robots and computers is a new research field, thus, the creation of

---

[2]https://www.aans.org/en/Publications/The-Journal-of-Neurosurgery.

scientific societies and launching the publication of J4–J6 was delayed compared with J1–J3.

There are some limitations to this approach. First, the full text is analyzed only in J1 among the six journals, whereas only the abstracts are analyzed in the other five journals. The reason is that obtaining data in full text is very expensive. Second, for J1, where the full text is analyzed, pdf files and HTML files were converted to text data files. It is inevitable that errors such as garbled characters occur in the process of this conversion work as well as technical mistakes occurring during the processing of line feed codes and so on. Therefore, it is difficult to obtain data that completely match the original text. Furthermore, for J1, the author selects only academic theses among all articles contained in the journals; thus, it is inevitable that arbitrariness enters into this selection. It is considered that these limitations have no significant effect on this analysis and its results.

8.3   Current State of Development of Surgical Robots

Surgical robots, like industrial and various service robots, rarely fall under the current accepted definition of "robot." The basic criterion to be considered a robot is "autonomy" (Chinzei 2015). Autonomy is the ability to recognize and avoid obstacles in the patient's body independently and without human intervention. However, a precise definition of a surgical robot in this way would mean that there are very few surgical robots currently available, including the most widely used surgical robot, "da Vinci." Therefore, in this article, surgical robots are viewed in the broad sense; thus, tools with primitive autonomy, such as some endoscopes, a portion of computer control systems, and so on that provide a technical basis for acquiring true autonomy in future, are included. Hence, the term "robot" and a portion of both "endoscope" and "computer" appearing in J1–L6 are viewed as robots because these apparatuses represent a technological base toward autonomy in future.

Surgical robotics is expected to be a promising new industry. The reasons are as follows:

(1) Advances in image diagnostic technology have been seen based on deep learning since about 2012 and signs of increasing the number of new drugs approved by the FDA since around 2010 are beginning to be seen, but technological breakthrough for improving medical technology is not still fully available, as shown in the fact that the application of genetic engineering to clinical practice has not advanced much. Under such circumstances, medical robots, especially robots used in surgery, can be expected to contribute to improvements in medical technology.

(2) With rapid advances in computer and control technologies, there is a possibility that various elemental technologies developed in robotic R&D activities can be translated to the medical field also, where robotization has been slow but is increasing.

However, it is not easy for a company to bring surgical robotics into a profitable business. The da Vinci Surgical System, which was approved by the US FDA in 2000 and is widely accepted throughout the world, is one of the few successful examples. NeuroMate, launched by Integrated Surgical Systems in 1987 to support stereotaxic surgery, is continuing to evolve and is still used in many hospitals as described below. However, most of the currently popular surgical robots have not developed the level of autonomy required to be considered a true "robot."In addition, powered suits as cyborg type robot are expected to be utilized in rehabilitation after illness, but it will take time to disseminate it.

The reasons why surgical robots have not achieved sufficient results are as follows.

(1) Industrial robots only process objects of uniform shape and material by repeated operation and their operation is not required to be fail-safe. On the other hand, in the case of surgical robots, the object to be processed is the human body, with different shapes and materials (i.e., organs); thus, standardization of the robot's actions is not possible. In addition, surgical robots also require high reliability without mistakes. Therefore, the computer and control technology used must be far more reliable than for industrial robots.
(2) In general, developing technologies requires matching technical seeds with needs; thus, collaboration between developers familiar with elemental technology and surgeons as users familiar with the functions required for surgical robots is indispensable. However, it is difficult for engineering researchers engaged in robotic technologies to understand highly specialized and unique surgical procedures (Iseki et al. 2009).

To overcome these difficulties, it is required to grasp the global trends in surgical robotics, clarify the technologies required for practical applications, and discover ideas that lead to a technological breakthrough. For this purpose, pinpoint focus text mining is expected to play an important role.

8.4   Perspective of Surgical Robot Technology

8.4.1   Bird's-Eye-View Based on Time Series Data

As noted above, when mining for detailed knowledge, it is necessary to begin extracting the necessary parts from the text data as a whole, guided by the work of grossly catching the whole of the text. Then, analysts gradually narrow down the important points to focus on the deeper knowledge. This is described below.

One indicator as to how robotics interacts with surgery and neurosurgery is the time series change in the number of theses.

Figure 8.1 shows the number of theses published in three surgical robotics journals (J4, J5, and J6). Almost all theses in these journals appear to be directly or indirectly related to surgical robotics, with J4 being dedicated to this field. All three journals were launched in the 1990 and 2000s, suggesting that robots for surgery increased in their availability during the early 2000s. For each of the three neurosurgery journals (J1–J3), the following data are collected: (1) the number of papers containing the term "robot" in full text or abstract, (2) the number of theses containing the term "computer" that were related to robotization and promoted it, and (3) the number of theses containing the term "endoscope," which is considered the technical foundation of surgical robotics technology (Fig. 8.4).

Of the papers in J1, 30–50 theses contain the term "robot," "computer," or "endoscope," or the like since around 2000, which represents 7–9% of all theses published in the journal. In the 1990s, there were no theses containing the term "robot";

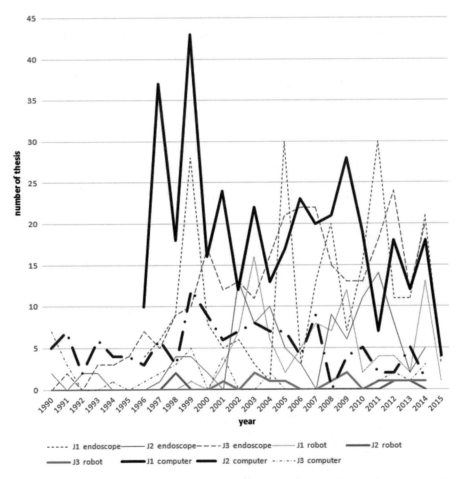

Fig. 8.4 Number of thesis related to surgical robot by year. *Source* J1 = Neurosurgical Focus, J2–3 = JDream3

however, the proportion of theses containing "robot" had been greatly fluctuating between 1 and 8% in the 2000s. Although robotics research for neurosurgery has been active since around 2000, the clinical adoption of these technologies has been limited. This seems to be consistent with the fact that although movement toward the development and practical application of surgical robots was found, such as the da Vinci Surgical System in the 2000s, their uptake has not been strong. Moreover, the number of theses containing the term "computer" is not increasing and the number containing the term "endoscope" is increasing only slightly.

8.4.2 Change in Numbers of Theses by Disease in Neurosurgery

Next, we examine time series changes in the appearance of disease terms exclusive to neurosurgery to understand what kinds of diseases are being studied.

Figure 8.5 shows the changes in the names of major diseases in J1 by four periods (1996–2002, 2003–2007, 2008–2012, and 2013–2017). According to the figure, the largest number of cases is cerebrovascular disorders such as "cerebral stroke" and "brain infarction." Among them, cerebral stroke has a large increase in

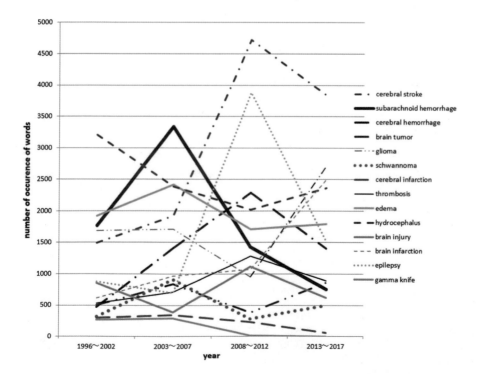

Fig. 8.5 Number of occurrence of disease words in neurosurgery (J1). *Source* Neurosurgical Focus

the number of cases while "subarachnoid hemorrhage" has not increased. The increased rate of brain infarction is also remarkable; thus, it can be concluded that cerebral infarction is the most important research topic in neurosurgery. The number of articles concerning "hydrocephalus" makes it the second most research topic after cerebrovascular disorder, although the number of studies has been decreasing, whereas "epilepsy," with only a few cases, shows an increasing trend.

There are fewer theses on brain tumor-related diseases compared with cerebrovascular disorders. While there is an increase in the term "glioma," the terms "brain tumor" and "schwannoma" do not show a marked increase.

The term "brain injury" has few cases and it has not increased.

These findings indicate the types of surgical robots that are most required by surgeons and, therefore, should be candidates for core competence in robotics companies.

8.4.3 Bird's-Eye-View Based on a Sketch of Technology and Function

To move from a bird's-eye-view type to a pinpoint focus type text mining, it is necessary to extract essential parts from the texts based on an overall understanding. For this purpose, it is necessary to add new terms that are indispensable for discovering knowledge. However, because truly important terms can be obscured in a huge group of terms, appropriate guidelines or sketches suitable for the knowledge to be found must be set in advance to remove unnecessary terms ("noise") and make important terms easier to discover. This article aims to discover the technological trends and the functions necessary for surgical robots; therefore, we will systematize the knowledge in advance, which will lead to the discovery of important terms, as shown in Table 8.1.

Table 8.1 Overall picture of neurosurgical robot technology

(1) Systems/devices
endoscope, master-slave surgical robot, intelligent surgical robot, intelligent operating room
(2) Parts/element technologies
arm, manipulator (multi-joint), navigation technology, image processing (three dimensional image recognition), sensors (image sensors, tactile sensors, speed sensors), intelligent control (feedback control, feedforward control, decision support)
(3) Required functions
minimally invasive, autonomous, accurate, miniaturized, automated, complex and flexible movement of arm, improvement of degree of freedom, expansion of motion range, rotation, articulation, speed control, movement that human fingers can't do, firm grasping of organs, image processing (image recognition of wide range including hidden organs, sharp image recognition, real-time image recognition), sterilization, ease of maintenance, low cost, cost performance improvement

Source Same to Fig. 8.1

First, from the viewpoint of the total system, the surgical robot starts with a surgical endoscope as its precursor device and evolves into a master–slave robot; ultimately, it is expected to evolve into a robot with higher intelligence. Next, from the viewpoint of elemental technologies, navigation technology, image processing technology, sensors, etc. are necessary.

Among the required functions, technologies that contribute to minimally invasive surgery are required. For that purpose, miniaturizing systems and components are required for both image recognition and navigation technologies in order to reach accurately the physical target of the operation. In addition, it is required to flexibly perform complicated movements by speed control and feedback control of multiarticular arms. Furthermore, a mechanism necessary for firmly gripping without damaging organs or tissues is required. Therefore, various elements technologies, such as a navigation method for moving the arm/manipulator, image recognition technology, and a pressure sensor are required.

In addition, it is also required to overcome the current situation where the equipment must be maintained and repaired for each operation, and the integrity of the sterilization procedures must be ensured. Cost reduction is also a crucial requirement.

8.4.4   Basic Trends in Surgical Robotics Technology

With reference to Table 8.1, nouns/compound words with high frequency of occurrence are extracted from J4 and J6, in which a lot of various detailed terms related to robots appear. This makes it possible to view the entire surgical robotics technology in more detail. The results are shown in Table 8.2.

What is understood from Table 8.2 is as follows.

First, the number of occurrences of terms relating to imaging techniques is overwhelmingly large. There are 132 cases of "image" contained in J4 in 11 years, 88 cases of other compound terms including "image," and similarly in J6. This "image" may include many technologies, such as MRI, which is not directly related to the surgical robot; however, even taking this into consideration, the fact that this number is large is undeniable. It is understood that image processing technology is the most important and it is inferred that the significance of rendering objects as three-dimensional graphics in particular is great. In J4, the number of "three-dimensional visualizations" remains one, whereas in J6 "11" in "3-D" and "3-D model." Although the meaning of this number must be evaluated carefully, there is no doubt that one of the major trends in technological progress of surgical robots is image processing technologies.

Table 8.2 Number of Occurrences of Surgical Robot Words (J4, J6)

J6	Number of occurrence	J4	Number of occurrence
invasive surgery	22	invasive surgery	101
minimization	5	minimization	3
shift minimization objective function	2	miniaturization	6
speed performance	5	conflict minimization	1
flexibility	4	miniature robotics	6
decision making	12	automation	6
automation	7	automatic technology	13
automatic segmentation	21	decision-making process	2
automatic method	7	feedback	46
automatic detection	7	slave manipulator	5
automatic image	23	freedom	26
freedom	13	free motion	2
degrees of freedom	9	degrees of freedom	10
computational complexity	11	flexibility	3
navigation	60	flexible endoscope	5
neuronavigation	1	feedback system	4
arm	25	feedforward	1
arm system	11	feedforward scheme	1
manipulation	18	complexity	10
microgripping manipulator system	13	navigation	86
joint	7	neuronavigation	4
multiple joint	10	navigation system	45
roll	0	neuronavigation sysrem	3
sensor	24	intraoperative navigation	3
position sensor	3	arm	42
tracking sensor	3	robot arm	12
image	274	articulating arm	1
video	16	manipulator	51
picture	4	telemanipulator	7
video Image	10	micromanipulation	0
endoscopic video	3	micromanipulation application	3
image processing	56	telemanipulator system	6
endoscopic Image	7	joint	10

(continued)

Table 8.2 (continued)

J6	Number of occurrence	J4	Number of occurrence
other compound words including "image"	253	joint mechanism	16
3-D	7	articulating joint	1
3-D model	4	distal rolling	1
microgripping	1	speed	16
gripper	1	velocity	8
force feedback	4	speed control	9
touch	7	velocity pattern	5
gripping force	1	sensor	41
slippery environment	1	biosensor	3
tactile sensor	1	multisensor	1
force sensor	2	neurosensor	1
		biosensor	2
		other sensor	15
		video	12
		image	132
		image guidance	10
		MRI, CT imaging, X-ray imaging	23
		oreoperative imaging mordalities	3
		intraoperative ultrasound imaging	1
		real time image	1
		force impulse imaging technique	1
		other compound words including "image"	88
		3-D	5
		three-dimensional model	2
		three-dimensional visualization	3
		other three-dimensional	9
		visual feedback	10
		gripper	3
		slippage	1
		gripping end	1
		touch	6
		touch signal	2
		touch feedback	1

(continued)

Table 8.2 (continued)

J6	Number of occurrence	J4	Number of occurrence
		tactile feedback	4
		force feedback	12
		force sensor	11
		tactile sensor	4
		haptic feedback	10

Note Each term contains related analogous words. For example, "navigation" includes "navigator", "neuron navigator", etc., and "manipulation" includes "micromanipulation" and the like.
J4 = 2004–2014, J6 = 2009–2014.
Source JDream3

Second, it is noticeable that the cases of "navigation," "navigation system," "arm," and "manipulator" is large compared with image-related technologies. This is an inevitable result as it is an essential function of the surgical robot for its arm to be controlled optimally.

In addition, navigation technology is aimed at flexible operation with multi-joint arm and control software. The relatively high frequency of "degree(s) of freedom," "joint," and "feedback" in Table 8.3 can be interpreted to correlate with this fact.[3]

Third, in order to make use of image processing and navigation control technology with high importance as described above, various sensors for acquiring a lot of information, including those related to image, position, speed, and the like are indispensable. Various sensors are one of the most important core technologies of surgical robots.

Fourth, "automation" and "automatic technology," which are terms that embody the automation technology related to research on surgical robots, are also frequent in appearance.

Fifth, a technique to grip organs firmly without damaging them is essential. However, the terms "gripper" and "gripping end" contained in J4 are only 3 and 1, respectively. "Touch," "touch signal," and "touch feedback" are 6 cases, 2 cases, and 1 case, respectively. The occurrence of "slippage," which is related to sliding when grasping organs, is only one. However, as for sensors, "force sensor" and "tactile sensor" are 12 cases in total. It is understood that the tactile sensing is one of the most important research fields studied intensively, although not as much as image processing and navigation technology. In particular, "force feedback" appears more frequently, accounting for 12 cases. It can be found that tactile information is beginning to be used for feedback control of intelligent surgical robots. Intelligent techniques for controlling such mechanical forces are essential

---

[3]For the importance of a highly flexible manipulator, see Kobayashi et al. (2005).

Table 8.3 Number of occurrences of neurosurgical robot words (J1)

Year	Total				Endoscope				Robot			
	1996–2012	2003–2007	2008–2012	2013–2017	1996–2012	2003–2007	2008–2012	2013–2017	1996–2012	2003–2007	2008–2012	2013–2017
noninvasive method	30	40	35	40	0	0	0	0	0	0	0	0
noninvasively	55	50	80	60	4	0	0	0	0	0	0	0
minimal invasiveness	20	20	25	20	10	3	0	2	0	0	0	0
neurovascular	326	771	786	733	40	22	57	51	0	0	0	12
flexible endoscope	58	45	45	21	21	23	24	10	0	0	0	0
master slave	0	0	0	0	0	0	0	0	0	0	0	0
intelligence	143	130	209	73	0	4	0	0	0	0	0	12
intelligent	15	15	15	80	0	0	0	4	5	10	0	21
microsurgery	260	428	590	285	44	37	85	29	0	0	0	5
miniaturization	5	25	15	23	5	0	5	0	0	0	0	3
navigation	499	376	671	2195	18	64	31	76	0	0	11	116
neuronavigation	241	314	901	1029	30	34	86	62	0	0	38	16
robot arm	0	5	0	38	0	0	0	11	0	0	0	16
manipulation	1035	476	1001	730	50	10	58	36	0	4	0	9
joint	1064	2131	1064	1100	42	11	4	3	0	4	0	21
speed	203	202	357	285	11	0	9	10	0	0	0	19
velocity	1072	378	466	147	2	0	2	0	0	0	4	6
flexibility	141	366	198	175	10	0	9	0	0	0	3	1
complexity	279	390	333	716	15	4	0	10	0	4	0	8
feedback	223	226	451	410	8	3	2	7	0	10	0	3

(continued)

Table 8.3 (continued)

Year	Total				Endoscope				Robot			
	1996–2012	2003–2007	2008–2012	2013–2017	1996–2012	2003–2007	2008–2012	2013–2017	1996–2012	2003–2007	2008–2012	2013–2017
tactile feedback	5	35	30	25	5	5	11	0	0	0	0	5
feedforward	16	0	0	0	0	0	0	0	0	0	0	0
automation	10	10	15	16	0	0	0	0	0	2	0	15
adhesion	321	320	276	342	5	0	4	1	0	0	0	0
anastomosis	427	338	719	153	0	0	5	0	0	0	0	5
microanastomosis	0	10	31	0	0	0	0	0	0	0	0	0
fixation	2430	2031	1622	2804	18	26	28	29	0	7	9	28
precise location	51	30	33	40	10	3	0	0	0	0	0	0
precise localization	20	49	90	75	0	0	0	0	0	0	0	0
precise mechanism	0	0	0	0	0	0	0	0	0	0	0	0
removal	4614	3646	4408	3407	259	152	300	143	0	0	0	21
microsurgical removal	55	30	55	55	6	0	2	0	0	0	0	0
endoscopic removal	89	15	31	5	33	4	6	4	0	0	0	0
clipping	326	40	388	390	10	0	0	0	5	0	0	0
precision	50	110	70	103	0	0	1	0	0	5	5	9
coil placement	25	193	50	10	0	0	0	0	0	0	0	0
accuracy	956	888	1419	2782	33	30	20	12	15	22	18	376
accurate localization	20	50	50	20	0	5	0	5	0	0	0	0

(continued)

Table 8.3 (continued)

Year	Total				Endoscope				Robot			
	1996–2012	2003–2007	2008–2012	2013–2017	1996–2012	2003–2007	2008–2012	2013–2017	1996–2012	2003–2007	2008–2012	2013–2017
accurate placement	29	25	41	88	10	0	0	0	0	0	0	9
roll	91	78	60	60	91	78	60	60	0	0	0	0
high-speed drill	239	122	176	169	15	3	8	25	0	0	0	0
tactile	35	70	95	55	0	0	7	10	0	0	0	2
touch	95	184	167	147	0	0	0	0	0	0	0	9
grasp	35	65	15	40	5	3	5	0	0	0	0	0
grasper	35	0	20	21	25	0	10	8	0	0	0	0
grasping force	0	0	5	0	0	0	0	0	0	0	0	0
grip	45	20	33	61	8	0	0	0	0	0	0	0
grip strength	96	15	15	5	0	10	0	0	0	0	0	0
slippage	319	87	40	55	0	0	0	0	0	0	0	0
tremor	0	0	0	2	0	0	0	10	0	0	2	2
sensor	116	66	83	66	0	0	5	15	0	0	2	2
force sensor	0	10	0	9	0	0	0	0	0	0	0	0
image	4081	2521	3786	4300	186	117	104	142	12	19	18	120
video	859	665	841	929	164	121	46	68	0	0	0	18
3-D	513	0	0	0	57	0	0	0	0	0	0	0
three dimensional	609	343	149	56	44	29	0	3	5	0	0	0
image-guidance	85	30	34	41	0	0	0	4	0	0	5	0
decision making	150	237	440	771	0	0	0	4	0	0	0	0
force feedback	5	24	0	0	5	0	0	0	0	0	0	0

(continued)

Table 8.3 (continued)

Year	Total				Endoscope				Robot			
	1996–2012	2003–2007	2008–2012	2013–2017	1996–2012	2003–2007	2008–2012	2013–2017	1996–2012	2003–2007	2008–2012	2013–2017
tactile feedback	5	35	30	25	5	5	11	0	0	0	0	5
visual feedback	0	0	25	15	0	0	5	0	0	0	5	0
haptic feedback	0	0	5	31	0	0	0	0	0	0	0	10
cost	815	416	1372	3212	27	12	12	9	0	2	0	39
low cost	10	25	55	65	5	0	0	5	0	0	0	0
cost effectiveness	63	25	73	62	10	0	0	0	0	0	0	0

Source Neurosurgical Focus

for the introduction of robots into surgical operation and further promotion of this technique is considered to be necessary.[4]

Sixth, surgical robotics aims for higher intelligence robots of the master–slave type as symbolized in the term "feedback" appearing as many as 46 times in J4. In J6, "feedback" is rare, but "decision-making" is common as well as terms such as "automation," "automatic method," and "automatic detection," which represent automation.

It has also been pointed out that the intelligence of surgical robots should not be used as a robot alone, but it should be targeted to ensure that the entire operating room is intelligent as a system, including sterilization and the placement of equipment (Iseki et al. 2003). However, there seem to be few occurrences of terms showing such signs.

Seventh, the number of occurrences of terms related to speed such as "speed," "velocity," and "speed control" are not small. This can be interpreted as reflecting the fact that it is necessary to precisely perform grasping, resectioning, peeling, anastomosis, and the like by controlling the motion speed of the manipulator and the arm.

8.4.5 Time Series Data Analysis Based on a Sketch of a Neurosurgical Robot

Next, this section focuses on neurosurgery rather than surgery in general and clarifies the robotic techniques and functions required in neurosurgery in particular compared with surgery.

In general, the following functions and techniques are considered inherent in neurosurgical procedures.

First, neurosurgical surgery requires techniques for exfoliating, resecting, and anastomosing finely, accurately and precisely. The brain is the most complex organ, with innumerable neurons; thus, a precise control technology is needed. The miniaturization of equipment is also required. At the same time, the robot must be supple and tough against external forces.

Second, minimally invasive techniques are required in brain neurosurgery more than other organs. One of the functions required for neurosurgery is that the surgical tools are placed in deep positions in the brain. It is necessary to reach the surgical target position without damaging the normal tissue. For that reason, the space available to reach the target is narrowed, thus, extremely precise procedures are required. The damaging to tissue greatly affects the prognostic quality of life of the

---

[4]The importance of tactile techniques for promoting minimally invasive procedures is mentioned in Onishi (2009), Kobayashi et al. (2015).

patient. For example, it is necessary to develop a system in which the endoscope itself is soft, but with a tip that is sufficiently rigid (Tanaka et al., home page). It is also necessary to increase the range of motion. For that purpose, improvement in the degrees of freedom of arm rotation, multi-articulation, etc. are required.

Third, for precise surgery, a system to prevent the operator's hand trembling is needed. This is achieved, for example, by a incorporating a support for the operator's arms (Goto 2014).

Fourth, devices used in neurosurgical procedures often require maintenance for each operation. Therefore, a mechanism to facilitate maintenance is necessary.

8.4.6 Research Topics Found from Neurosurgery Theses

Table 8.3 shows the number of occurrences of terms that represent treatment methods for disease in the four articles in J1. These terms are extracted based on a sketch of functions and elemental technologies required for neurosurgery as described in the previous section.

About 10 pages of theses published in J1 are mined in full text. Therefore, the "number of occurrences" shown in Table 8.3 is not the "number of theses" but rather the total number of times a term appears. The more occurrences of a term in a single thesis the greater the importance of the term; therefore, "number of occurrences" can be considered to be a better indicator than "number of theses."

However, the following precautions must be taken. Table 8.3 shows not only the total appearance cases of a certain term but also the co-occurrence relationship between a certain term and "endoscopy" or "robot" in a TB. The reason for using TB to analyze the co-occurrence relation is as follows. The analysis of one thesis as a mining unit results in extracting many co-occurrence relationships between a term and other terms unrelated to the former, which are only noise for the mining purpose. On the other hand, if co-occurrence relations are discovered on a sentence-by-sentence basis, then a lot of co-occurrence relations between a term and other terms with important semantic associations are lost. About five sentences are considered to be the most appropriate mining unit to discover co-occurrence relationships. However, as with this result, the following attention is necessary. That is, the appearance frequency of one term is increased five times compared with the original text data. In addition, the closer the distance in the texts between two co-occurring terms the more the co-occurrence frequency of terms and terms is increased in comparison with the original text data.

First, we examine the terms related to surgery in general.

Surgical robots continue to evolve from the master–slave type and are oriented toward system intelligence; however, as seen in Table 8.3, there were no cases of "master–slave." "Intelligence" occurs more than 500 times among all theses over the entire period. However, when these are limited to articles related to surgical robots, only 10–20 cases of "intelligence" or "intelligent" appear.

Focusing on feedback techniques that embody the specific content of intelligence, the number of occurrences of "feedback" is very high and has further increased since 2008; however, the number of co-occurrences with "robot" has not increased. In addition, "decision-making" also has a large number of occurrences but few co-occurrences with "robot." This fact may indicate that the intelligence of surgical robots has not progressed smoothly. In particular, "feedforward" only appeared in 15 cases during 1996–2002 and has not appeared since then. Acquiring higher intelligence capability, such as feedforward, does not seem to be straightforward. Therefore, for example, it may be correct that navigating actions should not be hastily targeted for full automation; instead, a step-by-step approach to developing intelligent automation should be taken (Masamune et al. 1999).

Noninvasive or minimally invasive surgery is ideal. However, an increasing trend for these terms is not found; i.e., the numbers of occurrences of "noninvasive method" are 30, 40, 35, and 40, respectively, during the period under study. To determine whether endoscopes and robots contribute to noninvasive surgery, we examine the relationship of "noninvasive method," "noninvasively," and "minimal invasive" to "endoscope" and "robot" and find that the number of co-occurrences is not large. Especially the former's relationship with "robot" can be seen. There is no indication that surgical robots are being developed for noninvasive procedures. Regarding the basic procedures of surgery such as removal, exfoliation, fixation, dissection, anastomosis, and clipping, "removal" is the most frequent in 16,075 cases during the years under review, "fixation" appears in 8,887 cases, and "anastomosis" in 1,637 cases. There are 1,259 occurrences of "adhesions," similar to "exfoliation," and 1,144 occurrences of "clipping." In addition, the terms "grasp," "tactile," and "touch," which embody the meanings of the procedure of touching and grasping the target organs, are about 150–550 cases, which is not large compared with other terms, but it is undeniable that these are important research themes. As can be understood that its number of cases is large and its number of co-occurrences with "robot" increased from 0 before 2012 to 21 in 2013–2017, "removal" among these procedures is the most relevant to endoscopes and robots. It turns out that the robotization of neurosurgery is progressing in a direction that emphasizes the resection function.

"Anastomosis" has five co-occurrence relationships with "endoscope" in 2008 to 2012 and five co-occurrence relationships with "robot" in 2013–2017. From this fact, it is suggested that anastomosis has begun to be a major movement in endoscopic and robotic surgery. That surgical robots, in general, have been used for anastomosis as seen in the previous section, suggest it to be valid in neurosurgery as well. For example, the surgical robot MM1 aims to anastomose the deepest microvessels in the brain, which is a technically difficult surgery.[5]

No co-occurrence is shown between "fixation" and "robot" or "endoscope," and no co-occurrence is shown between "clipping" and "robot" or "endoscope" also except for "1996 to 2012."

---

[5]http://plaza.umin.ac.jp/~ikourenk/department/project_06/index.html.

In addition, it is essential for surgeons to grasp and securely fix the target position (e.g., organ) so that it does not slip. From the fact that the co-occurrence relationship between "grasp" and "endoscope" is shown, and that the co-occurrence relationship between "tactile" or "touch" and "endoscope" or "robot" appears since 2008, it can be concluded that tactile sense is an important research theme for the introduction of endoscopes and robots.

Furthermore, comparing "clipping" with "coil placement," the latter has decreased significantly while the former is increasing, indicating that clipping is more important than coil embolization for the treatment of cerebral arteries.

Next, in surgery, it is required for surgeons to perform complicated tasks flexibly and quickly; therefore, "flexibility," "speed," "velocity," and "complexity," terms related to these tasks, are often found. This may mean that the importance of dealing with complexity is increasing more than speed for surgery.

However, paying attention to the co-occurrence relationship between these terms and "robot," the co-occurrence between "speed" or "velocity" and "robot" is stronger than "complexity" or "flexibility" and "robot" since 2008. In spite that flexibility is an important function for a surgical robot, it may be understood that a key theme of a surgical robotics development is speed improvement. This may indicate that it is not easy for surgical robots to achieve the flexible and complex movements required for surgery. Indeed, although the NeuroMate robot, with a multi-articulated arm to enable six degrees of freedom and used for brain stereotactic surgery, is capable of performing highly flexible movements (Kajita et al. 2013). This research seems not to be sufficient yet.

In addition, there were no cases of co-occurrence between "microanastomosis" and "endoscopes" or "robot," indicating that mechanization has not contributed to precise anastomosis.

Next, we examine "accuracy," "precise location," etc. as terms that embody the accuracy of mechanical motion, the most important function for surgery. Significantly, "accuracy" is extremely large, with 6,045 cases in total. Moreover, as can be seen from the fact that the growth rate has increased by nearly three times from period 1 to period 4, it is understood that its importance is increasingly being emphasized. The number of occurrences of "precision," "accurate localization," and "accurate placement" is also not small, and the numbers of "precise location" and "precise localization" also are the same. In particular, the number of "precise localization" occurrences was markedly increased by about four times. Accuracy is the most important requirement for surgery, especially neurosurgery.

Furthermore, when limiting the search to co-occurrences with endoscopes or robots, "accuracy" is most strongly related to "endoscope" and "robot." In particular, in 2013–2017 "robot" co-occurred extremely strongly with "accuracy" (376 cases) and "robot" co-occurred in nine cases with "accurate placement." It turns out that accuracy is one of the most important research themes of surgical robotics. In contrast, the relationship between "precise location" or "precise localization" and "robot" is not strong. This meaning will be described later.

Interestingly, the occurrence of "tactile feedback" is relatively large (95 in total) and "force feedback" appeared considerably in 2003–2007. In addition, the co-occurrence of "tactile feedback" and "robot" has become visible since 2013 (5 cases). These data may suggest that work to skillfully touch and grasp the target organs is important, and so is the effectiveness of introducing feedback control system for this purpose.

One of the most important core technologies for surgical robots is navigation. The number of occurrences of "navigation," "neuronavigation," "robot arm," and "manipulation" as terms related to navigation, as well as the number of co-occurrences between these terms and "endoscope" or "robot," has increased from 1996 to 2005.

Another interesting fact is the remarkable increase in occurrences of "cost," showing a big change from 815 cases to 416 cases, 1372 cases, and 3212 cases, respectively, over a four-period period. Similarly, occurrences of "low cost" increased from 10 to 25, 55, 65 cases, and that of "cost effectiveness" increased from 53 to 25, 73, and 62 cases in the same period. The co-occurrence of "cost" and "robot" remarkably increased to 39 cases in the 4th period, although it was nil in the third period. This may reflect a tendency for the introduction of surgical robots to lead to high-cost medical care or it could reflect conflicting demands for reduced medical costs common to developed countries. This can be interpreted as reflecting the reality of increasing medical costs in order to enhance the functioning of surgical robots and to achieve low invasiveness and high accuracy.[6]

The above analyses show that the technological development of surgical and neurosurgical robots has focused on advances in image-related technologies, and that tactile sensation, accuracy, and flexible processing of complex movements also are important issues. However, it was also proven that this development and popularization is not advancing smoothly enough yet. Therefore, it is required to know more about the techniques needed to facilitate the development and diffusion of surgical robots. To do this, the pinpoint focus type text mining technique, in which a narrow-range of technology, such as imaging and tactile sensing are extracted and focused on in detail instead of describing the overall picture of the surgical robot, is required. This technique is considered below.

8.5   Pinpoint Focus Type Text Mining Method

8.5.1   Discovery of Key Words Based on Co-occurrence Relationship

One of the difficulties in narrowing down to a specific topic from all the text data is that the texts are enormous and contain mostly "noise" that is unrelated to the topic

---

[6]It has been reported that the cost of endoscopic forceps sets is 10 times higher than that of conventional forceps sets and thatthe cost of surgical robots is 30 times higher than that of conventional forceps sets (Inaki 2008).

to be analyzed. For this reason, the terms that are essential for discovering the meaning of the topic are buried in an enormous amount of noise, making it difficult to extract specific topics and analyze them using pinpoint methods.

In many cases, gaining insight into the meaning of a term is made possible by terms with a low frequency of occurrence rather than words a high frequency of occurrence. This is because terms with a low frequency of appearance may express a prediction of the future and embody innovative, promising, and emerging technologies. Discovering these terms makes it possible to know future profitable business for the company and to obtain useful technological ideas necessary for developing new products. Rather, it is not unusual that valuable technical information that is only known to a few people due to its novelty also appears at a low frequency in texts. Thus, even a single occurrence of a term may represent a very important technique to do so and so it is imperative to search for important terms with a low frequency.

One approach to overcoming the difficulty of finding these key terms with low frequencies of occurrence is using association rule analysis. This paper uses the "correlation value" in the IBM Watson Explorer, in which association rule analysis has been introduced. In this tool, the correlation value is defined as $(A \cap B/A)/(B/D)$, where A and B are two document sets including the terms a and b, respectively, and D is the whole document set. Furthermore, when the number of cases is small, the correlation value tends to be higher than the actual condition; thus, the correlation value of words with few cases may be corrected to be lower. Therefore, by paying attention to terms with a large correlation value, it becomes possible to find a key term b that is strongly related to the word a despite word b being inconspicuous due to a small number of occurrences.

As an example of finding important terms with a small number of occurrences based on the correlation value, the author extracts terms that are strongly co-related to "robot" or "endoscope" in J5. The results are shown in Table 8.4.

According to Table 8.4, the terms having a large correlation value since 2008 and having a value larger than the correlation value before 2007 are "CyberKnife Robotic Radiosurgery system," "da Vinci," "complete elimination," and the like. In addition, image-related terms, such as "visual acuity," "visual navigation," "preoperational imaging," and "video feedback," which are not found in the top rank before 2007, first appeared in the top rank since 2008, suggesting that imaging technology has become an important research issue. In addition, although its number of occurrences is only one, the occurrence of "orifice procedures" since 2008 may be a sign of increasing research focusing on minimally invasive endoscopy and robotics. This may show that in order to promote minimally invasive surgery, advanced control systems, where the arm is manipulated using both hands, and feedback control technology enabling complicated manipulation are implemented, etc. in addition to the extension of the existing technology of laparoscopic technology, have become important since around 2010.

Furthermore, although not shown in Table 8.4, there are 23 types of navigation-related technologies before 2007, including "navigation system," "hybrid navigation," and "neuronavigation" and their correlation was 0.29 at

Table 8.4 Words co-occurring with "robot" or "endoscope" (J5)

Before 2007			After 2008		
Word	Number	Correlation value	Word	Number	Correlation value
robot system	6	3.95	CyberKnife Robotic Radiosurgery system	2	2.77
robotic system	7	2.64	surgical robotic	2	2.77
virtual endoscopy	2	2.07	visual acuity	2	2.77
neuroendoscopic procedure	2	2.07	bimanual robot	2	2.77
endoscope tip	2	2.07	surgical robotics	2	2.77
endonasal sinus surgery	2	2.07	surgical robot	3	2.55
Endonasal sinus surgery	2	2.07	da Vinci	2	1.78
endoscopic lens	2	2.07	invasive surgery	4	1.49
position measurement system	2	2.07	robotic assistance	2	1.31
robotic device	3	1.36	video Feedback	1	1.15
endoscopic procedure	2	1.33	existing laparoscopy	1	1.15
endoscopic image	2	1.33	orifice procedure	1	1.15
needle tip	3	1.07	complete elimination	1	1.15
real time	7	1.16	peritoneal cavity	1	1.15
			metal ball target	1	1.15
			navigational software	1	1.15
			visual navigation	1	1.15
			joint movement	1	1.15
			photometric calibration method	1	1.15
			preoperative imaging	1	1.15
			laparoscopic procedure	2	1.04

Source JDream3

maximum. On the other hand, since 2008, the number of term types has decreased to 14 and their correlation values have also decreased to 0.03 at the maximum. This may reflect the fact that neuro-techniques are not being adopted smoothly in arm navigation techniques as one of core technologies for surgical robots.

8.5.2 New Prediction of Surgical Robot Technology Viewed from Less Frequent Terms

The same points as in the previous section will be analyzed in this section using J4. Although the results cannot be listed in tabular form, "robot" correlates well with "operational time" (correlation value 1.27) and "blood loss" (1.21). However, the two terms are often used (50 and 36 cases, respectively), indicating that they are important skills even if not viewing the correlation value.

Despite the lack of significance due to the small number of events, the following are noteworthy terms with high correlation values.

First, although the occurrence frequency of "capsule endoscope" and "wireless power" are only 4 and 3, respectively, the correlation values between these two terms and "endoscope" are 4.60 and 3.22, and in the top 2nd and 3rd places, respectively,. This can be interpreted as a precursor both to the widespread use of capsule-type endoscopes and the increased importance of a wireless power supply technology capable of doing so. It should be interpreted that these technologies be considered for long-term availability, even if not applicable in the short-term to surgical robotic technology development.

Second, in terms of "anastomosis," there were 14 cases of co-occurrence with "robot" (correlation value 0.80) while there was only one case (0.01) of co-occurrence with "endoscope." It is presumed that the evolution from the endoscope to the robot has been effective for anastomosis. The co-occurrence between "adhesion" and "robot" was 4 cases (0.37), suggesting that the surgical robot is about to be mainly used for anastomoses rather than for the dissection of adhesions.

8.6 Neurosurgery Robot

8.6.1 Evolution of the Neurosurgery Robot and Its Future: The Capsule Endoscope

In the preceding paragraph, it was described that using association rule analysis raises the likelihood of extracting meaningful terms for a specific topic. However, as in the previous section, based on only the height of the correlation value between "robot" and "endoscope," it is impossible to find sufficiently important key terms. For example, "SpineAssist miniature robotic device," "optimal miniaturization," "effective miniaturization," "functional miniature," and "device miniaturization" are important technologies that will make it possible to miniaturize surgical robots. However, because the correlation value (0.02 or less) as well as the number of cases is small, it is hard to find. To discover such key terms, it is necessary to continue searching for other important terms having a low appearance frequency, starting from already discovered important terms.

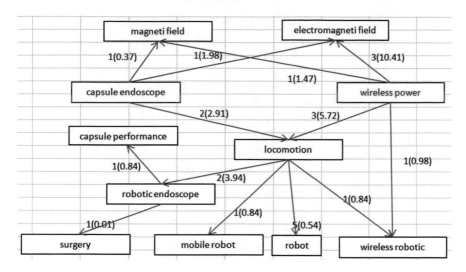

Fig. 8.6 Network of "capsule endoscope". *Note* Numbers are frequency of occurrence. Numbers in parentheses are correlation values.
Source Neurosurgical Focus

Several examples will be described below. For that purpose, the three important surgical robot technologies discussed in the previous paragraph will be described in detail: the capsule endoscope, accuracy, and tactile sensing.

First, the correlation values between "capsule endoscope" or "wireless power" and "endoscope" are large. Thus, mining other terms co-occurring with these two words reveals that "wireless power" is correlated strongly with "locomotion" (5.72). Thus, when searching for terms co-occurring with "locomotion," it can be found that it co-occurs with both endoscope-related technologies, such as "capsule endoscope" and "endoscope," and robot-related technologies such as "robotic endoscope," "mobile robot," "wireless robotics," "robot," etc. Further, searching for terms that coincide with "robotic endoscopes" or "wireless robotics", "capsule performance" and "surgery" can be found. This is illustrated in Fig. 8.6. It can be understood that capsule endoscope technology and robot technology are linked together by "locomotion," which is located in the hub. Indeed, incorporating a compact actuator is expected to enable the capsule endoscope to self-propel in the body and allow various tasks. Even though capsule endoscopy is unlikely to be applied as a surgical robot in the near future, it will become available in the relatively far future (Moglia et al. 2007) and, therefore, may be interpreted as an important current research theme.

As is widely known, many excellent algorithms for natural language processing have already been developed and research results have been obtained as a way of networking multiple nodes (Mima 2012; Kushima et al. 2017). In this article, however, the author adopts the technique of stretching links of nodes and creating a network of terms by appropriately filtering individual terms of interest by using analysts' knowledge (part of background knowledge) rather than simply using the algorithm-dependent approach. The reason is that by making both nodes and links only based on the algorithm, a lot of meaningless terms (noise) are extracted, important key terms are eliminated, and it is often difficult to express the correct meaning. In other words, to focus on the terms of greatest importance with less frequent occurrence, it is difficult to rely solely on computer algorithms.

8.6.2   Evolution of the Neurosurgery Robot and Its Future: Accuracy and Precision

One of functions especially required in neurosurgery robots is accurate movement. As stated in the previous section, two different meanings are contained in the phrase "accurate movement": accuracy, which means the closeness between the value measured by the sensor and the true value, and precision, which means the variation in robot movement when the same operation is repeated. Surgical robots should aim to improve the *precision* of repetitive movements based on improved *sensingaccuracy*. For example, it is a prerequisite in surgery that the precision of the repetitive operations of the master–slave system is high; for this reason, further improvement on the sensing accuracy is required. As seen in Table 8.3, the less frequent the occurrence of terms, e.g., "precision" compared with "accuracy" may indicate the current evolutionary direction of master–slave surgical robots.

Figure 8.6 shows the network of terms originating from "accuracy" and "precision," clarifying the research trend of surgical robotics. There are 56 theses that contain both "accuracy" and "precision." When searching for terms with large correlation values with these 56 cases, "feedback" is discovered in 13 cases. This may indicate that the feedback mechanism plays an important role in achieving both sensing accuracy and operation precision. Therefore, when searching for terms with large correlation values for these 13 theses including all three terms ("accuracy," "precision," and "feedback"), the most prominent terms are "fuzzy logic" and "robotics." It can be inferred that surgical robotics research is promoted with an emphasis on realizing accuracy and precision by introducing the new advanced algorithm. Of interest is that all 13 papers including all three terms also contain a large number of terms embodying real-time characteristics such as "real-time imaging,," "real-time," "real-time radiography," and "real-time target." From this, it can be inferred that speeding up the information processing is essential for realizing both accuracy and precision.

Furthermore, "accuracy" and "precision" co-occur directly or indirectly with "tactile feedback," and both "real-time imaging" and "image quality" co-occur with "force feedback." This fact may indicate that cooperation between image processing technology and tactile processing technology is indispensable for achieving sensing accuracy and operation precision. Moreover, the fact that "tactile feedback" and "visual feedback" co-occurred with "parallel" or "parallel structure" may imply the importance of parallel processing technology.

While improving real-time performance may be achieved in a variety of directions, such as improved information processing/communication speed and the evolution of sensor technology, the evolution of algorithms seems to be one of the most important innovations.

8.6.3 Evolution of the Neurosurgery Robot and Its Future: Tactile Sensing and Imaging

As described above, it is necessary for a surgical robot to grip and fix various organs without injuring them; thus, tactile sensing technology is expected to play an extremely important role for surgical robots. However, tactile technology is still in an immature phase and, thus, relies heavily on image processing technology. Therefore, in this section, the relationship between tactile-related technology and image processing-related technology will be clarified.

Figure 8.8 shows how image and tactile technologies are related to the evolution of surgical robots. Searching for image/visual- and tactile-related technologies co-occurring directly or indirectly with "endoscope" or "robot," as shown in Table 8.4, reveals an overwhelmingly large number of occurrences of image-related technology and many types of image-related terms. However, as shown by the fact that "tactile feedback" co-occurs with many types of terms related to many images, it is not sufficient to manipulate the surgical robot with image data alone: it is also necessary to combine tactile data with image data. In particular, the co-occurrence relationship between "tactile feedback" and "removal" or "anastomosis" suggests that research aimed at linking image information with tactile information is being conducted more for resection than for anastomosis. It can be interpreted further that this coincides with the co-occurrence of "force feedback" and "view-only software" (one case). This implies that it is difficult to manipulate only with image processing data and that image data has to be used in combination with pressure data to control a surgical robot.

To grip and fix various organs without hurting them, it is essential to prevent hand trembling by the surgeon operating the surgical robot as much as possible. However, unlike tactile sense, it was difficult to extract terms for hands movement. The reason for this is that the surgeon's hand movement is expressed by terms such as "tremor," "tremble," "handshaking," "movement," etc., most of which are used for tremors of the body due to the disease in the thesis; thus, it is not possible to be distinguish their meaning. Therefore, these terms are not included in Fig. 8.8.

Interestingly, "force feedback" co-occurred with "rigid body" (correlation value 17.38) and "tactile feedback" co-occurred with "rigid instrument" (0.38). This indicates that rigid materials for robot arms are required to increase accuracy and precision because the surgeon's hand tremble is amplified by the arm of the device.

Tactile-related terms co-occur with "angle," "angle sensing," "screw," and "screw insertion." This may indicate that robots that prevent the surgeon's hand tremble movements to control correctly the insertion angle are required. For that reason, a mechanism to feedback tactile information for controlling the drill angle is an extremely important research issue (Masamune et al. 2003). This can be seen in the fact that the terms "angle" and "screw" are included in Fig. 8.8.

Furthermore, "tactile feedback" is strongly correlated with "learning curve" (correlation value 2.81). It cannot be understood from Fig. 8.5 whether this implies the necessity for feedback mechanisms or the necessity for surgeons to acquire endoscope and robot operation skills. Or it may have both meanings.

8.7 Sophistication of Knowledge by Linking TBs

8.7.1 Linking TBs Based on Key Terms

As seen in the previous section, we can obtain deeply detailed knowledge in narrowly focused fields by discovering important terms with less frequent key terms and discovering the network of these terms using association rule analysis. In other words, it is possible to approach pinpoint focus type text mining from a bird's-eye-view type by doing so. However, in order for companies to know the profitability of their business in future and to select a business among a lot of candidates vying for limited management resources, it is necessary to acquire more deeply detailed knowledge about what technology is required for a new project and how to overcome technical bottlenecks. For that purpose, it is essential to understand meaning based on linking TBs by similarity and relevance of context rather than simply being bound by term co-occurrence relationships.

Figures 8.6, 8.7, 8.8 show the relationships among terms discovered by association rule analysis. The important meanings are understood as follows. For example, the high correlation between "tactile feedback" and "needle tip" or "ideal position" shown in Fig. 8.6 suggests that the technique of processing tactile information with the feedback system plays an important role in guiding and securing the needle tip in the correct position. However, it is not known whether this is the correct interpretation; even if it is true, it is impossible to obtain a detailed knowledge about tactile feedback techniques from Fig. 8.6. In order to know this, it is necessary to discover relevant TBs and link them together to understand their meaning.

Therefore, this section describes a method for exploring and extracting TBs containing required knowledge and linking the TBs together accurately. However, it is not easy to acquire this method because given the current level of natural language processing it is impossible to understand the meaning and context of sentences correctly. Until natural language processing technology can understand the context correctly in the future, there is no choice but to satisfy with using

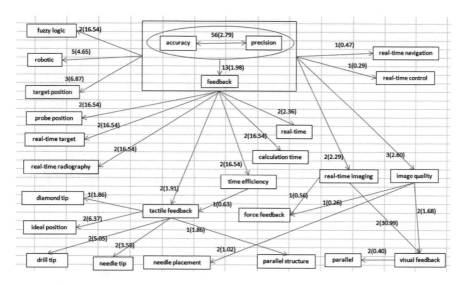

Fig. 8.7 Network of "accuracy" and "precise". *Note* Numbers are frequency of occurrence. Numbers in parentheses are correlation values.
Source Neurosurgical Focus

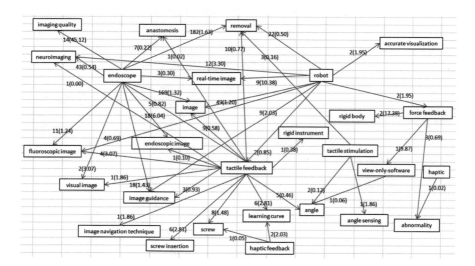

Fig. 8.8 Network of "tactile" and "image". *Note* Numbers are frequency of occurrence. Numbers in parentheses are correlation values.
Source Neurosurgical Focus

insufficient techniques as elaborately as possible and approaching the truth little by little. For this purpose, this section illustrates how TB can be linked using two criteria: (1) the commonality of terms and (2) the similarity in the context of TBs.

In this section, the procedure of linking TBs on the basis of the commonality of the terms contained in TBs will be considered. Taking an example of the strong co-occurrence between "tactile feedback" and terms related to image information, such as "fluoroscopic image," which are found in Fig. 8.8.

It is presumed that TBs that contain both terms related to tactile sensing and to images are more likely to have a narrative of knowledge about the connection between touch and image. Therefore, TBs containing "tactile feedback" and "image" are extracted. As a result, 14 TBs are extracted. Among them are the following TBs (Table 8.5).

Table 8.5 TBs including both "tactile feedback" and "image"

<Text Block 109743> "3,9 Evaluation requires imaging. Intraoperative imaging during the revision surgery is important for accurate localization of instrumentation and its relationship to vital neurovascular structures, the achievement of adequate decompression, and maintaining structural integrity within the spinal column. Currently, many centers routinely use the C-arm (fluoroscopy); however, fluoroscopy is limited in resolution and depth of field. Improved CT-IGS navigational functions of systems like the O-arm provide 2D and 3D resolution of structures in real time to ensure proper instrument placement and alignment. 11 Neuronavigation reduces the risk that tactile feedback cues from previously violated or weakened bone or dense scarring will mislead the surgeon, and also provides the option of acquiring an intraoperative postinstrumentation CT image to confirm correct placement of screws."

<Text Block 109,746> "... Our retrospective study found that CT-IGS provided safe screw placement in both revision (98.6%) and primary (98.7%) spine surgeries."

<Text Block 350,816> "It was more difficult to locate the curved ventral margin of the sacrum than the anterior cortex of a vertebral body with lateral fluoroscopy. ... Another key difference is that the cancellous bone in the sacrum is much softer and less dense than in a vertebral body, resulting in limited tactile feed-back to the surgeon during cannula placement. Balloon pressures were generally low, and there was very little resistance to PMMA injection. The combination of difficulty with fluoroscopic imaging and reduced tactile feedback made it hard to tell whether PMMA might be extruding into the soft tissues of the buttock or pelvis during injection."

<Text Block 350,820> "... For these reasons, we elected to use BrainLAB image guidance for the third patient, and found that this modality significantly improved the accuracy of cannula placement and PMMA delivery. With the BrainLAB system, it was possible to navigate instruments and inject PMMA directly into the fractures identified on the preoperative CT scan. Although further work will be needed, BrainLAB image guidance may offer some advantages over conventional fluoroscopy with respect to instrument placement and PMMA delivery to the fracture site. A potential concern with any method of sacral augmentation is medial migration of cement, which would compromise the exiting sacral nerve roots and sacral spinal canal."

Note Numbers in parentheses following TB are ranked with similarity of context in height.
Source Neurosurgical Focus

First, from TB 109,743, the more detailed knowledge that CT-IGS navigation systems like O-arms using tactile feedback makes it possible to accurately position the screw of the endoscope, reducing the risk of injuring the organs, is obtained. Further linking this TB with TB 109,746 provides further sophisticated knowledge that its safety is 98.6–98.7%.

It is also stated in TB 350,816 that acquiring tactile information from soft tissues, such as cancellous bone, are not easy and the difficulty in establishing a drainage tube during surgery increases. Thus, the need for progress in tactile technology is understood. By linking this TB with TB 350,820, knowledge that using the Brain LAB image guidance image diagnostic system enables accurate induction of PMMA delivery into soft tissue, although conventional fluoroscopy and tactile feedback have technological limits, is obtained. In this way, it will be possible to gain valuable knowledge to help solving the technical bottleneck.

As can be understood from the above, by evolving a method to extract co-occurring relationships among terms obtained by association analysis into the method of linking TBs, it is possible to acquire precise knowledge at a pinpoint level.

8.7.2 Linking TBs Based on Deep Learning

However, as described in the previous section, searching based on term commonality alone would miss many of the important TBs because natural languages often have the same meaning and context even when they are written in different terms; conversely, when they are used in the same terms they often have completely different contexts. The ideal method for finding the necessary TBs is based on the context similarity and not on the commonality of the appearing terms.

Methods for finding similarities in the meanings of sentences include potential semantic analysis, K-means method, etc. and based on machine learning algorithm and deep learning techniques such as Doc2vec, which has been drawing much attention recently. However, despite such excellent ambitious attempts, extremely ambiguous and polysemantic natural language is impossible to be understood with computer algorithms alone. Therefore, it is indispensable that the reader has background knowledge in order to understand the correct meaning. The necessity for background knowledge is especially significant in the case of technical documents described in technical jargon. Moreover, most of these approaches are aimed at clustering myriad documents and few discuss similarities between two extracted documents. However, for the pinpoint focus type text mining method, it is necessary to discover similarities between two documents rather than roughly clustering all of the documents. Doc2vec can elucidate the ranking of similarities between two documents and so this is used here.

Of course, Doc2vec, like other tools, is far from fully appreciating correctly the similarity between documents. Documents that are not quite similar are often regarded as similar. However, from documents that are too enormous for humans to read, it is often possible to find closely related documents that would otherwise not have been discovered. Thus, by accepting the technological limitations of the current state of deep learning methods, valuable documents are expected to be found. Moreover, the potential for the future development of deep learning technologies is undeniably large and the significance of using them is expected to increase.

To learn more about how to use tactile techniques and imaging information to achieve accurate device guidance by using Doc2vec, it is useful to search for TBs with a context similar to the TB described for this technique. Therefore, from all the TBs (511,449 cases in total), the top 300 TBs with high similarity to TB 179,043 are extracted. In doing so, 14 TBs that can be regarded as having similarity to TB 179,043 in the meaning that a system for guiding devices using tactile information and imaging information is described in those TBs are found. This is shown in Table 8.6.

Table 8.6 TBs, context of which is similar to that of TB179,043 (Doc2vec)

<TB202,422> (97) "... Functional MR imaging, for example, is capable of revealing the locations of lower CN nuclei.31 The complementary technique of diffuser tensor imaging may resolve the positions of axonal tracts within the brainstem and can be used to guide the surgical approach to a CM.28,59 While the highly conserved nature of brainstem anatomy across individuals usually renders more exhaustive neuroimaging in any particular case unnecessary, these diagnostic studies might be useful because of their potential to reveal the true positions of structures displaced by larger lesions; knowing whether a motor tract is displaced medially or laterally by a lesion in the medulla, for example, could help determine the most advantageous pial point of entry. The integration of these imaging techniques with intraoperative neuronavigation technologies will likely improve safety during the resection of CMs juxtaposed to white matter tracts and nuclei."

<TB258,124> (2) "... Neuronavigation Image-guided neuronavigation uses preoperative, intra-operative, or real-time imaging to allow the surgeon to understand spatial relationships within the brain that are not visible by line-of-sight. ... Similarly, intraoperative CT or MR imaging can be used in this fashion to shorten the temporal relationship between obtain-ing and utilizing the imaging information, as well as to obtain updated information at specific time points during the case. Lastly, intraoperative ultrasonography and fluoroscopy can actually provide real-time imaging and hence eliminate inaccuracy due to the movement of structures following image acquisition."

<TB418,353> (66) "Although the selection of reference points in cases of anterior spinal surgery is limited by the relative lack of prominent bone landmarks on the anterior aspect of the spinal clumn, the degree of accuracy required is less than that needed in most posterior screw fixation procedures. This degree of accuracy, termed "clinically relevant accuracy," will change according to the procedure being performed. ...For example, insertion of a C1-2 transarticular screw demands a higher degree of clinically relevant accuracy than placing an anterior fixation screw across a large thoracic or lumbar vertebral body. In both cases image-guided navigation provides clinically relevant accuracy more consistently than fluoroscopy alone."

<TB45,834> (44) "The benefit of navigated pedicle screw placement is particularly apparent in cases where there is deformed anatomy and anatomical landmark-based freehand screw placement is challenging. Jin et al. used fluoroscopic cone-beam navigation to facilitate

(continued)

Table 8.6 (continued)

placement of pedicle screws in patients with dystrophic neurofibromatosis Type 1-associated scoliosis and reported 79% accuracy compared with 67% in the freehand
group. Intraoperative use of cone-beam fluoroscopy may be especially valuable in helping to localize lesions on the thoracic spine, where localization can be notoriously difficult with conventional fluoroscopy...."

<TB325,222> (21) "...These systems can allow precision to within millimeters; 10 however, they have some drawbacks. The visualization of the anat-omy is confined to the state it was in when the radiographic images were obtained; most neuronavigation systems cannot yet compensate for changes in this anatomy."

<TB462,954> (71) "The quality of the eventual 3-D reconstruction depends on the mode and resolution of image acquisition. Both slice thickness and image matrix size are basic determinants of final resolution. In soft-tissue imaging of the head, neck, and spine, MR imaging has been and likely will continue to be the imaging modality of choice. Continuing refinements in MR imaging sequence protocols, field strength, and gradient coil design have been important in the improvement of final image quality...."

Note Numbers in parentheses following TB are ranked with similarity of context in height.
Source Neurosurgical Focus

The term "tactile" rarely appears in the 14 TBs. Most of the terms that appear are related to "image," such as "fluoroscopy." This fact may show that imaging information seems to be more important than tactile information for device navigation in surgical robots. The number of occurrences of "accuracy," "intraoperative," "real-time (real time)," and "neuronavigation" in 14 TBs are 7, 5, 2, and 5, respectively. This suggests that it is important to properly guide the device in real-time during surgery and that the neuro technique may be effective for doing so. Therefore, it will be necessary to acquire from these 14 TBs more detailed and specific knowledge about the technique and solution methods for accurately guiding the equipment in real-time by using image information.

Because it was found that the use of image information is common for the guidance of equipment as mentioned above, further detailed knowledge about image information processing technology is required next. Reading the TB 202,422 (rank 97) for this purpose shows that the introduction of the complex technique of diffuser tensor imaging can increase the resolution of the position of axial tracts with the brain stem. In doing so, one candidate of the technology used for the image processing technology improvement is discovered.

In addition, technical solution methods for improving the image processing technology are required. Reading TB 258,124 (rank 2) makes it possible to find that using ultrasonography and fluoroscopy together is also effective.

In particular, it can be understood from TB 202,422 that the safety of surgery can be enhanced by combining navigation and image processing techniques. Reading TB 418,353 (rank 66), the more concrete example showing the effectiveness of the integration of navigation technology and image information processing technology will be obtained not only using fluoroscopy alone, but also combining it with image-guided navigation to improve the quality of positioning and fixing the screw. From TB 45,834 (rank 44), it is possible to obtain knowledge that using fluoroscopic cone-beam navigation rather than conventional fluoroscopy makes possible more accurate positioning and fixation of the pedicle screws. By linking similar TBs together in this way, we can learn a number of examples of techniques useful for the evolution of image information processing, allowing us to refine and deepen our knowledge.

The significance of introducing neuro technology into the navigation of equipment also is understood from reading TB 258,124.

It should be noted, however, that the introduction of neuro technology into surgical robots is not easy as understood from the description in TB 325,222 (rank 21). Indeed, as described in Sect. 8.5.1, the introduction of neuro technology into robotics navigation seems to have been unsteady.

Furthermore, as seen in TB 109,746 in the previous section, it is difficult to obtain tactile information of soft tissue. According to TB 462,954 (Rank 71), MR imaging is used for image diagnosis of soft tissues such as heads, necks, and spines; thus, it is understood that (1) MR imaging sequence protocols, (2) field strengths, and (3) gradient coil designs play important roles in improving this quality.

Thus, by using a five-sentence TB as the mining unit, it is possible to locate important information from the enormous amount of the technical documents written in esoteric jargon and to extract precise and detailed knowledge from them.

Of course, such a method is not yet satisfactory, and many insufficiencies remain to be solved in future. For example, the size of the TB is also controversial, such as, whether five-sentence or ten-sentence TBs are preferable. Algorithms for finding similarities between TBs are also still not sufficient. Overcoming these issues is a future task.

8.8 Conclusions

It is essential for companies to know the global trends in technological advances and acquire the ideas needed for overcoming bottlenecks to technological advances in order to accurately design business and R&D strategies. To do so successfully, it is insufficient to rely solely on knowledge stocks that managers and engineers have accumulated through past experiences. In the era of newly emerging innovation paradigm characterized by big data, dramatic improvements in computerized information processing speed and the digitization of text data through the dissemination on the Internet have led to the creation of text mining techniques for obtaining knowledge from unstructured natural languages, and it has begun to be used for corporate strategy planning as well.

However, conventional text mining has a major focus on grasping the overall tendency, taking a crude bird's-eye-view of enormous texts as a whole. Such a method may be effective for obtaining knowledge expressed in daily language, including market needs. This is because the analyst has a certain amount of relevant background knowledge as common sense in advance, so the correct solution is relatively easy to obtain by combining the results of analysis with the analyst's background knowledge. On the other hand, because knowledge concerning technology is generally less commonly possessed by analysts, it is required that the results of analysis themselves are highly accurate. This requires a pinpointing where knowledge is written by using a pinpoint focus type text mining method.

As mentioned in this article, detailed knowledge can be provided by creating TBs consisting of five sentences, reaching a point where the desired solution may be described, and further linking TBs with similar contexts.

This article takes the example of neurosurgical robots to describe how text mining methods, especially the pinpoint focus method of text mining, contribute to reasonable decision-making on projects where companies intend to input their limited management resources and plan an effective R&D strategy. It is expected that text mining will be increasingly used for planning corporate strategy.

References

Abe, H., Hirano, S., & Tsumoto, S. (2005). Byojo keika chisiki chushutsu no tameno tekisuto mainingu (Text mining for extraction of pathology knowledge). The 19th Annual Conference of the Japanese Society for Artificial Intelligence: 2005: 1–2 (in Japanese).

Chesbrough, H. (2003). *Open innovation* (p. 227). Boston: Harvard Business School Press.

Chinzei, K. (2015). Iryo kiki kokusai kikaku no saikin no doukou (Recent trend of international standard of medical equipment). *The Japan Journal of Medical Instrumantation, 85*(5), 530–534. (in Japanese).

Christensen, C. (1997). *The innovator's dilemma* (p. 253). Boston: Harvard Business School Press.

Fujimoto, T. (2007). Architecture-based comparative advantage: A design information view of manufacturing. *Evolutionary and Institutional Economics Review, 4*(1), 55–112.

Goto, T., Hongo, K., Yako, T., Hara, Y., Okamoto, J., Toyoda, K., et al. (2013). The concept and feasibility of EXPERT: Intelligent armrest using robotics technology. *Neurosurgery, 72*, A39–A42, 2013/01: 39–42.

Goto, T. (2014). Nou shinkei geka ryoiki ni okeru robotto shujutsu no genjou (Current status of robotic surgery in the field of neurosurgery). *The Shinshu Medical Journal, 62*(1), 68–69. (in Japanese).

Gupta, V., & Lehal, G. S. (2009). A survey of text mining techniques and applications. *Journal of Emerging Technologies in Web Intelligence, 1*(1), 60–76.

Inaki, N. (2008). Atarashii mekanizumu no tajiyuudo kanshi "radius surgical system" wo mochiita naishikyou geka shujutsu (Newly designed mechanical manual manipulator-radius surgical system-for endoscope surgery). *Journal of Japan Society for Endoscopic Surgery, 13*(6), 723–728. (in Japanese).

Iseki, H., Muragaki, Y., Nakamura, R., Hori, T., Takakura, K., Sugiura, M., et al. (2003). Interijento opeshitsu MRI yudo shujutsu taiou shisutemu (Intelligent operating theater and MR-compatible operating system). *MEDIX, 39*, 11–17. (in Japanese).

Iseki, H., Muragaki, Y., Maruyama, T., Suzuki, T., Ikuta, S., & Akimoto, J. (2009). Iryo kiki kaihatsu to iryo kiki no ishi shudo chiken (Medical device development and medical device doctor-initiated clinical trial). *The Journal of Japan Society for Laser Surgery and Medicine, 30*(1), 64–67. (in Japanese).

Japan Patent Office, home page, https://www.jpo.go.jp/shiryou/gidou-houkoku.htm.

Japan Patent Office. (2014). Tokkyo shutugan gijutu doukou chosahoukokusho—robotto—(Patent application technology trend survey report—robotics—), Japan Patent Office: 1–341 (in Japanese).

Japan Patent Office. (2015). Tokkyo shutugan gijutu doukou chosahoukokusho—naishikyo— (Patent application technology trend survey report—endoscope—), Japan Patent Office: 1–561 (in Japanese).

Kajita, Y., Mori, K., Hayashi, Y., Wakabayashi, T., & Yoshifa, J. (2013). Nabigeshon noushinkei geka shujutsu no genjo to tenbo (The current status and perspective of navigation neurosurgery). *Japanese Journal of Neurosurgery, 22*(7), 510–518. (in Japanese).

Kaneko, S., & Ootake, H. (2010). Raifu saiensu jisho kara kurinikaru infomatikusu e (From a life science dictionary to clinical informatics: Toward the knowledge finding from medical texts). *Journal of Information Processing and Management, 53*(9), 473–479. (in Japanese).

Kataoka, H., Masamune, K., Sakuma, I., & Dohi, T. (1999). Nou moderu no chosei ni yoru jutsu chu nou henkei no henkei keijou suitei (Shape estimation of brain shift by model adjustment). *Journal of Japan Society of Computer Assisted Surgery, 22*(1), 30–38. (in Japanese).

Kim, G., & Bae, J. (2016). A novel approach to forecast promising technology through patent analysis. *Technology Forecasting and Social Change, 117*(April), 228–237.

Kobayashi, N., Miyamoto, U., & Ooyama, K. (2005). Teishinshu shujutsu shien sisutemu "Navoit" nokaihatsu (Development of Naviot for minimally invasive surgery). *Journal of the Robotics Society of Japan, 23*(2), 22–25. (in Japanese).

Kobayashi, E., Mitsushima, I., & Oonishi, K. (2015). Shujutsu robotto wa hito no bisaina rikishokkaku jouhou wo saigen dekiruka (Can a surgical robot reproduce human's fine tactile information). *Medical Torch, 11*(1), 28–33. (in Japanese).

Koike, A. (2007). Tekisuto maining ni okeru senzaiteki chisiki no hakken shien (Support for potential knowledge finding based on text mining). *Journal of Information Processing, 48*(8), 824–829. (in Japanese).

Komoda, F. (2011). Tango setto no sakusei to shinka ni motozuku tekisuto maining shuho (Text mining technique based on creation and evolution of 'word set'). *Information and Documentation, 54*(9), 568–578. (in Japanese).

Kurata, M., & Takigawa, K. (2010). Nihon no igaku ronbun ni okeru seitai kan-ishoku no hatten katei (Development process of living liver transplantation in Japanese medical papers). *Journal of Nursing, Shiga University of Medical Science, 8*(1), 26–29. (in Japanese).

Kushima, M., Araki, K., Suzuki, S., Araki, S., & Nikama, T. (2012). Denshi karute nyuuin kanja kiroku no tekisuto maining (Text data mining of in-patient nursing records within electronic medical records). The 26th Conference of the Japanese Society for Artificial Intelligence; 2012, 2–1, 3K2-NFC-3-1: 1–2 (in Japanese).

Kushima, M., Araki, K., Yamazaki, T., Araki, S., Ogawa, T., & Sonehara, N. (2017). Text data mining of care life log by the level of care required using keygraph. In *Proceedings of the International MultiConference of Engineers and Computer Scientists* (Vol. I). http://www.iaeng.org/publication/IMECS2017/IMECS2017_pp327–331.pdf.

Masamune, K., Nakamura, R., Kobayashi, E., Sakuma, I., Dohi, T., Iseki, H., et al. (1999). Kudou bunrigata teiinou shujutsu shien manipyureta sisutemu (Development of sterilizable manipulator for neurosurgery with separated drive mechanism). *Japan Society of Computer and Surgery, 1*(1), 24–29. (in Japanese).

Masamune, K., Ooshima, K., Nagano, H., Kuribayashi, S., & Nakajima, S. (2003). Teishinshuu sekitsui geke shujutsu no tameno nabigeta robotto sisutemu oyobi sousa intafeisu no kenkyu (Study on user interface system of the navigation robot for minimally invasive spine surgery). Annual Report, Research Institute for Technology, Tokyo Denki University, 23: 119–124 (in Japanese).

Mima, H. (2006). Shizen gengo shori niokeru tan-i no settei (Word unit setting in natural language processing). *Gekkan Gengo (Monthly Language), 35*(10), 56–64. (in Japanese).

Mima, H. (2012). Kindai bunken no dejitaru akaibu to tekisuto maining (Digital archiving and text mining of modern-style japanese literatures using iwanami shoten's journal "shisou" (thoughts)). IPSJ SIG Technical Report, 2012-CH-95-4: 1–8 (in Japanese).

Moglia, A., Menciassi, A., Schurr, M. O., & Dario, P. (2007). Wireless capsule endoscopy: From diagnostic devices to multipurpose robotic systems. *Biomedical Microdevices, 9*, 235–243.

Nakatsuji, T., & Y. Hashijume, M. (2008). Gekai no motomeru robotto hando (The robot hand required by the surgeon). *Journal of the Society of Biomechanisms, 32*(3), 125–129 (in Japanese).

Muragaki, Y., Iseki, H., et al. (2000). Opun MRI wo mochiita "real-time" navigation no kaihatsu (Development of "Real-Time" navigation system updated with intraoperative MR imaging for total removal of glioma). *Japan Society of Computer and Surgery, 2*(3), 213–214. (in Japanese).

Netzer, O., Feldman, R., Goldenberg, J., & Fresko, M. (2012). Mine your own business: Market-structure surveillance through text mining. *Marketing Science, 31*(3), 521–543.

Nishiyama, R., Takeuchi, H., Watanabe, H., & Nasukawa, T. (2009). Shin gijutsu ga motsu tokucho ni chumoku sita gijutsu chosa shien tsuru (Technology survey assistance tool focusing on their advantages). *Journal of the Japanese Society for Artificial Intelligence, 24*(6), 541–548. (in Japanese).

Nonaka, I. (1991). The knowledge creating company. *Harvard Business Review, 69*(6), 96–104.

Onishi, K. (2009). Teishinshu geka shujutsu robotto no rikishokkaku fiidobakku (Force haptic feedback of minimally invasive surgical robot). *Journal of Japan Society of Computer Aided Surgery, 11*(2), 57. (in Japanese).

Onishi, K. (2015). Iryou-you rikishokkaku tsuki kanshi (Forceps with medical force tactile sense). *Medical Torch, 11*(1), 30–32. (in Japanese).

Ozawa, S. (2006). *Chansu hakken no deta bunseki (Data analysis for chance discovery)* (pp. 1–273). Tokyo: Denki University Press (in Japanese).

Ozawa, S. (2009). Shujutsu shien robotto no kenkyu-kaihatsu to shorai tenbo (Research and development of surgical robots and perspective). *Journal of the Robotic Society of Japan, 27*(3), 284–286. (in Japanese).

Patel, F. N., & Soni, N. R. (2012). Text mining: A brief survey. *International Journal of Advanced Computer Research, 2*(4), 234–239.

Porter, M. E. (1980). *Competitive strategy: Techniques for analyzing industries and competitors* (pp. 1–396). New York: Free Press.

Rzhetsky, A., Seringhaus, M., & Gerstein, M. (2008). Seeking a new biology through text mining. *Cell, 134*(1), 9–13.

Tanaka, K., Watanabe, T., & Yoneyama, T. (home page). Iryou kiki no hensei sutoresu hyouka (Denaturing stress evaluation of small medical devices". http://www.altairhyperworks.jp/html/ja-JP/PDF/AOP/1509_kanazawa.pdf (in Japanese).

The Japan Research Institute, Robotic Society of Japan, The Japanese Society for Artificial Intelligence, & Japan Ergonomics Society. (2008). Robotto bunya ni okeru akademi rohdo-mappu houkokusho (Academy road map in robot). Ministry of Economy and Industry: 1–35 (in Japanese).

Veugelers, M., Bury, J., & Viaene, S. (2010). Linking technology intelligence to open innovation. *Technology Forecasting & Social Change, 77,* 335–343.

Dr. Fumio Komoda (Ph.D., Kyushu University) is Professor Emeritus of International Business at the Faculty of Economics, Saitama University, Japan, and visiting professor of International Business at the Faculty of Economics and Business Management, Saitama Gakuen University. His main research interests is developing text mining methods that can contribute to building a company's technology strategy. He received his PhD in Economics from Kyushu University. He has published widely works and academic papers in the domain of management of technology and text mining.

Dr. Yoshihiro Muragaki (Ph.D., TWMU) is Professor of Faculty of Advanced Techno Surgery (FATS) at Tokyo Women's Medical University (TWMU). He is also a professor of Department of

Neurosurgery of TWMU and he has been conducting brain glioma surgeries including awake craniotomy, and achieved excellent performance. Recent his work is developing the intelligent operating room named SCOT and as a leader he is gaining global attention. He graduated from Kobe University, received his PhD in Medicine from TWMU, and also in Biomedical Science from Waseda University, Tokyo.

Dr. Ken Masamune (Ph.D., University of Tokyo) is Professor of Cooperative Major in Advanced Biomedical Sciences, Joint Graduate School of Tokyo Women's Medical University and Waseda University. He is an excellent computer science and robotics researcher. His main research interests is developing intelligent surgical robot with artificial intelligence. He received his PhD in Engineering from the University of Tokyo.

Chapter 9
Paradigm Change in the History of the Pharmaceutical Industry

Sarah Edris

Abstract The history of the pharmaceutical industry is told within three major paradigms, each of which arises through a continuous interaction between science, technology, business organizations, national institutions, and the wider growth of economic and social developments. The first paradigm begins in the mid-19th century, when the influence of chemistry on medicinal research had reached a degree of maturity, and ends with the outbreak of WWII. The second paradigm requires a better understanding of organizational features that differ from the old-line pharmaceutical companies that emerged in the 19th century. These companies were largely influenced by new institutions, restrictive environments, and turbulent decades. The third paradigm reflects geographic shifts of specialization in pharmaceuticals and changes in the composition of organizations involved; its setting is more collaborative and networked than previous paradigms. The shift between paradigms in this industry is therefore associated with the character of international business efforts and the context for which these efforts occur, with implication for firm responses to new developments in science and technology, and theories in international business, strategy, innovation, and economic geography.

9.1 Introduction

The history of the pharmaceutical industry as told in this chapter is focused on the limits to growth of the sciences upon which the industry relies on, the changes in the nature and structure of the largest companies active in the industry, and the knowledge networks between those companies and various organizations. It also tries to make the connection between organization-level developments and macroeconomic trends as well as with the institutional structure that affect or encourage linkages between the major companies and universities, hospitals, research institutes, government organizations, even companies in other industries.

S. Edris (✉)
Rutgers Business School, Rutgers University, Newark, NJ, USA
e-mail: sarah.edris@rutgers.edu

© Springer Nature Singapore Pte Ltd. 2019
J. Cantwell and T. Hayashi (eds.), *Paradigm Shift in Technologies and Innovation Systems*, https://doi.org/10.1007/978-981-32-9350-2_9

In doing so, I maintain that paradigm change can only arise through a continuous interaction between science, technology, business organizations, national institutions, and the wider growth of economic and social developments. I demonstrate that the technological progress in pharmaceuticals cannot be fully understood where any of these are taken in isolation.

I build on several strands of research to identify and conceptualize paradigm change in the pharmaceutical industry. The first parallels the historical development theorized by Freeman and Louca (2001). This work has directed attention to the interdependencies of economic and social movements, which include technological and scientific innovations, within the framework of institutional settings. It does so without assigning primacy in causal relationships. The second adopts the three epochs, or major paradigms suggested by Henderson et al. (1990), and implicitly recognized in the specialist literature on pharmaceuticals. The third builds upon the systemic perspective suggested by Cantwell et al. (2010), who proposed that international business co-evolves with its institutional environment. The objective is to connect the three major paradigms in the history of the pharmaceutical industry to global developments, since in the evolutionary process of substantial change that has characterized this sector, firm behavior and institutions related to both science and public policy have strongly interacted with one another.

Indeed, the pharmaceutical industry is driven by advances made in science, high research and development (R&D) investments and long development cycles. It has produced the majority of new medicine, and in many ways, changed the character of the medical community. The evolution of the industry has also been significantly shaped by the environment for which it has taken part. To understand the continuous historical transformation of the industry, I therefore analyze the leading area of interaction with the external environment, differentiating contexts within which different views about modes of organizing business activity have taken place. However, in studying the emergence and evolution of the pharmaceutical industry, the relationship between science and technology are central, and central to this chapter.

While theoretical and empirical studies have paid attention to the structure of the health care system, the institutional arrangements surrounding health-related research, and the role of intellectual property protection in affecting the processes of innovation, scant attention has been paid to the interaction between the shift in the technological paradigm in this industry from chemistry to the life sciences, the changes in organizational composition over time, as well as the context within which the interactions between various organizations have taken place. I examine the changes in corporate structure which reflect paradigm change in the industry, the varying types of investments, and increasing cooperation of the major companies with various types of organizations, and within the wider context of shifting paradigms. The history that lies behind the term 'paradigm change' in the industry under study is therefore an intriguing mix of accounts of science, technology, institutions, political and social trends, and conceptions of chance and history.

The chapter is structured as follows. First, in Sect. 9.2, I conceptualize the nature of paradigm change as a shift from preceding patterns of commercializing pharmaceutical technologies by relating scientific advances to organization-level developments as well as macroeconomic trends. This provides the analytical structure for which the histories of the pharmaceutical industry are told in Sect. 9.3: Paradigm 1 (1850 to 1939), Paradigm 2 (1940 to 1989), and Paradigm 3 (1990 to date). Section 9.4 interprets the process of paradigm change in the history of the pharmaceutical industry, and summarizes the major developments in science, technology, types of business organization, national institutions, and political and economic trends since the mid-19th century. It also predicts potential challenges and opportunities major pharmaceutical firms will confront in the near future. Section 9.5 suggests future research directions and potential sources of data to investigate proposed questions. A listing of consolidated companies can be found at the end of this chapter. Section 9.6 concludes.

9.2   Understanding the Nature of Paradigm Change in the Pharmaceutical Industry

Paradigm change in the pharmaceutical industry can be perceived broadly, dating back to Ancient Egyptian prescription records, or narrowly, to the first specialized pharmaceutical plant in Germany. Put differently, a broad lens would include all efforts of drug discoveries over time, much of which took place in isolated societies that progressed at much slower speed with simpler forms of technology. Paradigm change in the pharmaceutical industry as seen through a narrow lens focuses on activities that necessarily entail a strong interaction between technological efforts and scientific advances, each of which have been influenced by the tighter and more complex form of interdependencies of recent history. From this dichotomy, industry specialists, business historians, as well as scientists in the field of biology, have taken a narrow perspective of development in the pharmaceutical industry, where the science and technology linkages are central. Within such narrow perspective, the relative autonomy of evolutionary developments in science and technology justifies independent consideration.

Over the last few decades, economists have paid attention to Schumpeter's claims, particularly that technological innovation revolutionizes the economic structure from within, destroying old formations of markets and incumbent advantages (Christensen 1993; Tripsas 1997), allowing for new entrants unencumbered by firm history; though the modern evolutionary literature on technological change has tended to follow the Penrosean tradition (Cantwell 2002), which accounts for the cumulative nature of corporate learning (Helfat and Raubitschek 2000; Zollo and Winter 2002), i.e. deeply entrenched capabilities that accumulate and coevolve with products and markets, as well as the complexities that arise in different institutional contexts.

The parallel observation made in technology holds for science; the most discussed account of scientific revolutions is Thomas Kuhn's. In his work, scientific paradigms displace the old, and are so radically different that they can no longer be compared with the preceding paradigms in guiding future research. Nelson and Winter (1977, 1982) had drew attention to the relative autonomy of developments along a technological trajectory, and the possibility of noncumulative, conceptual and practical changes in subsequent epochs. Dosi (1982) further developed this work by drawing analogies to Kuhn's paradigms in science for technology; though, unlike Kuhn, Dosi's ideas didn't speak to a particular community, and for this reason, can be seen cited in the wider social sciences (Tunzelmann et al. 2008), as well as higher levels of analysis, such as the evolution of an industry. Nonetheless, there is evidence that suggests, as the sciences upon which technological knowledge relies on shifts, so does the locational specialization of the technological knowledge; and though science and technology may have their paradigms, more needs to be understood about the extent to which patterns of science have affected technological developments, and the limits to growth science has to offer for commercializing new drugs.

Put differently, while developments in technology and science can be seen as relatively independent, as is reflected in their social reorganization, methodological standards, content and goals, it is essential to take into account their interdependencies, and their reliance on institutional and economic developments for the purpose of understanding paradigm change, emergence, and evolution in the pharmaceutical industry. Freeman and Louca (2001) have suggested a coevolution of social subsystems (science, technology, politics, economics, culture) to provide insight into the process of paradigmatic shifts. This work builds on earlier theorizing by Derek De Solla Price (1984), Nathan Rosenberg (1969, 1974, 1976, 1982a, b), Keith Pavitt (1995), which demonstrated the way in which systemic features of scientific and technological developments have interacted, and with the wider economic and institutional environment. Recent traditions in management and economic geography already reflect the perspective of technological change as an evolutionary process, and take into account the wide range of institutions that coevolve with technology at the micro-level of the firm. A long discussion in the international business literature has, in many ways, rethought what is meant by 'evolutionary' processes, drawing attention to sociological, wider ranged approaches of technological accumulation, the interaction between production, institutions, and governance structures, as well as the influences the expansion of the firm has itself had in helping to stimulate change in the environment (Cantwell et al. 2010). There is also work in the strategy and organizational theory literature, which has also begun to suggest that organizational analysis should be more about how individuals aggregate to the collective level, capturing a more complex social interaction and interdependence with institutional entities.

In what follows, I assume the definition of paradigm change as put forth by Dosi (1982): discontinuous change is associated with the emergence of a new technological paradigm, defined as a pattern of solutions selected and established by the interplay between scientific advances, various economic and institutional variables. Once a technological paradigm has been selected among new, competing paradigms, in relation to the long-run patterns of social development, the path selected and established shows a momentum of its own (Nelson and Winter 1982; Rosenberg 1969), which is consistent with classical theories of social evolution, most notably Hegel and his contemporaries. Furthermore, technological knowledge is cumulatively developed (Cantwell 1989), and new technologies rely on novel combinations of prior knowledge, product portfolios and adaptive organizational capabilities (Arthur 2009; Zott and Amit 2008; Helfat and Raubitschek 2000; Teece et al. 1997). In this way, while a discontinuity reflects a break with the past, the past is necessarily synthesized in subsequent paradigms. To be sure, a trajectory is defined as the direction of advance within the boundaries of an established paradigm (Dosi 1982), whereas the existence and nature of paradigm change is induced at the crossing between science, technology, institutions, political and economic trends, and for this reason, needs to be explained historically.

As previously mentioned, I also parallel the co-evolutionary developments presented in Freeman and Louçã (2001) with adjustments to the subsystems as would be applicable to the process of paradigmatic shifts in the context of the pharmaceutical industry. Revised definitions of these subsystems are the following. In particular, I've replaced culture with institutions, since I regard culture as a partial feature of what is meant by institutions in the relevant literature, (Cantwell et al. 2010; North 2005) and is consistent with prior work on institutional change and social and physical technologies (North 2005; Nelson and Sampat 2001). Furthermore, developments which occur within each paradigm require understanding and reviewing the feature of subsystems which, taken together, are profoundly different in subsequent paradigms.

(1) **Science** is the generalizable and replicable knowledge of nature, usually resulting from basic research and represented by refereed and published papers. The science involved in drug discovery, through which potential new medicines are identified, include biology, chemistry, and pharmacology.
(2) **Technology** is a collective capability or knowledge of production (products, processes and services), which is not easily replicable. The technologies involved in drug discovery in the current paradigm include various computational tools.
(3) **Business organization** is a form within which application of science to production and practitioner knowledge of technology take place in pursuit of commercial interest. Firms in the pharmaceutical industry were historically integrated. The industry now relies on networked based structures.
(4) **National institutions** are such things as laws, scientific regulations, educational institutions, national innovation policies, culture, etc., which mold and differentiate the nature of innovative activity across countries.
(5) **Economic, political, and social trends** reflect the wider economic and social growth and demand, political developments, or industry-related crises.

In summary, although each of these subsystems have distinctive features and have their own paradigms, their interdependencies and interaction express the process of paradigmatic shifts of the industry. First, for a coherent paradigm to take hold, each of the aforementioned subsystems need to be harmoniously connected. Thus: a paradigm is a social system with interconnected parts. Changing one part cannot be done without changing others. There may be paradigms within each of the sub-systems identified: e.g. as argued by Kuhn for science. However, in a science-based industry where the developments of science and technology are well connected, the reciprocal influence of the developments that occur within each of the aforementioned subsystems need to be taken into account (de Solla Price 1984; Rosenberg 1969, 1974, 1976, 1982a, b; Pavitt 1995). Moreover, while the emergence of a new paradigm causes conflict with older paradigms, as reflected in the nature of linkages between scientific, business and government institutions, old paradigms can be seen as retained and embedded in a subsequent paradigm. Put differently, science and technologies are cumulatively developed, new institutions embody old formations, and political, economic and social contradictions are ultimately resolved and integrated in practice.

9.3   Histories of the Pharmaceutical Industry

The history of the pharmaceutical industry is told within the three major paradigms implicitly recognized in the specialist literature, and explicitly, though roughly divided in Henderson et al. (1990), due to its' usefulness for examining the evolution of the modern pharmaceutical industry. The shift between each paradigm is due to a number of key features that are so profoundly different and merit descriptions. The first paradigm receives a longer period of time through which prescriptions were customized and formulated at local laboratories in various societies. It begins in the mid-19th century, when the influence of dye chemistry and analytical chemistry on medicinal research had reached a degree of maturity, and ends with the outbreak of WWII. The second paradigm requires a better understanding of organizational features that differ from the old-line pharmaceutical companies that emerged in the 19th century. These companies were largely influenced by new institutions, restrictive environments, and turbulent decades. The third paradigm reflects geographic shifts of specialization in pharmaceuticals and changes in the composition of organizations involved; its' setting is more collaborative and networked than previous paradigms. Needless to say, the ability to link the research concerns of markets, and the interaction with the emergence of external research institutes as well as connection to university science and engineering for background knowledge and training has been critical, and so much of the focus will therefore be on the current paradigm.

9.3.1 First Paradigm (1850 to 1939)

In the mid-19th century, the science of chemistry was providing new learning and generating new product opportunities. During this time, Germany and Switzerland were leaders in the science of chemistry and in the synthetic dye industry. It was thus initially Swiss and German chemical producing enterprises, such as Roche, Ciba, Sandoz, BASF, Bayer, and Hoechst that exploited their technical competencies and knowledge accumulated in organic chemicals and dyestuff to commercialize drugs based on synthetic dyes. A number of British and American firms, such as Wyeth (later, American Home Products), Eli Lilly, Squibb, Upjohn, Pfizer, Merck, Abbott, SmithKline, Warner-Lambert, and Brothers-Wellcome emerged as specialized in pharmaceuticals in the later part of the 19th century. These firms would eventually become old line pharmaceutical companies which grew internally then via merger and acquisitions to overcome barriers to entry, and pioneer technical capabilities needed to produce and market prescription drugs for national and international markets.

To be sure, in the US, the industry evolved largely in response to the advent of modern transportation and communication (Chandler 2009). The major US core companies were rather wholesaler/producer enterprises with strong marketing capabilities and distribution channels—e.g. use of new radio networks to reach mass markets—operating in commercial cities (Liebenau 1987). They relied on German and Swiss firms to supply new prescription drugs based on revolutionary organic chemical technologies, and merely processed, packaged, and marketed a variety of existing over-the-counter drugs derived from natural resources. In general, large firms tended to grow in the most technologically advanced and dynamic centers, and so the new transport and communication technologies of the 19th century was a precondition for the administrative coordination of production decisions, and were therefore regulated by the potential for economies of scale and scope. In fact, up until the end of the second paradigm, scale and scope were closely linked. Achieving scale depended upon bringing together the production of related products and common distribution networks.

However, strategies of diversification and internationalization required an evolution in the organizational structure that would later support the new sciences, and capture returns on new technologies. First, an evolution from the hierarchical to a managerial structure (Chandler 1986) was required to provide the institutional support for in-house corporate R&D, which would become essential for large firm survival in the second paradigm. This evolution can be traced back to the early years of the 20th century, when three major German companies had begun to integrate upstream into the production of essential raw materials, expanding into other areas of chemistry, pharmaceuticals, nitrogenous fertilizers, plastics, photographic products and synthetic materials (fibers), by extension from organic chemistry (Cantwell 2004). Following the formation of IG Farben in 1925, intensification of research was in the hands of professional managers than it was influenced by top-level decision-makers, thus becoming the "world's first truly

managerial industrial enterprise (Chandler)". Likewise, corporate R&D needed to
be divisionalized, with a central facility that commanded a strategic overview
function led by technically trained managers. It was therefore the central labs which
were likely to incorporate some basic research activities with the more localized
support of divisions, which had their own R&D facilities. In this way, the firm was
not only narrowly focused on the demands and narrowing research agendas of each
division, but with the wider firm and industry within which these demands belong.

   Second, the more restricted environment that took hold toward the end of the
first paradigm and through to the second paradigm obliged firms to jump barriers by
investing in markets abroad. Thus, German chemical firms, such as Merck and
Schering, established subsidiaries in the US, whereas firms whose innovative diffusion of home markets led them to replicate their activities abroad included
companies like Roche, which would later become a world leader in genetic engineering. The openness which characterized the first paradigm allowed American
firms to effectively incorporate European, largely Swiss and German, development
capabilities, and build research laboratories to create the necessary production
facilities, and commercialize new prescription drugs on their own. This cumulative
experience would have great implications for large firm survival in subsequent
paradigms.

   However, the science of chemistry began to peter out by the early 20th century,
as the science of biology created new learning and opportunities for commercialization. In addition, the institutions that supported the efforts that were driven by
chemistry did not represent suitable platforms for the newly emerging drug research
that had become increasingly guided by pharmacology and clinical sciences.
During this time, the US was relatively specialized in biology, becoming more
sensitive and aware to public health concerns, and the need the contents and sale of
pharmaceutical products (the history of the FDA requires a chapter in its own right).
But it wasn't until the US ceased importing German products during WWI that
American firms began to investigate throughout the interwar years how to develop
the technical capabilities needed to commercialize prescription drugs based on
newer pharmaceutical technologies. After the discovery of penicillin and other
antibiotics, pharmaceutical companies either established departments of microbiology and fermentation units, or exploited their microbiological capabilities to find
drugs.

9.3.2   Second Paradigm (1940 to 1989)

The institutions created for drug research and development in the second paradigm,
which has its' roots during WWII, led to the formation of the pharmaceutical
industry as we know it, where mass production required greater linkages with
universities, medical schools, hospitals, research institutes, and/or national laboratories, depending on the way in which funding is administered nationally, and the
nature of the linkages between research and practice locally. In the US,

government-sponsored crash programs provided the financing required to build research labs and the necessary production facilities focused on commercial production techniques and chemical structure analysis. The increasing dialogue between microbiologists, biochemists, pharmacologists, and chemists resulted in substantial advances in physiology, pharmacology, enzymology, and cell biology, as well as a revolution in prescription drugs, in antibiotic drugs and then in other therapeutic areas. In addition, the industry's transition to an R&D intensive business increased the likelihood of large firm survival by increasing the capacity to learn and adapt to changing environments. Though not all German firms adopted the managerial form structure, they did have centralized science based strategies, benefited from research institutes, and pioneered industry-university cooperative relationships, which would later become critical in the third paradigm.

With the outbreak of the second world war, pharmaceutical companies were faced with a "target rich" environment, but had very little detailed knowledge about the causes, much less of the biological underpinnings of specific diseases. These companies relied on what has become known as "random screening", a method for finding new drugs, by which the specific biochemical and molecular roots of many diseases were not well understood. The advances made in traditional biology (physiology, pharmacology, enzymology, and microbiology) and use of enzyme systems as screens led to enormous progress in the medical understanding of both, the chemical reactions of existing drugs, as well as diseases for which no drug therapy existed (Gambardella 1995; Henderson 1994). Several companies in the US, UK, and Switzerland were among the first to design significantly more sophisticated and sensitive screens to screen a wider range of compounds that were previously available in either small quantities or difficult to evaluate due to the complex mixture of reactions in living animals (Henderson and Cockburn 1994; Scannell et al. 2012; Lipinski and Hopkins 2004). French, Italian, other European, and Japanese firms were slower in absorbing the concepts introduced by biochemistry, such as enzymes and receptors. This is partially due to a number of factors, including a comparative weakness in scientific research functions in addition to absence from the global industry (with the exception of Takeda). These differences would have significant implications for firm response to the advent of molecular biology as well as the technological specialization of countries in pharmaceuticals.

To be sure, the introduction of genomic sciences, rapid DNA sequencing, combinatorial chemistry in the 1970s required specially trained scientists, innovative approaches to drug discoveries, and a fresh set of ingredients and services— that is to say, the institutional and organizational support that distinguishes the US environment and explains the success of biotech in the US. First, the Diamond versus Chakrabarty Supreme Court patent decision, allowing the patenting of novel living organisms and their DNA in 1980, removed barriers to biotechnology. Second, a federal statute enacted in 1983, known as the Orphan Drug Act, gave tax benefits and granted a 7-year monopoly to enterprises that commercialized drugs to treat relatively uncommon life threatening conditions, which would otherwise be uneconomic to discover and bring to market. Major pharmaceutical firms had been investing their efforts on best-selling drugs (with sales of $1 billion or more), and neglected marketing efforts behind smaller, existing products, many of which could have become best sellers, and benefit a wider array of patients than originally anticipated (as was the case with AIDs, asthma, etc.). Third, changed social relationships within universities attracted the entrepreneurial interest of numerous professors who knew how to use advances made in genetic engineering to enhance the productivity of discovery in addition to benefiting from niche drugs large pharmaceutical firms neglected in favor for those in demand. Inevitably, a number of startups emerged in the late 1980s, quickly built integrated learning bases, and created from scratch the functional capabilities needed to, and succeed in commercializing products from new technology based on the new discipline of molecular biology (Chandler 2009). Together, the new startups created the infrastructure of a new industry—biotechnology, that marked the transition to the third paradigm.

However, the evolution of biotechnology, and ultimate success of the biotech industry was unclear for some time, so the use of molecular biology hadn't yet been adopted as a research tool until the third paradigm. Moreover, the 1970s was an instable decade, characterized by stagnation, recession and high unemployment, in addition to high prices, decreasing rate of successful new product introduction, and the breakdown of the postwar system of international monetary exchange (Chandler 2009). During this period, pharmaceutical companies were merely investigating the transition to guided discovery, and what it would take to incorporate the new technical knowledge and procedures with the advent of molecular biology. However, no momentum appeared in light of the new technology and no successful biotech firm emerged until the late 80s, with the exception of Genentech, founded in 1976, now a subsidiary of Roche as of 2009.

Following the chemical crisis between 1979 and 1982, chemical firms had also begun to redefine product lines to commercialize, mostly by buying and selling business units from one another (Chandler 2009). The coming of new pharmaceutical technologies based on a new science (molecular genetics) presented a dramatically different pattern of growth to which pharmaceutical firms had no inherent advantage over chemical firms (Kenney 1986). To increase profits, there was a trend among US pharmaceutical companies to diversify into related consumer chemicals (household and personal goods), food and drink, as well as medical instruments and devices—an industry related in terms of markets, but not technology. However, these firms quickly realized that the less diversified their company the better their financial performance, and the stronger positioned they were to recognize the importance of biotech in addition to exploiting the new learning in traditional biology, especially microbiology.

In the meantime, chemical companies began a large-scale restructuring of their product portfolios, first by strengthening their core competencies followed by the drive into pharmaceuticals—that is, developing products based on the new science of biology than chemistry. For example, Monsanto was the first to successfully grow capabilities in traditional biology and biotechnology, and in trying to commercialize chemically based and genetically engineered agricultural products.

However, not all chemical firms were particularly successful. Du Pont's response to the crisis was to diversify into pharmaceuticals by acquiring a small drug company, Endo, then by forming a joint venture with Merck, through which Du Pont would learn how to develop the necessary functional capabilities. Unable to compete with industry rivals, Du Point sold its' pharmaceutical division to Bristol-Meyers Squibb in 2001. Dow had also quickly failed, abandoning its attempt to enter through acquisitions, including its' 1989 Marion Laboratories, which it later sold to Hoechst 5 years later. While these companies had strong marketing capabilities, they failed to build ties with startups, universities, and research institutes, which would later become especially important in the third paradigm. European chemical firms were more successful in shifting their focus from chemicals to pharmaceuticals by quickly selling or spinning off their chemical businesses: e.g. following the merger of Ciba-Geigy and Sandoz in 1995 that formed Novartis, the firm quickly sold off its' chemical businesses to focus on pharmaceuticals; Hoechst merger with Rhone-Poulenc Rorer to form Aventis, led the firm to spin off its chemical businesses; Britain's ICI spun off its' pharmaceutical division as a separate enterprise (Zeneca).

In sum, large pharmaceutical firms were fully integrated from drug discovery through clinical development, regulations, manufacturing, and marketing. Within this paradigm, these firms produced a range of drugs, prescription and over-the-counter, relying on a few best sellers, each with annual sales in excess of $1 billion. Drug discovery had been conducted in-house, where large random-screening programs were used with, as mentioned earlier, limited knowledge about the underlying physiological processes. However, this in-house capacity had begun to rely upon collaborations with other organizations, including universities, research institutes, and government-funded institutes.

9.3.3   Third Paradigm (1990 to Date)

The advent of molecular biology attracted a number of firms from different technological traditions, and had a great impact on a number of other fields, including chemicals, diagnostics and agriculture. Needless to say, the new science had its greatest impact on human therapeutics, and the organizational structure of the pharmaceutical industry. First, the potential to understand disease processes at the molecular (genetic) level and to determine optimal molecular targets for drug intervention presented new concepts of drug discovery. In the current paradigm, the search for a new drug begins with a therapeutic area of interest, and analysis of existing drugs and patents. Scientists mine the scientific literature to identify a drug target—a protein (e.g. enzymes, receptors) or nucleic acid (e.g. DNA, RNA) involved in a disease to which a drug is directed to change its' behavior or function. With information on drug targets, genes, and the biological mechanisms responsible for the disease, medicinal chemists comb through a what is known as a "chemical space" to design and synthesize compounds that can bind to the target. Leading candidate compounds are then selected for advancement into clinical trials. This approach to drug discovery is known as reverse pharmacology or target-based drug discovery (Drews 2000).

By the mid-1990s, firms were quick to adopt a novel technique called "combinatorial chemistry" and replaced synthetic and medicinal chemists with powerful computation tools. These tools lowered the cost per molecule achieved, by allowing chemists to create millions of related compounds in a single step, and model 3-dimensional structures of gene targets with those chemical structures. Around the same time, the first draft of the human genome was introduced by the Human Genome Project, which enhanced coordinated search efforts in drug discovery by providing a clearer picture of the target landscape. In other words, the availability of the human genome map was a complementary innovation for target-based drug discovery (Hoang and Rothaermel 2010); the map was therefore mainly beneficial to firms experienced in target-based strategies, which were either large, diversified, and research intensive firms or smaller specialized biotech firms with unique intellectual properties (Zucker et al. 1994).

To be sure, the challenge for long-established core pharmaceutical companies—which had continued to rely on the support of the existing nexus that had been so enlarged by the post-WWII therapeutic revolution (Chandler 2009)—was to incorporate new technical knowledge and procedures and transition to the "drug discovery by design" (from a process of trial and error). In addition, these companies had to maintain the new sub disciplines of biology (including microbiology, enzymology, and biochemistry). Pharmaceutical companies were the first to shift towards the biotech model for R&D, and therefore develop strong ties to various scientific institutions (universities and government research laboratories) which were heavily supported by governments with funding for research in the new science that could aid the private sector's R&D. Countries therefore sought to create the optimal financial arrangements for public-private partnerships. In the US, funds are administered through the government-owned National Institute of Health (NIH), funding more than one-third of biomedical research. The Foundation of the NIH (FNIH) was an additional government agency that supported research and educational programs as well as foster collaborative relationships between the NIH, firms active in pharmaceuticals, universities, and non-profit organizations. The nature of these collaborations were specified in formal contracts (including cost sharing and rights to patents). In the UK, biomedical research is funded by the Department of Health, the Medical Research Council, and private foundations such as the Wellcome Trust (UK); France and Germany, biomedical research is performed directly in government research laboratories, such as CNRS, INSERM, Deutsche Forschungsgemeinschaft and the Max Plank Gesellschaft.

Pharmaceutical companies therefore gained advantage in becoming the first industrial group to learn and recognize the importance of molecular biology, rDNA, and biotech through a number of intangible benefits, including membership to scientific networks, where interactions improve the learning activities of participants, i.e. company scientists (Kenney 1986; Pavitt 1991). And so were also the first to redefine their strategic boundaries and acquire the research tools and production technology the new sciences had to offer. However, the maturing of biotechnology as a research and manufacturing technique brought an increasing number of MNCs into the industry and increased competition. As discussed previously, chemical firms sought to reverse stagnation, using the products of biotech as a tool to create new products for agriculture.

The synergies between the skills developed in applications of biotech to agriculture and medicine meant that pharmaceutical companies had no inherent advantages over chemical firms in producing drugs (Kenney 1986), and so the technical barriers between chemical and pharmaceuticals, both producers and marketers of molecules, eroded. Put differently, the use of molecular biology as a production tool was a competence destroying innovation (Henderson et al. 1999), particularly for firms that had little experience with target-based strategies. While pharmaceutical companies had the advantage in becoming the first industrial group to recognize the importance of biotech, and take the lead in their investments, chemical firms were larger financial entities and increasingly research-oriented (Kenney 1986). Thus, chemical executives saw pharmaceuticals as an ideal area to expand into, given the technical similarities (e.g. strong screening programs), and especially due to the relative importance of owning propriety rights to molecules in the new era; whereas pharmaceutical companies met the competitive entry of the leading chemical companies, resulting in a decade of international mergers. The purpose of these mergers was to enter markets where the barriers to entry were too high due to an increase in patenting new disease related genes. The only way to combine strengths in different geographical markets, capture economies of scale and scope, and expand potential products in the pipeline was through a merger (Chandler 2009).

By the 1990s, every large chemical and pharmaceutical firm had adopted more than one strategy to ensure a strong position in the bio-revolution. Investments have ranged from the development of in-house research capacity, linkages with universities to compensate internal resources and infrastructure for conducting R&D, capital investments in biotech start-ups, to a combination of strategies in accordance with the particular market position of the firms. Prior investments in R&D (i.e. genome technologies, combinatorial chemistry tools and specific diseases) have shown to influence the firm's exploratory search efforts and ability to absorb new scientific knowledge (Rosenberg 1990). This 'preadaptation' to exploit the opportunities (Cattani 2005) provided by the new science also explains the uneven distribution of exploratory search efforts across firms located near universities.

To be sure, whereas the world's largest industrial firms account for 60% of total patenting across all fields, large firms active in the pharmaceutical sector only account for 37% of pharmaceutical patents. This is likely due to the fact that there are other important actors in the industry, including smaller and specialized biotech firms, universities, hospitals, research institutes, and government agencies and national laboratories. For such smaller actors, particularly smaller pharmaceutical firms, and where biotech firms emerged, knowledge flows tend to be geographically localized. Indeed, according to one line of research, the economic benefits of basic research are not really widely and freely available (D'Este Guy and Iammarino

2013; D'Este and Iammarino 2010; Jaffe et al. 1993). For the most part, these benefits are geographically (and linguistically) localized, since they are embodied in institutions and individuals, and transmitted principally through personal face-to-face contacts. However, it has also been shown, though not as subject to empirical testing, that large pharmaceutical firms search globally for the best university science-based knowledge (de Solla Price 1965).

The international business literature in particular has highlighted the role of cross-border external knowledge acquisition for innovation strategy (Monteiro 2015; Monteiro and Birkinshaw 2017; Cano-Kollmann et al. 2016; Cantwell 2017), and the relative significance of local versus global knowledge sourcing in technological development. The establishment of common social communities across the subunits of the international MNE network (Kogut and Zander 1993), the collaboration between those subunits (Berry et al. 2014; Cantwell and Piscitello 2015), and deliberate effort of the MNE to coordinate the activities performed by those subunits, particularly those which develop a knowledge embeddedness within the MNE group (Asakawa et al. 2018; Cantwell and Piscitello 2014) and search for excellence within their core domain globally (Turkina and Van Assche 2018; Scalera et al. 2018), has been shown to be critical for the development of new combinations of knowledge around the core business expertise of the corporate group. Moreover, because the context of the pharmaceutical industry involves a community of firms from different technological traditions, large firms are increasingly motivated by cooperation in learning processes across locationally dispersed centers (Sachwald 1998; Mowery et al. 1998; Rosenberg 1982a, b). This is due to (1) the growing significance of basic science within technological knowledge, (2) rising technological interrelatedness and technology fusion, (3) emergence of broader technological systems, and (4) rising costs of R&D.

The ability of large firms to develop new knowledge by combining their internal (specialized) knowledge base with complementary areas of knowledge developed by other actors, such as firms in the same industry, as well as those in a different industry, and especially universities and research institutes (which provide basic science), and hospitals (which undertake basic and applied research), has therefore been critical for a successful innovation strategy (Veugelers and Cassiman 1999; Cohen and Levinthal 1990). Together, internal research, collaborations with scientific institutes, in-licensing, etc., increase absorptive capacities (Cockburn and Henderson 1998; Gittelman and Kogut 2003). As shown in the context of biotech, inter-firm agreements for technology exchange result in a more focused profile of technological specialization, which varies depending on whether agreements extend to cooperative learning or restricted to a simple exchange of knowledge (Hagedoorn 1990; Hagedoorn and Schakenraad 1992). Cooperative research ventures can therefore be seen as a compliment to in-house development whereby the firm's own problem-solving and learning sets the agenda for what is usefully searched when monitoring the external environment.

In summary, the environment within which the major pharmaceutical, biotech, and chemical firms operate in the current paradigm has become more focused and business networked, where various organizations interact more directly, and within which the gap between academia and large pharmaceutical companies has bridged and strengthened. "Networking has become more essential than ever in scientific and technical activities, as can be demonstrated by the rapid growth of collaborative research, joint ventures, consultancy, various types of licensing and know-how agreements, joint data banks, and, of course, innumerable forms of tacit informal collaboration (Freeman and Louca 2001, p. 327)."

9.4   Interpreting the Process of Paradigm Change in the History of the Pharmaceutical Industry

The received structure of the pharmaceutical industry is best represented in summary of the three paradigms. Table 9.1 summarizes the major developments in science, technology, types of business organization, national institutions, and political and economic trends since the mid-19th century.

The predominant drug discovery strategy in the earlier paradigm was non-target based, or phenotypic drug discovery (Scannell et al. 2012; Lipinski and Hopkins 2004), a method of modifying bioactive compounds without a clear understanding of the drug target or underlying disease, and testing them for efficacy in animals. This method is also known as non-target based, or trial and error learning approach, which had been long established in pharmaceutical firms (Gittelman 2016). Target-based strategies were resource intensive and required deep background knowledge of gene targets, and the adoption of computational tools and screening capabilities. The historical coincidence between World War II crash programs, the introduction of new organizational routines, and the relative strength of American and British positions in the science of biology differentiated the pattern of development of pharmaceutical activities in the English-speaking world. These countries witnessed the birth of specialized pharmaceutical producers who leveraged on the technical experience and organizational capabilities accumulated through wartime efforts to develop antibiotic drugs. By contrast, German and Swiss companies, which had previously dominated the world's prescription drug markets, as well as French producers, were preoccupied with wartime pressures. The advent of molecular biology in the 1970s and entry of chemical firms following the crisis in the chemical industry created many challenges and increased competition. Economic, political and social trends in the second paradigm continued to disturb the environment for which long-time pharmaceutical producers operated.

Drug discovery in the current paradigm is an inter-disciplinary endeavor (Fleming and Sorenson 2004); it involves the recombination of new and existing component technologies (Henderson and Clark 1990; Fleming and Sorenson 2001) as well as scientific knowledge to explore the technological landscape and work with smaller sets of possible combinations without full experimentation. In addition, the organization of drug discovery involves more actors and requires new firm capabilities (genomics, combinatorial chemistry, computational sciences), and so

Table 9.1 Major developments in the history of the pharmaceutical industry

	Paradigm 1 (1850–1939)	Paradigm 2 (1940–1989)	Paradigm 3 (1990–date)
Science	• Chemistry	• Pharmacology • Physiology • Enzymology • Microbiology	• Molecular biology • Genetics
Technology	• Chemical synthesis • Fermentation	• Transition from phenotypic to target-based discovery	• Combinatorial chemistry and computational tools
Business organization	• Hierarchical organizations/ functional specialization • Scale and scope/vertical integration • International expansion	• Managerial structure organizations • In-House R&D • Fully integrated • Diversification via mergers	• Network organizations • Equity/Research contracts; JVs; Licensing • Divestures; Mergers/ Acquisitions
Institutional environment	• Germany/Switzerland: Strong university training in chemistry • No connections with science • Food and Drug Act	• Diamond versus Chakrabarty Supreme Court patent decision • Bay-Dohle Act, and similar national policies • Orphan Drug Act • Loose connections with science	• US: Venture capital; Entrepreneurial interest of Professors • Scientific maps • Worldwide: intimate connections with science
Political and economic trends	• International networks (including cartels) • New transport/ communication technologies • Energy intensity (oil based)	• Centralization/ metropolitan centers • Nationalistic policies, world agreements and confrontation • Wartime investments • Crisis in the chemical industry	• Global and local connectivity • Information intensity (ICT) • External versus Internal cooperation (clusters)

Source Author's own analysis

the setting of the third paradigm is more collaborative and networked than previous paradigms. The strong university-industry interactions in the US, for example, specifically the openness of American universities to entrepreneurial activity on the part of their researchers (Mazzoleni and Neslon 2007), further encouraged the developments that opened up a new route for the American pharmaceutical industry, and in gaining a leading position in the new biotech sector. With several thousand biotech firms launching by the end of the second to the beginning of the third paradigm, drug discovery became more science-intensive, and the importance of firm-university relations, and relations with publicly funded institutes increased.

Finding, characterizing, and developing medicines has become so complex that new technical and institutional instruments are being generated to apply new scientific advances to the solution of societal problems. While areas like cancer have benefited from strong academic and corporate research (several anti-cancer drugs achieved blockbuster status by the early 2000s, including MabThera, Gilvec, Eloxatine, Gemzar, Casodex, Taxotere and Zometa), other areas, like neurology and neuroscience have not benefited from target-based research to the same extent. Health issues in less developed countries have also received less attention than the US, Western Europe and Japan, which consumed most of the world's total production of pharmaceuticals. Advances made in technology, artificial intelligence in particular, is continuing to change drug discovery science, and appropriate strategies, i.e. developing optimal research relationships across countries.

The paradigms identified herein therefore exhibit a change that is subsequently synthesized in light of new interconnections between scientific, business and government institutions. As argued in Sect. 9.2, old paradigms can be seen as retained and embedded in subsequent paradigms due to the cumulative nature of its subsystems. Science and technology are cumulatively developed, new institutions embody old formations, and political, economic and social contradictions are ultimately resolved and integrated in practice. The implication for large pharmaceutical companies developing new strategies to deal with new challenges and opportunities is to take into consideration various differences between countries, long governmental approval processes, R&D projects and collaborative research efforts, and participation in supporting a biotech sector that appears to be in financial disarray. Various new technologies, including AI, also seem to be changing patterns of drug discovery science, and so pharmaceutical companies will need to adopt the new developments in technologies.

9.5 Future Research and Potential Uses of Data

This chapter has assessed changes over time in the drivers of paradigm change in the pharmaceutical industry, including the effects of scientific and technological innovation, business organization and governance structures, and background economic and social developments. The histories told herein do not provide a complete synthesis of events, but rather highlight the major developments that explain shifts between each paradigm, emphasizing a systemic interpretation, not just the developments of its' parts per se. The chapter relies on a general framework that draws attention to issues that relate science, technology, business organization, institutions, national boundaries, and the interconnectedness of the global economy. Consequently, the chapter can be used as a starting point for understanding when new paradigms emerge, and how, in terms of the processes discussed herein. The discussions also help facilitate a more in-depth analysis of the shifts in science-technology relationships in the course of paradigm change. In fact, a series of questions have been raised in the course of this research, and which are worth noting for future research.

First, what is the structure of the knowledge networks on which a population of firms in the pharmaceutical, biotechnology, and chemical firms have relied internationally? What are the preferred sources of knowledge in the pharmaceutical industry in leading countries for innovation? *Second*, when organized by nationality, how does the structure of knowledge networks compare to the structure of knowledge sources on which firms rely in their research elsewhere in the world? In particular, what is the organizational and geographic composition, and the relationship between localized geographic proximity and global excellence at a distance? How should internal and external knowledge sources be combined, how important are hospitals compared to universities as knowledge sources for the industry, and to what extent are these relationships geographically localized? *Third*, what are the implications of technological change as a corporate learning process for technology-based alliances? What is the relative significance of public and private linkages? *Fourth*, what are the pressures that seem to lead pharmaceutical firms to relocate their activities from one city to another within countries? How does the composition of innovation, and the linkages between science and technology in particular be responsible for an evolution in the geographic profile and nature of investments? *Finally*, what is the role of cross-border networks in the transmission of knowledge within the firm? What are the implications for managers and public policy?

Given the difficulty and secrecy involved in obtaining firm R&D investments, patents of large firms and corporate groups active in the pharmaceutical industry provide an interesting empirical setting for future research to examine these questions. A patent is a legal document and a set of exclusionary rights granted by an authorized governmental agency (Griliches 1990). The right embedded in the patent can be assigned by the inventor to somebody else, usually a corporation, and/or sold to or licensed for use by somebody else. In the past, inventors contracted with firms. But in view of changing organizational structure, this relationship became adapted within the firm, as specified in the employment contract, where the inventor is an employee of a firm, and assigns the invention to a firm, if such was invented in a corporate lab. In this way, the patent is granted to the inventor (inventor-team). And at the time of grant, the inventor then assigns the patent to an organization. The dimensions covered for the use of patent statistics include: (i) the year of grant and application; (ii) type of technological activity, derived from historically consistent patent class system; (iii) city or town and country of the inventor's residence (host country); (iv) the organization to which patent has been assigned. In turn, the assignee organization can be identified with an owner, such as a corporate group (corporate consolidation requires an extensive search into the history of those firms), along with its' sectoral and home country code (derived from external sources), as well as the nature of the background context; (v) citations, essentially knowledge building, because of the dependence on earlier (interdependent) increments of knowledge.

In fields like chemicals and pharmaceuticals, a large part of the inventions is codified in patent applications, because they provide firms with key, inimitable resource (Gambardella 1992), the ability to "patent block" their rivals from entering product markets or disease areas, as well as forcing rivals those rivals into negotiations (Cohen et al. 2000; Ziedonis 2004; McGrath and Nerkar 2004). Table 9.2 provides a listing of the top patenting corporate groups and private companies in pharmaceutical technologies. Future research can examine the ownership structure and identify the subsidiaries of each firm and corporate group involved in pharmaceutical research through an extensive search into their history (including merger and acquisition activity, sectoral and home country codes) using the D&B Who Owns Whom directories, Bloomberg, public announcements made by those companies, company websites, and via informal interviews with contacts employed at those companies. It can then extract from the USPTO websites all patents granted to those firms. US data offer a disaggregation by cross-country, cross-firm, structural and historical dimensions on a scale that is not achievable through other sources.

Together, firms reported in Table 9.2 account for about 37% of pharma patenting in the US between 1976 and 2016, implying there are other important actors in the industry, such as smaller and specialized biotech firms, universities, hospitals, research institutes, and government agencies and national laboratories.

Table 9.2 Listing of major companies

Company	Nationality	Current product lines	Date of foundation
Abbvie	US	Pharmaceuticals and biologics	2013
Abbott Labs	US	Pharmaceuticals, diagnostics, Medical devices	1888
Bristol Myers Squibb	US	Pharmaceuticals and biologics	1887, merged with Squibb in 1989
Eli Lilly	US	Pharmaceuticals	1876
Johnson & Johnson	US	Pharmaceuticals, medical devices, consumer health	1886
Merck & Co.	US	Pharmaceuticals	1891 as subsidiary of Merck; 1917 as independent
Pfizer	US	Pharmaceuticals	1849
Valeant Pharmaceuticals	CA	Pharmaceuticals	1859
AstraZeneca	UK	Pharmaceuticals and biologics	1999 by merger of Astra & Zeneca
GlaxoSmithKline	UK	Pharmaceuticals, vaccines, healthcare	2000 by merger of Glaxo & Smith Kline
Allergan	IE	Pharmaceuticals	2013
Sanofi Aventis	FR	Pharmaceuticals and biologics	2004 by merger of Sanofi & Aventis
Bayer	DE	Pharmaceuticals, diagnostics, women health, plant biotechnology	1863
Merck Group	DE	Biologics	1668
Hoffman-La Roche	CH	Pharmaceuticals, diagnostics	1896

(continued)

Table 9.2 (continued)

Company	Nationality	Current product lines	Date of foundation
Novartis	CH	Pharmaceuticals, consumer Health, animal health	1996 by merger of Ciba-Geigy and Sandoz
Novo Nordisk	DK	Pharmaceuticals	1923
Teva Pharmaceuticals	IL	Pharmaceuticals	1901
Astellas Pharma	JP	Pharmaceuticals	2005 by merger of Yamanouchi & Fujisawa
Daiichi Sankyo Co. Ltd	JP	Pharmaceuticals, medical equipment	2005 by merger of Daiichi & Sankyo
Ono Pharmaceuticals	JP	Pharmaceuticals, diagnostics	1717
Otsuka Pharmaceuticals	JP	Pharmaceuticals	1964
Shionogi	JP	Pharmaceuticals, diagnostics, medical devices	1878
Takeda Chemical Industries	JP	Pharmaceuticals	1781
Amgen	US	Biologics	1980
Biogen	US	Biologics	1978, by merger
Celgene	US	Biologics	1986, spinoff of Celanese
Gilead Sciences	US	Biologics	1987
Immunomedics	US	Biologics	1982
Incyte Pharmaceuticals	US	Biologics	1991
Ionis Pharmaceuticals	US	Biologics	1989
Regeneron Pharmaceuticals	US	Biologics	1988
Rigel Pharmaceuticals	US	Biologics	1996
Vertex Pharmaceuticals	US	Biologics	1989
Colgate Palmolive	US	Household and personal, healthcare supplies	1806
Dow Chemical	US	Chemicals, plastics, paints, agrochemicals, gas and oil	1897
Du Pont	US	Chemicals, plastics, paints, agrochemicals, gas and oil	1802
Monsanto	US	Agrochemicals	1901 until 2018, acquired by Bayer
Procter & Gamble	US	Personal health/consumer care	1837

(continued)

Table 9.2   (continued)

Company	Nationality	Current product lines	Date of foundation
Reckitt Benckiser	UK	Household and personal care, healthcare, pharmaceuticals	1999, merger of Reckitt & Colman and Benckiser
Syngenta	CH	Agrochemicals	2000
BASF	DE	Chemicals, plastics, paints, agrochemicals, gas and oil	1865
Novozymes	DK	Biologics	2000
AkzoNobel	NL	Chemicals, paints	1994, merger of Akzo and Nobel
Mitsubishi Chemical Holdings	JP	Chemicals	2005, merger of Mitsubishi Chemical and Mitsubishi Pharma

Source Authors' own analysis

These firms also reflect recent consolidation, including Pfizer's acquisition of Pharmacia, the merger of Sanofi and Aventis to form Sanofi-Aventis, and so on. Network analysis can complement traditional empirical methodologies to understand the innovation landscape using patent citations, which, by constructing nodes (organizations) from the patents and the links between them using citations, reveals the social structure of the network, identifying actors and their connections. From this perspective, the focus of attention is the wider system or the structure of knowledge flows, not the individual actors per se. To examine the organizational character of the firm's knowledge network (using SNA methods), the patents cited by the top corporate groups can also be extracted from the USPTO. Further consolidation of the knowledge sources (the organizations being cited by those firms, e.g. universities, hospitals, research institutes, firms outside of the industry, etc.) is then required. All pairs of citing and cited patents identified can be grouped according to whether the implied knowledge flow was intra- versus inter-organizational, and whether it was localized versus global, etc., depending on the research question.

9.6   Conclusion

In this chapter, I tell the history of the pharmaceutical industry by focusing on the relationship between science and technology, the changes in the nature and structure of the largest companies active in the industry, the connection between organizational level developments and macroeconomic trends, and the institutional structure that affect the organizational and locational character of the industry's knowledge network. I conceptualize the nature of paradigm change as a shift from preceding patterns of commercializing pharmaceutical technologies by relating scientific advances to organization-level developments as well as macroeconomic trends. This provides the analytical structure for which the histories of the pharmaceutical industry are told: Paradigm 1 (1850 to 1939), Paradigm 2 (1940 to 1989), and Paradigm 3 (1990 to date), within which the reciprocal influence of the developments that occur within each of the aforementioned subsystems are harmoniously connected.

However, this chapter is not without its limits. The accounts of the three paradigms told herein, and the shifts between them, are brief, as it was not my immediate intention to synthesize all events for event sake. I merely distinguish three paradigms to conceptualize the way in which path-dependent and mainly incremental changes to previous patterns of development led to a qualitative transformation that induced a movement from one epoch to another. I therefore hope this chapter can be used as a starting point for researchers interested in the industry, and able to provoke an interest in the questions proposed herein or inspire new questions to be explored, through which tentative predictions about firm and knowledge networks, and the evolution of the industry and geographic setting can be made.

As previously noted, while various studies have paid attention to the structure of the health care system, the institutional arrangements surrounding health-related research, and the role of intellectual property protection in affecting the processes of innovation, scant attention has been paid to the interaction between the shift in the technological paradigm in this industry from chemistry to the life sciences, the changes in organizational composition over time, as well as the context within which the interactions between various organizations have taken place. As maintained herein, the process of paradigm change in the pharmaceutical industry cannot be fully understood where any of these developments are treated in isolation of one another.

References

Arthur, W. B. (2009). *The nature of technology: What it is, and how it evolves*. New York: Free Press.

Asakawa, K., Park, Y., Song, J., & Kim, S.-J. (2018). Internal embeddedness, geographic distance, and global knowledge sourcing by overseas subsidiaries. *Journal of International Business Studies, 49*(6), 743–752.

Berry, H., Guillé, M. F., & Zhou, N. (2014). Is there convergence across countries? A spatial approach. *Journal of International Business Studies, 45*(4), 387–404.

Cano-Kollmann, M., Cantwell, J. A., Hannigan, T. J., Mudambi, R., & Song, J. (2016). Knowledge connectivity: An agenda for innovation research in international business. *Journal of International Business Studies, 47*(3), 255–262.

Cantwell, J. A. (1989). *Technological innovation and multinational corporations*. Oxford: Basil Blackwell.

Cantwell, J. (2002). Innovation, Profits and Growth: Penrose and Schumpeter. In Pitelis, *The growth of the firm: the legacy of Edith Penros* (pp. 215–248). Oxford: Oxford University Press.

Cantwell, J. A. (2004). An historical change in the nature of corporate technological diversification, chapter 10. In J.A. Cantwell, A. Gambardella & O. Granstrand (Eds.), *The economics and management of technological diversification*. New York: Routledge.

Cantwell, J. A. (2017). Innovation and international business. *Industry and Innovation, 24*(1), 41–60.

Cantwell, J., Dunning, J. H., & Sarianna, M. L. (2010). An evolutionary approach to understanding international business activity: The co-evolution of MNEs and the institutional environment. *Journal of International Business Studies, 41*, 567–586.

Cantwell, J. A., & Piscitello, L. (2014). Historical changes in the determinants of the composition of innovative activity in MNC subunits. *Industrial and Corporate Change, 23*(3), 633–660.

Cantwell, J. A., & Piscitello, L. (2015). New competence creation in multinational company subunits: The role of international knowledge. *World Economy, 38*(2), 231–254.

Cattani, G. (2005). Preadaptation, firm heterogeneity, and technological performance: A study on the evolution of fiber optics, 1970–1995. *Organization Science, 16*(6), 563–580.

Chandler, A. D. (1986). Technological and organizational underpinnings of modern industrial multinational enterprise: The dynamics of competitive advantage, chapter 2 in A. Teichova, M. Lévy-Leboyer and H. Nussbaum (Eds.), *Multinational enterprise in historical perspective*, New York: Cambridge University Press.

Chandler, Jr., A. D. (2009). *Shaping the industrial century: The remarkable story of the evolution of the modern chemical and pharmaceutical industries*. Cambridge: Harvard University Press.

Christensen, C. M. (1993). The rigid disk drive industry: A history of commercial and technological turbulence. *Business History Review, 67*(4), 531–588.

Cockburn, I., & Henderson, R. M. (1998). Absorptive capacity, coauthoring behavior, and the organization of research in drug discovery. *The Journal of Industrial Economics, 46*(2), 157–182.

Cohen, W. M., Nelson, R. R., & Walsh, J. P. (2000). Protecting their intellectual assets: Appropriability conditions and why U.S. manufacturing firms patent (or not). National Bureau of Economic Research, Working Paper No.7552.

Cohen, W. M., & Levinthal, D. A. (1990). Absorptive capacity: A new perspective on learning and innovation. *Administrative Science Quarterly*: 128–152.

D'Este, P., Guy, F., & Iammarino, S. (2013). Shaping the formation of university-industry research collaborations: What type of proximity does really matter? *Journal of Economic Geography, 13*(4), 537–558.

D'Este, P., & Iammarino, S. (2010). The spatial profile of university-business research partnerships. *Papers in Regional Science, 89*(2), 335–350.

de Solla Price, D. (1965). Is technology historically independent of science? A study in statistical historiography. *Technology and Culture, 6*, 553–568.

de Solla Price, D. (1984). The science-technology relationship. *Research Policy, 13*(1), 3–20.

Dosi, G. (1982). Technological paradigms and technological trajectories. *Research Policy, 11*(3), 147–162.

Drews, J. (2000). Drug discovery: A historical perspective. *Science, 287*(5460), 1960–1964.

Fleming, L., & Sorenson, O. (2001). Technology as a complex adaptive system: Evidence from patent data. *Research Policy, 30*(7), 1019–1039.

Fleming, L., & Sorenson, O. (2004). Science as a map in technological search. *Strategic Management Journal, 25*, 909–928.

Freeman, C., & Louca, F. (2001). As Time goes by: From the industrial revolutions to the information revolution. Oxford: Oxford University Press.

Gambardella, A. (1992). Competitive advantages from in-house scientific research: The US pharmaceutical industry in the 1980s. *Research Policy, 21*(5), 391–407.

Gambardella, A. (1995). *Science and innovation in the US pharmaceutical industry*. Cambridge: Cambridge University Press.

Gittelman, M. (2016). The revolution re-visited: Clinical and genetics research paradigms and the productivity paradox in drug discovery. *Research Policy*.

Gittelman, M., & Kogut, B. (2003). Does good science lead to valuable knowledge? Biotechnology firms and the evolutionary logic of citation patterns. *Management Science, 49*(4), 366–382.

Griliches, Z. (1990). Patent statistics as economic indicators: A survey. *Journal of Economic Literature., 28*(4), 1661–1707.

Hagedoorn, J. (1990). Organizational modes of inter-firm cooperation and technology transfer. *Technovation, 10,* 17–30.

Hagedoorn, J., & Schakenraad, J. (1992). Leading companies and networks of strategic alliances in information technologies. *Research Policy, 21,* 163–191.

Helfat, C., & Raubitschek, R. (2000). Product sequencing: Co-evolution of knowledge, capabilities and products. *Strategic Management Journal, 21*(10–11), 961–979.

Henderson, R. (1994). The evolution of integrative competence: Innovation in cardiovascular drug discovery. *Industrial and Corporate Change, 3*(3), 607–630.

Henderson, R. M., & Clark, K. B. (1990). Architectural innovation: The reconfiguration of existing product technologies and the failure of established firms. *Administrative Science Quarterly*, 9–30.

Henderson, R., & Cockburn, I. (1994) Measuring competence? Exploring firm effects in pharmaceutical research. *Strategic Management Journal, 15,* 63–84. Winter special issue.

Henderson, R., Orsenigo, L., & Pisano, G. P. (1999). The pharmaceutical industry and the revolution in molecular biology: interactions among scientific, institutional, and organizational change. In D. C. Mowery & R. R. Nelson (Eds.), *Sources of industrial leadership* (pp. 267–311). New York: Cambridge University Press.

Hoang, H., & Rothaermel, F. T. (2010). Leveraging internal and external experience: Exploration, exploitation, and R&D project performance. *Strategic Management Journal, 31,* 734–758.

Jaffe, A., Trajtenberg, M., & Henderson, R. (1993). Geographical localization of knowledge spillovers, as evidenced by patent citations. *Quarterly Journal of Economics, 58*(3), 577–598.

Kenney, M. (1986). *Biotechnology: The university-industrial complex.* New Haven: Yale University Press.

Kogut, B., & Zander, U. (1993). Knowledge of the firm and the evolutionary theory of the multinational corporation. *Journal of International Business Studies, 24*(4), 625–645.

Kuhn, T. (1962). *The structure of scientific revolutions.* Chicago: University of Chicago Press.

Liebenau, J. (1987). *Medical science and medical industry: The formation of the americal pharmaceutical industry.* Macmillan Press.

Lipinski, C., & Hopkins, A. (2004). Navigating chemical space for biology and medicine. *Nature, 432*(7019), 855–861.

Mazzoleni, R., & Neslon, R. R. (2007). Public research institutions and economic catch-up. *Research Policy, 36,* 1512–1528.

McGrath, R. G., & Nerkar, A. (2004). Real options reasoning and a new look at the R&D investment strategies of pharmaceutical firms. *Strategic Management Journal, 25*(1), 1–21.

Monteiro, L. F. (2015). Selective attention and the initiation of the global knowledge-sourcing process in multinational corporations. *Journal of International Business Studies, 46*(5), 505–527.

Monteiro, L. F., & Birkinshaw, J. (2017). The external knowledge sourcing process in multinational corporations. *Strategic Management Journal, 38*(2), 342–362.

Mowery, D. C., Oxley, J. E., & Silverman, B. S. (1998). Technological overlap and interfirm cooperation: Implications for the resource-based view of the firm. *Research Policy, 27*(5), 507–523.

Nelson, R. R., & Sampat, B. N. (2001). Making sense of institutions as a factor shaping economic performance. *Journal of Economic Behavior and Organization.* 31–54.

Nelson, R. R., & Winter, S. G. (1982). *An evolutionary theory of economic change.* Cambridge, Massachusetts, USA: The Belknap Press of Harvard University Press.

Neslon, R., & Winter, S. G. (1977). In search of a useful theory of innovation. *Research Policy, 6* (1), 36–76.

North, D. C. (2005). *Understanding the process of economic change.* Princeton, NJ: Princeton University Press.

Pavitt, K. L. R. (1991). What makes basic research economically useful? *Research Policy, 20*(2), 20–109.

Pavitt, K. (1995). Academic research and technical change. In J. Krige & D. Pestre (Eds.), *Science in the 20th Century* (pp. 58–143). Amsterdam: Harwood Academic.

Rosenberg, N. (1969). Directions of technological change: inducement mechanisms and focusing devices. *Economic Developments and Cultural Change, 18,* 1–24.

Rosenberg, N. (1974). Science, inventions and economic growth. *Economic Journal, 100,* 725–729.

Rosenberg, N. (1976). *Perspectives on technology.* New York: Cambridge University Press.

Rosenberg, N. (1982a). Inside the Black Box: Technology and Economics. Cambridge: Cambridge University Press.

Rosenberg, N. (1982b). Technological interdependence in the American economy, chapter 3 in Inside the Black Box: Technology and Economics, New York: Cambridge University Press.

Rosenberg, N. (1990). Why do firms do basic research (with their own money)? *Research Policy, 19,* 165–174.

Rosenberg, N., & Nelson, R. R. (1994). American universities and technical advance in industry. *Research Policy, 23*(3), 323–348.

Sachwald, F. (1998). Cooperative agreements and the theory of the firm: Focusing on barriers to change. *Journal of Economic Behavior & Organization, 35*(2), 203–228.

Scalera, V. G., Perri, A., & Hannigan, T. J. (2018). Knowledge connectedness within and across home country borders: Spatial heterogeneity and the technological scope of firm innovations. *Journal of International Business Studies, 49*(8), 990–1009.

Scannell, J. W., et al. (2012). Diagnosing the decline in pharmaceutical R&D efficiency. Nature reviews. *Drug Discovery* 11(3) (pp. 191–200). Tripp, Simon, and Martin Grueber.

Teece, D. J., Pisano, G., & Shuen, A. (1997). Dynamic capabilities and strategic management. *Strategic Management Journal,* 509–533.

Tripsas, M. (1997). Unraveling the process of creative destruction: Complementary assets and incumbent survival in the typesetter industry. *Strategic Management Journal,* 119–142.

Tunzelmann, N., Malerba, F., Nightingale, P., & Metcalfe, S. (2008). Technological paradigms: past, present and future. *Industrial and Corporate Change, 17,* 467–484.

Turkina, E., & Van Assche, A. (2018). Global connectedness and local innovation in industrial clusters. *Journal of International Business Studies, 49*(6), 706–728.

Veugelers, R., & Cassiman, B. (1999). Make and buy in innovation strategies: Evidence from Belgian manufacturing firms. *Research Policy, 28*(1), 63–80.

Ziedonis, R. H. (2004). Don't fence me in: Fragmented markets for technology and the patent acquisition strategies of firms. *Management Science.* 50(6) 804–820.

Zollo, M., & Winter, S. (2002). Deliberate learning and the evolution of dynamic capabilities. *Organization Science, 13,* 339–351.

Zott, C., & Amit, R. (2008). The fit between product market strategy and business model: implications for firm performance. *Strategic Management Journal, 29*(1), 1–26.

Zucker, L. G., Darby, M. R & Brewer, M. B. (1994). Intellectual capital and the birth of US biotechnology enterprises. No. w4653. National Bureau of Economic Research.

Sarah Edris is a Ph.D. Candidate at Rutgers University. She is interested in the evolution of MNE networks and the changing geographical distribution of pharmaceutical innovation. Specifically, she is interested in how the structure of knowledge sources evolve over time, within and between firms, universities, and other organizations; and across geographical space, and technological fields. She is also interested in understanding the institutional rationale that shapes organizational search, and learning from failure. Prior to joining Rutgers, she attended CUNY, where she earned a B.Sc. in management and finance; her second major was in philosophy.

Chapter 10
Knowledge Transfer and Creation Systems: Perspectives on Corporate Socialization Mechanisms and Human Resource Management

Tamiko Kasahara

Abstract This study explores the role played by corporate socialization mechanisms (CSMs) and human resource management (HRM) practices in inter- and intra-unit knowledge flows and creation systems in multinational corporations (MNCs), especially focusing on professional service firms (PSFs). Drawing on a longitudinal case study of Cambridge Technology Partners (CTP), which was established during a paradigm shift in the consulting industry arising from the advancement of information technology (IT), the findings of this study suggest that CSMs are incorporated into HRM practices. When CTP was a global consulting company, CSMs worked as an infrastructure for transferring knowledge from the headquarters (HQs) to a focal subsidiary. HRM practices were designed to enhance and practice the HQs' goals of achieving global competitiveness. Specifically, training and development practices played a role in transferring knowledge from the HQs to a focal subsidiary at the corporate level, while tacit and explicit knowledge were transferred from senior to junior consultants at the individual level. At present, sharing the corporate philosophy within CTP has generated a high project success rate in Japan and laid the foundation for further knowledge creation for CTP. HRM practices are designed to enhance CTP's business model, methodology, and corporate culture and to create new knowledge. Training and development practices enhance the creation of new knowledge within the company. Additionally, performance appraisals and incentive practices are designed to encourage consultants to create new knowledge.

Keywords Corporate socialization mechanism · Human resource management · Knowledge transfer · Knowledge creation system · Cambridge Technology Partners

---

The original version of this chapter was revised: Correction in Table 10.1 has been updated. The correction to this chapter is available at https://doi.org/10.1007/978-981-32-9350-2_12

T. Kasahara (✉)
School of Information and Management, University of Shizuoka, Shizuoka, Japan
e-mail: kasahara@u-shizuoka-ken.ac.jp

10.1   Introduction

Research in the area of strategy and management indicates that the primary source of competitive advantages for multinational corporations (MNCs) lies in their ability to create and transfer knowledge within their intra-organizational networks (Ghoshal and Bartlett 1990; Birkinshaw and Hood 1998). An MNC is a globally distributed and integrated network in which knowledge is created in various units and transferred within the MNC (Hedlund 1986; Bartlett and Ghoshal 1989). This perspective has triggered research on the factors that influence inter-unit knowledge creation and transfer within a differentiated network (Zander and Kogut 1995; Szulanski 1996; Gupta and Govindarajan 2000).

Previous research has argued that human resource management (HRM) practices have a significant impact on knowledge transfer in MNCs (Minbaeva et al. 2003, 2014; Kaše et al. 2009; Mäkelä and Brewster 2009; Yamao et al. 2009; Peltokorpi and Vaara 2014). Since human resources-embodied knowledge is viewed as the foundation of a firm's core capabilities (Grant 1996; Argote and Ingram 2000), the resource-based view (RBV) argued that managing knowledge stock such as human resources, can be the source of a firm's value creation (Kang et al. 2007). Recent research has examined how HRM architecture can manage knowledge stocks and flows within various employee groups (Lepak and Snell 1999, 2002; Kang et al. 2007); how human resource management (HRM) practices and organizational governance mechanisms, such as corporate socialization mechanisms (CSMs), affect inter-unit knowledge transfer (Björkman et al. 2004; Minvaeba 2005; Kaše et al. 2009); and how HRM practices contribute to facilitating or hindering the absorptive capacity within MNCs (Minbaeva et al. 2003).

Research has also indicated the existence of barriers to knowledge transfer due to lack of knowledge of the source's motivation and the recipient's absorptive capacity (Szulanski 1996). To overcome these barriers, researchers have addressed the role of CSMs (Bartlett and Ghoshal 1989; Gupta and Govindarajan 2000; Björkman et al. 2004) and HRM practices, which can enhance inter-unit knowledge transfers and facilitate the absorptive capacity between the knowledge sender and receiver (Gupta and Singhal 1993; Minvaeba 2005).

HRM practices and knowledge transfer are associated concepts, but their link does not explain the underlying mechanisms at play. The theoretical work focusing on the causal connections between HRM and knowledge processes needs qualitative research for inductive inquiry and theory development (Minvaeba 2005). In this chapter, I explore the role played by CSMs and HRM practices in inter-unit knowledge flows and creation systems in MNCs. In particular, I focus on professional service firms (PSFs) because knowledge is a crucial source of competitive advantage for PSFs and their business and business models have changed from traditional operations to offering software and advisory services to promote business using IT, given the paradigm shift driven by IT advancement.

Although the interaction between coworkers and the transfer and internal creation of knowledge have been shown to be pivotal (Collins and Smith 2006;

Reed et al. 2006), PSFs have received much less research attention in the literature than manufacturing MNCs.

10.2   Knowledge Transfer Within MNCs

MNCs are integrated network organizations that generate knowledge across differentiated units to achieve a global competitive advantage. This view of MNCs indicates the existence of inter-unit knowledge transfer within their organizations, and researchers have investigated how MNCs manage this process (Hedlund 1986; Bartlett and Ghoshal 1989; Szulanski 1996; Gupta and Govindarajan 2000). Knowledge transfer is seen as "an unfolding process of dyadic exchange of organizational knowledge between a source and a recipient unit consisting of four stages: initiation, implementation, ramp-up, and integration" (Szulanski 1996, p. 28). The initiation and implementation stages comprise all events that lead to the knowledge transfer decision and resource flows between the recipient and the source. The ramp-up and integration stages comprise a phase in which the recipient starts using the transferred knowledge and a stage in which the recipient obtains satisfactory results from the transferred knowledge, which gradually becomes routinized. Recipients need to acquire useful knowledge from a source and utilize it in their units by adapting the new knowledge to their business circumstances (Minbaeva et al. 2014). Nonaka and Takeuchi (1995) depicted knowledge creation as the result of different modes of knowledge conversion: socialization, externalization, combination, and internalization. Socialization refers to the process in which tacit knowledge is transferred through social contact, such as communication and interaction through discussions and sharing experience among organizational members. Externalization is defined as the process in which tacit knowledge is converted into explicit knowledge in the form of concepts, metaphors, hypothesis, descriptions, and models. Combination is a process wherein knowledge is transformed from explicit knowledge to more complex, explicit knowledge by merging, categorizing, reclassifying, and synthesizing existing explicit knowledge using a database. The internalization process focuses on transforming explicit knowledge into tacit knowledge through application and practice and is shared across the organization. Nonaka and Takeuchi indicated that knowledge can be created and, then, shared, improved, and justified through social (collaborative) and individual cognitive processes in an organization.

Although inter-unit knowledge transfer has been considered a source of competitive advantage for MNCs, previous studies showed the difficulties of knowledge transfer within MNCs. One complication is related to the characteristics of certain types of knowledge, such as idiosyncratic, tacit, and non-codified knowledge. These types of knowledge are difficult to formalize in the absence of direct social interaction (Nonaka 1994; Zander and Kogut 1995; Hansen et al. 1999). Moreover, making a clear distinction between tacit and explicit knowledge can also be difficult due to their interrelatedness (Tsoukas 1996). When transferred knowledge has tacit components, these require the sender and recipient units to have intimate relationships (Szulanski 2000). Other issues are related to the knowledge sender's motivation, the knowledge recipient's absorptive capacity, and relationships between source and recipient units. Knowledge senders' motivation problems arise when a sender does not feel adequately rewarded for sharing knowledge and fears losing ownership or a position of privilege by sharing crucial knowledge with others (Szulanski 1996). Björkman et al. (2004) showed a positive relationship of inter-unit knowledge transfer with the criteria used to evaluate subsidiary performance in terms of agency theory.

Absorptive capacity is defined as the "ability to recognize the value of new external information, assimilate it, and apply it to commercial ends" (Cohen and Levinthal 1990, p. 128). In addition, absorptive capacity has four dimensions: acquisition, assimilation, transformation, and exploitation (Zahra and George 2002). These characteristics differ across recipients depending on prior related knowledge and the extent of inter-unit homophily between recipient and sender units (Gupta and Govindarajan 2000). If recipient units do not possess the same relative absorptive capacity as senders, the knowledge transfer between them is not successful. In particular, when a sender's knowledge is created through relations with external organizations and internal development work, recipient units are required to engage the same experts and facilities to receive the shared knowledge (Björkman et al. 2004). Organizational knowledge creation requires an organizational mechanism and strategy, which facilitates learning, communication, trust, and motivation for its members (Senge 1990; Watkins and Marsick 1993; Nonaka and Takeuchi 1995). Research suggested that CSMs, which promote a higher degree of knowledge transfer between the sender and recipient in inter- and intra-organizational contexts, may act as facilitators.

10.2.1 Corporate Socialization

CSMs are shared goals, values, and beliefs (Nohria and Ghoshal 1994) for building interpersonal homophily, trust, and close interpersonal networks in an MNC. Corporate socialization is achieved through interpersonal interaction, such as transfers of executive managers between the HQs and a subsidiary, and joint-work in team task forces and committees (Ghoshal and Bartlett 1988). The more various distinct units socialize while being part of the MNC's goals and vision, the more likely they are to achieve open and dense communication, increase the number and quality of the communication channels, and enhance resource and knowledge exchange and trust relationships within MNCs (Ghoshal and Bartlett 1988; Szulanski 1996, 2000; Gupta and Govindarajan 2000; Björkman et al. 2004).

Empirical research on knowledge transfer and sharing pointed out that close interpersonal networks have a positive impact on the creation, adoption, and diffusion of knowledge within MNCs. For example, Ghoshal and Bartlett (1988) showed that normative integration through corporate socialization is positively related to an MNC subsidiary's knowledge creation, adaption, and diffusion. Their results also indicate that dense intra-inter unit communication is at least partially positively related to knowledge creation, adoption, and diffusion by MNCs' subsidiaries. Similarly, Szulanski (1996) showed that the arduousness of the relationship, which characterizes the ease of communication and the intimacy of the relationship between the source and recipient units, facilitates knowledge transfer within the firm. From the social capital perspective, Tsai and Ghoshal (1998) showed that social interaction ties between the knowledge sender and recipient have a positive impact on building trust and resource exchange while the existence of a shared vision has a positive impact on building trust. Tsai (2001) argued that intra-unit knowledge transfer, including learning practices, occurs in a shared social context in which different units are linked to one another. They also suggested that business units have a central network position in terms of intra-organization and absorptive capacity, thus showing that knowledge transfer positively affects their innovation capability. Gupta and Govindarajan (2000) examined the relationship between knowledge outflows from and knowledge inflows into the subsidiary as well as the existence and richness of the transmission channel, which comprises formal integration mechanisms such as task forces and permanent committees. They also addressed lateral socialization mechanisms including job transfers to peer subsidiaries, participation in multi-subsidiary executive programs, job transfers to the HQs, and participation in corporate mentoring programs. Their results showed that formal integration and socialization mechanisms are positively related to knowledge outflows from focal to peer subsidiaries, while only formal integration mechanisms are positively related to knowledge outflows from focal subsidiaries to the HQs. Knowledge inflows from peer subsidiaries into focal subsidiaries and from the HQs into focal subsidiaries are positively related to formal integrative mechanisms as well as lateral socialization mechanisms. Björkman et al. (2004) found that CSMs, such as inter-unit trips and visits, international committees, teams, task forces, and training involving participants from multiple units are positively related to outward knowledge transfer from focal subsidiaries.

As seen above, CSMs are believed to facilitate the development of interpersonal ties in MNCs, which, in turn, support knowledge transfer between the parties. HRM practices such as recruitment, training and development, performance appraisal, career development, and retention contribute to building and maintaining the firm's knowledge stocks (Lepak and Snell 1999; Minbaeva et al. 2003; Hatch and Dyer 2004; Cabrera and Cabrera 2005), thus facilitating knowledge flow, acquisition, transfer, and integration within companies.

10.3 Human Capital, Social Capital, and HRM Practices

HRM targets human capital, which is considered the foundation of a firm's core competencies. The role of HRM practices in knowledge transfer and sharing within inter- and intra-organizational networks has been explained through human capital and social capital perspectives. In terms of RBV, human capital, one of the three subcomponents of intellectual capital (together with social and organizational capital), refers to the aggregate set of skills and expertise possessed by employees and is presumed to contribute to sustainable competitive advantage. Human capital is, in fact, unique and firm-specific (Boxall 1996; Hatch and Dyer 2004). The premise of human capital theory is that the more employees acquire firm-specific knowledge through learning and training and development, the more they become unique and irreplaceable, thus enhancing firm performance (Lepak and Snell 1999; Hatch and Dyer 2004; Youndt et al. 2004). Although knowledge within firms may exist in various forms and functions, knowledge and skills are embodied in people (Grant 1996). Human capital is viewed as both a codified and tacit knowledge repository (Lado and Willson 1994). While codified knowledge mostly consists of explicit types of knowledge stored in documents and databases within firms and is at risk of imitation by rival companies, tacit knowledge resides in the understanding and skills of people and in the company's routines and relationships, and is acquired through repetition and experience. Therefore, it is difficult for competitors to imitate it. However, firms often risk losing employees who possess crucial and specific knowledge because companies cannot own human capital. Therefore, human capital theory argues that human resources are an asset that firms should invest in, and firm-specific human capital is essential for building knowledge stocks through learning and HRM practices (Lepak and Snell 1999; Hatch and Dyer 2004; Youndt et al. 2004; Kang et al. 2007).

Human capital theory argues that firms can increase their human capital through the following HRM practices: by acquiring highly talented individuals from the external labor market ("buying individuals"), through recruitment and selection practices, or by internally developing individuals through training and career development practices ("making individuals") (Lepak and Snell 1999; Youndt et al. 2004). Furthermore, firms need to consider how to retain individuals who possess firm-specific knowledge and skills through retention programs. Lepak and Snell (1999) proposed an HR architecture based on employment modes, employment relationships, and HR configuration (the bundle of HRM practices). This architecture was built by using two dimensions, namely the uniqueness and value of human capital, to show how firms manage different types of human capital that contribute in different ways to the firm's competitive advantage. The suitable HR configuration depends on the human capital type. For example, core employees, who possess highly valuable and unique (firm-specific) knowledge and, are a source of core knowledge for the firm, may be addressed by high-performance work systems, such as extensive training by commitment and team-based or skill-based pay for information sharing or learning. Hatch and Dyer (2004) found that selection, training and development, and deployment practices significantly enhance learning, which, in turn, improves firm performance. Therefore, HRM practices can be viewed as antecedents of human capital, as they contribute to increasing human capital characterized by firm specificity and decreasing imitability from rivals.

Along with human, and social capital, another subcomponent of intellectual capital is the resource embedded within, available through, and derived from a network of relationships (Nahapiet and Ghoshal 1998). All these elements contribute to creating a sustainable competitive advantage for firms since parties outside these relationships cannot access this knowledge network (Adler and Kwon 2002; Kostova and Ross 2003). According to Lengnick-Hall and Lengnick-Hall (2003), social capital has several benefits, such as improving cohesion within a group, facilitating inter-unit resource exchanges between parties, and reducing dysfunctional turnover when individuals have strong positive social connections with their colleagues. A relationship can be built within a particular group (relationships among employees within a firm) and with external stakeholders (relationships between a firm and external parties). In this study, I focus on the social capital embedded in MNC inter-units. At the individual level, employees can build social capital through teamwork with the firm's encouragement to learn from their colleagues and/or external stakeholders. At the organization level, companies can encourage employees to create a wide range of relationships by nurturing corporate culture (Lengnick-Hall and Lengnick-Hall 2003) through opportunities for employees to join inter-unit meetings, project groups, and cross-border teams for increased interaction (Mäkelä, and Brewster 2009). The concept of social capital is used to explain how firms manage CSMs to exchange knowledge in inter- and intra-organizational networks. As I mentioned earlier, this is done both at the individual level (Mäkelä and Brewster 2009) and at the organizational level (Nahapiet and Ghoshal 1998; Tsai and Ghoshal 1998; Björkman et al. 2004; Yamao et al. 2009).

Social capital and human capital are complementary (e.g., Coleman 1988; Burt 1992, 1997). For example, training and development programs may not only increase individuals' knowledge (human capital) but also help individuals build relationships with their colleagues (Youndt et al. 2004) and boost knowledge exchange, diffusion, and sharing within and outside firms, which may require the physical aspect of interpersonal connection, trust and norms, and shared interpretations and codes. Some scholars proposed three interrelated dimensions of social capital: (1) a structural dimension (i.e., the physical aspect of interpersonal connections), (2) a relational (affective) dimension, which comprises trust, norms, obligations, and expectations, and (3) a cognitive dimension, which refers to shared codes, understanding, and interpretations, including shared narratives and linguistic codes (Nahapiet and Ghoshal 1998; Tsai and Ghoshal 1998; Kang et al. 2007). Nahapiet and Ghoshal (1998) and Tsai and Ghoshal (1998) showed that the relationship between social capital and knowledge or resource exchange/combination between interacting parties enhances a firm's competitive advantage and value creation.

In the context of HRM, Kang et al. (2007) suggested HR configurations for managing relational archetypes. These comprise a cooperative archetype involving relationships between a firm's core employees and internal partners (colleagues and the like), and an entrepreneurial archetype for relations between a firm's core employees and external partners. Both of these archetypes enhance knowledge sharing between parties. The cooperative archetype, the interaction relationships within firms, is characterized by more structurally dense relationships based on generalized trust, organizational norms, and common architectural knowledge among employees. It is supported by HRM practices that include the following: job rotation for interdependent work structures, as a structural dimension; shared goals and values and team-based appraisal systems for engendering generalized trust, as a relational dimension; and selection based on organizational fit, training and development with socialization programs, and mentoring for broader skill development, as a cognitive dimension. In terms of affective and cognitive social capital, Mäkelä and Brewster (2009) showed that the four types of inter-unit interactions in MNCs have different consequences for knowledge sharing, thereby outlining a mediating role for social capital between inter-unit interactions and knowledge sharing. These were inter-unit meetings, project groups, cross-border teams, and expatriate and repatriate interactions.

Yamao et al. (2009) examined the association between subsidiary HRM practices (HR configurations), knowledge stocks (human capital and social capital), and subsidiary-to-HQs knowledge transfer. They showed that a developmental configuration that emphasizes training and development and related practices, such as incentive and performance appraisal (which encourage employees to develop their skills) have a positive impact on the development of human capital. Collaboration configuration, including a set of HRM practices that encourage teamwork, such as selection based on candidates' ability to collaborate with others, international training and development, performance appraisal, and incentives, has a positive impact on the development of social capital. As discussed above, HRM practices can be viewed as antecedents of social capital, as well as human capital. By using data from knowledge-intensive firms including PSFs, Kaše et al. (2009) examined the relationship between HRM practices, interpersonal relations, and intra-firm knowledge transfer. Their results indicate that HRM practices, work design (such as teamwork and project-based work for enhancing interaction with coworkers), and training and development for establishing and maintaining personal relationships among coworkers have a positive impact on interpersonal relations. They also showed that interpersonal relations mediated the relationship between HRM practices and intra-firm knowledge transfer.

Previous studies focused on CSMs and the role of HRM practices in knowledge transfer and sharing within MNCs in terms of human capital and social capital. To facilitate inter-unit knowledge transfer and sharing, social interaction ties are obtained through project groups, inter-unit meetings, and cross-border teams (Mäkelä and Brewster 2009). To this end, participation in international training, corporate programs, and corporate mentoring programs (Björkman et al. 2004; Tsai and Ghoshal 1998) is fundamental for MNCs. Although the content of HR configurations for enhancing social interaction differs across studies, training and development practices with related HRM practices, such as performance appraisal, incentives, and career development for encouraging employees to develop their knowledge and interpersonal network with colleagues, are identified as key drivers for MNCs (Kang et al. 2007; Kaše et al. 2009; Yamao et al. 2009). Based on the insights of previous studies, I introduce the relationship between social mechanisms and HRM practices for knowledge transfer and creation using a longitudinal case study that focuses on one PSF.

10.4 Method

Although the role of HRM practices and socialization mechanisms in knowledge transfer and sharing has often been examined using quantitative methods (e.g., Minbaeva et al. 2003; Minvaeba 2005; Kaše et al. 2009; Mäkelä and Brewster 2009; Yamao et al. 2009), in this study, I adopt a qualitative approach in the form of a longitudinal case study because theoretical work focusing on the causal connections between HRM and knowledge processes needs qualitative research for inductive inquiry and theory development (Minvaeba 2005). A qualitative approach has many advantages that derive from its ability to achieve in-depth and holistic understanding, thus mirroring the complexity of social life. Qualitative research has made a substantial contribution to interpreting relationships between variables and to explaining the factors underlying the broad relationships that are established (Punch 2005). By focusing on the historical perspective, a longitudinal case study may allow the researcher to consider the role played by HRM practices in enhancing social interaction among employees for inter-unit knowledge transfer and sharing in MNCs through a combination of retrospective and real-time analysis (Pettigrew 1990). Case studies on knowledge creation have been conducted with PSFs, such as McKinsey, a global consulting firm (Bartlett and Ghoshal 1989).

To this end, I selected Cambridge Technology Partners (CTP) for the case study. CTP was founded in 1991 in Cambridge, Massachusetts, as an IT consulting firm that provided comprehensive IT services. Two years after its founding, CTP was listed on NASDAQ. It attracted significant attention from academics and practitioners because of its rapid expansion in national and global operations. Until its takeover by Novell, Inc., CTP operated 55 offices in 19 countries and employed 4,300 individuals. CTP Japan, which was founded in 1997, carried on its operations with the brand after CTP was taken over by Novell. Eventually, Novell's mergers and acquisitions (M&A) strategy was unsuccessful, and the company formerly known as CTP virtually disappeared. Since 2006, CTP Japan has been operating independently as the "new-born" CTP affiliated with Nihon Unisys, Ltd. This study seeks to clarify how HRM practices enhance socialization among employees, as well as their role in knowledge transfer between the HQ and subsidiaries, and knowledge creation in a subsidiary. To this end, the proposed case study is framed in terms of three phases of CTP's development: the first phase (1991–2000) is referred to as the age of Global-CTP. The second phase (2001–2005) is referred to as the age of Novell-CTP; and the third phase (2006–Present) is referred to as Present-CTP. The reason for treating CTP and CTP Japan as a single case study is that they represent a rare instance in which the generation, development, extinction, and reemergence of a global company can be observed. Since Global-CTP and Novell-CTP no longer exist, I can openly investigate HRM practices, corporate socialization, and knowledge transfer and creation at CTP, as well as disclose the name of the company. Therefore, CTP is an optimal case for determining the relationships among HRM practices, corporate socialization, and knowledge transfer and creation over an extended period of time.

Data used in this study were obtained from multiple sources. Primary data were collected through semi-structured interviews with Mr. Tsutomu Suzuki (the current president of CTP Japan, and vice president of Novell in 2004), Mr. Masaru Shirakawa (the current vice president of CTP Japan), and Mr. Masami Saito (marketing director of Novell-CTP Japan at the time). A total of 15 interviews were conducted from 2004 to 2019, by me along with a colleague who has a connection with CTP. The interviews, which lasted 90 min each, on average, were recorded and transcribed word for word. At a later date, the author asked further questions by email to clarify unclear points. Secondary data were obtained from each company's website (and past versions of their website via Internet archives), press releases of Novell, and the CTP case study proposed by Harvard Business School (Ambile et al. 1995; Gompers and Catherine 1997a, b). In the case study, I address the transition of the HRM practices and socialization mechanisms at CTP, which utilized the same concepts and practices adopted within the group over its three development phases, although Present-CTP has not yet globalized its business. The company believes that these concepts and practices allow them to achieve a substantial competitive advantage. Before introducing the relationships between HRM practices, socialization mechanisms, and knowledge transfer/creation system at CTP, I will describe CTP's business model. Global-CTP was one of the pioneers in ICT-based consulting and helped create a paradigm shift in the consulting industry by introducing a novel business model. In addition, CTP globalized its business by using the same business model over three distinct phases; HRM practices and socialization mechanisms were used for guaranteeing the completion of projects on time and on budget and facilitating collaborations within the group.

10.4.1 Paradigm Shift in the Consulting Industry and the CTP Business Model[1]

PSFs are known as organizations that offer sophisticated knowledge or knowledge-based products (Alvesson 2004). In the past, the business of PSFs has been positioned as a local industry since these companies have primarily provided customized services for local clients' needs (Maister 2003; Lødendahl 2005). However, some PSFs, such as management consulting and ICT consulting, globalized their operations to address the needs of their multinational clients. Competitive advantages arose from the knowledge generated through global business experiences and collaborations within the group that led to their global

---

[1]In the case study, I use Global-CTP to explain the business situation in the Global-CTP era, while I use CTP for explaining the role of the HQs of Global-CTP and/or business activities, including the business model and the Cambridge Culture that Present-CTP inherited from the Global-CTP era. In addition, I use Global-CTP Japan as well as Novell-CTP Japan when explaining the Japanese subsidiary's business activities during the Global-CTP and Novell-CTP eras.

presence. A number of factors triggered the paradigm shift in the consulting industry, including the demand for integrated and "one-stop" professional service firms and the advancement of IT (Kubr 2002). Traditionally, consultants have been viewed as advisors who do not take responsibility for decisions concerning a client's business. Consultant remuneration reflected the time spent in providing advice. However, since the late 1990s, the advancement of IT led management consulting firms, IT consultancies, and certain manufacturing firms, such as IBM, to begin expanding their business into e-business consulting, offering software and advisory services for business promotion using IT. This sectoral restructuring made the distinction between traditional management consulting and IT consulting somewhat blurry. Clients began requiring integrated, one-stop services, including management consulting and the execution of services in combination with experience and know-how regarding the latest technologies. Although the e-business consulting boom peaked at the end of 2000, it created new business models and new ways of consulting in the industry. As a result, nowadays, consulting firms need to carefully consider the performance they promise to their clients. Consultants are now viewed as assistants, and their remuneration tends to be increasingly related to results.

CTP was born during this paradigm shift. It was a global consulting firm that helped clients achieve competitive advantage by implementing high-impact, technology-enabled business solutions in unprecedented time frames and provided consulting services in a consistent and systematic manner (both upstream and downstream). CTP originated in the Cambridge Technology Group, founded in 1984, which was an IT education seminar company that provided training services for UNIX. The group successfully developed the rapid prototyping of applications on UNIX and began providing applications development and system integration services. In 1991, CTP spun off from the group as a company offering consulting and software development services. The company focused on realizing a short cycle time for system installation (shorter than business cycle time). Therefore, CTP pursued a strategy of providing quick services through the implementation of optimal solutions for clients.

CTP's business model made maximum use of its distinctive consulting methodology, collectively called CTP's Rapid Application Development (CTP RAD), which could guarantee a unique business model based on a "Fixed Time/ Fixed Price approach" to CTP's clients on a global basis. CTP pioneered both the Fixed Time/Fixed Price approach and RAD, which is a software development methodology adopted for projects based on tight timelines. RAD uses prototyping and combines high-level development tools and techniques (Martin 1991). Many competitors used RAD but struggled with increasing the success rate of their projects. Using the RAD methodology does not necessarily lead to a higher success rate on projects. Initially, CTP RAD was one of the custom development methodologies for open operating systems; however, the system evolved CTP developed through the M&A of related companies (Table 10.1).

The Fixed Time/Fixed Price approach means that CTP guaranteed that projects would be completed on time and on budget as contracted with its clients. This

Table 10.1 History of CTP

Year	Events
1984	• Two MIT professors founded Cambridge Technology Group (CT Group), which provided training services for UNIX
1986	• CT Group succeeded in rapid prototyping for UNIX applications, which provided consulting services for applications development
1991	• Cambridge Technology Partners (CTP) was founded in Cambridge, Massachusetts as an independent consulting division spun out of CT Group
1993	• CTP went public on the NASDAQ. The number of employees was expanded to more than 300 people. CTP received high recognition from the market due to "Fixed Time/ Fixed Price" delivery. In addition, CTP successfully developed Rapid Application Development (RAD) • CTP implemented rapid expansion of branches all over the US, such as Atlanta, Chicago, Dallas, New York, and Seattle
1994	• CTP set up operations overseas (in 1995, accounting for 22% of the European area) including in North Europe: Sweden; West Europe: UK, Ireland; and Central Europe: Holland and Germany
1995	• CTP acquired The Systems Consulting Group, Inc., Miami and Axiom Management Consulting, Inc., in San Francisco, which added expertise in package software evaluation, implementation, and business process redesign
1996	• CTP continued to drive geographic expansion to Latin America and Oceania. • CTP merged with Ramos & Associates, Inc., a recognized leader in deploying Enterprise Resource Planning (ERP) solutions including financial and manufacturing/ distribution applications
1997	• CTP Japan was founded in Tokyo, which started to provide services of ERP, Customer Relationship Management (CRM), and E-business • New offices were opened in Australia and India
1998	• CTP's strength in developing innovative, state-of-the-art electronic commerce solutions was recognized by Forrester Research (a large and independent technology and market research company in the US)
1999	• Fortune selected CTP as a top e-consulting firm (e 50)
2000	• CTP (UK) won Computing magazine's IT Service Company of the Year award
2001	• Novell acquired CTP. CTP became a Novell subsidiary retaining its name • Novell-CTP Japan began to provide identity management services
2004	• Novell-CTP Japan began to provide change management services
2005	• Novell-CTP Japan began to provide meeting transformation services and published a book "How to Create an Efficient Meeting" (ASA Publishing co., LTD.) (in Japanese)
2006	• Novell-CTP Japan affiliated with Nihon Unisys, Ltd., retaining its brand and autonomy
2007	• Novell-CTP Switzerland and Hungary spun off Novell as an independent company
2008	• Present-CTP began to provide training programs for project facilitators
2016	• Present-CTP was nominated for the best company award from Great Place to Work® Institute
2017	• Present-CTP was nominated for the best company award from Great Place to Work® Institute
2018	• Present-CTP was nominated for the best company award from Great Place to Work® Institute

Source global-CTP's website (http://web.archive.org/web/20000815074522/ http://www.ctp.com/ inv/ar/), present-CTP website (https://www.ctp.co.jp/company/), and interviews with the CEO of Present-CTP

contrasted with many competitors who used per diem fees and tended to tack on fees for the implementation of their services in the case of a project delay. The project success rate[2] was approximately 90% during the Global-CTP era (1998–2000), compared with a success rate of 26% for IT projects in the US.[3] The success rate at CTP-Japan is 95.6% (as of June 2008, at an achievement value of 529 projects in Japan).[4] This success rate is staggering compared to the 52.8%[5] success rate of domestic IT projects in Japan. CTP operated its Fixed Time/Fixed Price business model by building and utilizing CTP RAD faster than its competitors. By 2000, CTP had earned high recognition from IT industries, as well as several awards from research firms and business magazines, such as Forrester Research, Fortune, and Computing. CTP achieved rapid growth and global expansion by executing its unique and reproducible business model. The company executed an identical business model on a global scale under the slogan "Building the New Economy," aiming to expand its business by focusing on offering advanced solutions, such as eBusiness, eCRM (customer relationship management), and eERM (enterprise risk management). This generated global competitive advantages for CTP. Local competitive advantages were also obtained by leveraging and maximizing the use of CTP RAD to expand in local markets. CTP Japan customized CTP RAD to Japanese market circumstances. Although CTP Japan contributed to sharing the solution development for the Japanese market within the group, Global-CTP's solutions were developed at the initiative of the HQs, which secured the degree of completion of each solution. During the Global-CTP era, a subsidiary was authorized to expand its business at its own will, but the HQs and other subsidiaries provided generous support.

In 2001, the second phase of CTP began (Novell-CTP: 2001–2005). CTP was acquired by Novell and became one of Novell's service divisions but retained its name, market presence, and independence to recommend best-of-breed solutions. Novell was the leading provider of the Net services software, which delivers services to secure and power all types of networks, including the Internet, intranets, and extranets.[6] The purpose of the acquisition was to expand Novell's ability to deliver consulting support to customers and other IT service companies. Novell thought of this acquisition as an opportunity to make a transition from a software development model to a solution business model, which was symbolized by the replacement of its top management. Although Novell acquired CTP, Jack Messman, the president and chief executive officer of CTP, was nominated

---

[2]"Project success" means that the project is completed on time and on budget, with all features and functions as originally specified (CHAOS: A Recipe for Success-Software and Systems Engineering:http://www4.informatik.tu-muenchen.de/lehre/vorlesungen/vse/WS2004/1999_Standish_Chaos.pdf).

[3]CHAOS: A Recipe for Success-Software and Systems Engineering (http://www4.informatik.tu-muenchen.de/lehre/vorlesungen/vse/WS2004/1999_Standish_Chaos.pdf).

[4]CTP website (https://www.ctp.co.jp/press/pr_20080707_01/).

[5]NIKKEI XTECH: Survey of IT projects 2018, 2018 March (1): 26–27.

[6]http://www.novell.com/news/press/2001/3/novell-to-acquire-cambridge-technology-partners.html.

president and CEO of Novell. Eric Schmidt (a former CEO of Google), formerly the company's CEO, continued to provide strategic counsel as chairman of the Novell board of directors.[7] Novell-CTP utilized the same business model used by the company in the age of Global-CTP. Joint projects between Novell and CTP Japan were carried out. In this phase, CTP had significant opportunities to share feedback about products and services within the Novell-CTP group, excluding CTP's consulting methodology. However, this acquisition caused confusion, which eventually undermined CTP's culture thereby leading to high turnover of industry-ready consultants. As a result, from approximately 100 consultants of Novell-CTP Japan in 2001, the companies retained about 40 consultants in 2005. Within five years, the relationship between Novell and CTP faced a crisis due to the difficulties of integrating Novell and CTP's business models and absorbing CTP RAD, which was grounded in the "Cambridge Culture" (Nishii 2013).

In 2006, the third phase of CTP (Present-CTP: 2006–Present) began. CTP Japan spun off from Novell and was affiliated with Nihon Unisys while retaining its brand name and continuing to provide original consulting services. Similarly, in 2008, some European subsidiaries, such as Switzerland and Hungary, decided to spin off from Novell to set up independent companies. While Present-CTP is no longer a global company, it is the only firm with unique characteristics inherited from the former Global-CTP, and it has transformed facilitation-driven consulting into an idiosyncratic consulting "methodology" and developed a comprehensive business model to benefit from its distinctive strengths. In particular, Present-CTP began to provide training and development programs for project facilitators and workshops on its methodology.

10.5 Results

10.5.1 Cambridge Culture as a Corporate Socialization Mechanism

Since the founding of Global-CTP, its members have emphasized its culture as the basis for executing CTP RAD. The "Cambridge Culture" features six core values, the code of conduct for Cambridge consultants, (i.e., FROGBB), and Cambridge Magic as the ground rule of project teams to globally apply CTP's consulting methodology, Cambridge RAD. This culture works as a CSM for enhancing collaborations within inter-intra-organizations and has been the foundation of CTP's business model over the three phases mentioned above. HRM practices were instituted in order to achieve CTP's business model based on its culture. Six core values lie at the basis of consultants' behavior and are the prerequisites for project

---

[7]http://www.novell.com/news/press/2001/7/novell-completes-acquisition-of-cambridge-technotech-partners-jack-messman-becomes-president-and-ceo-.html.

Table 10.2 Six core values of CTP

Take initiative	Understand what to ask Understand how uncertain matters should be handled Clarify and manage problems Proactively seek opportunities Explore to deliver performance better than expected
Dedication	Do a better job to complete projects Meet your commitments to issues Willingly take on work if it is good for the company Demonstrate an enthusiastic attitude toward others
Flexibility/openness	Help other people' in problem-solving Learn new things positively Cope with ambiguous and indeterminate matters Adapt to new circumstances quickly
Respect	Build rapport Be willing to make compromises Encourage yourself to learn from experienced persons Tune in with others who do not have experience and knowledge
Honesty	Face difficulties courageously Ask someone what he/she wants directly Share good and bad information Criticize constructively Share information openly
Trust	Keep a confident outlook Accept responsibility and accomplish tasks to the end Build a trust relationship with clients and team members

Source Present-CTP website (https://www.ctp.co.jp/company/) and interviews with the CEO of Present-CTP

success. Project team members can work together towards a common goal, sometimes in difficult conditions, because all Cambridge consultants share the same core values (Table 10.2).

Table 10.2 shows the outline of the six core values. Global-CTP defined the code of conduct for their consultants—the FROGBB—based on these values. In line with the Cambridge Culture, which aims at achieving Fixed Time/Fixed Price quick services, CTP believed that project success is not achieved by a consulting methodology, but rather by the consultants who execute the methodology. The term "FROGBB" was created by combining the first letter of the words Fast, Right, Open, Guaranteed, Business Case, and Behavioral Focus (Table 10.3).

Table 10.3 shows the contents of the FROGBB. The essence of the FROGBB was embodied in the company culture and was based on facilitation-driven consulting services. The CTP consultants were required to dutifully practice the FROGBB in order to support the clients' profits and the success of their business. The six core values and the FROGBB were shared within the group. These supported the company in the pursuit of their business model by encouraging cooperation among consultants in inter/-intra-organizations and their clients. Compared to traditional consulting firms, Global-CTP emphasized collaboration with their

Table 10.3 The FROGBB code of conduct for Cambridge consultants

Fast	We consider the time frames to be important. With our methodology, tightly integrated services, and developmental efficiencies, we can deliver results in unprecedented time frames
Right	To ensure that clients use the solutions we deliver to their full potential, we focus on more than their technological components—we also address people, behavioral, and organizational issues. We build consensus on objectives and functionality and gain buy-in from all parties—up front—to ensure that everyone is "on the same page" and committed to the project's success
Open	We commit to openness with our clients
Guaranteed	We commit to the success of our deployments with fixed time/fixed price contracts. Integrating technology-based solutions can be risky—many clients have been burned by "runaway" projects with significant time and cost overruns. Our approach lessens the risk to our clients and imposes a discipline to deliver projects on time and on budget
Business Case	We craft technology solutions that support our clients' business strategies and prioritize functionality to ensure that every component of the application creates business value. This strategy prevents clients from lengthening projects with "nice to have" features that will not create strategic advantage. It also enables us to deliver the return on investment promised at the outset of the project
Behavioral Focus	We act with integrity. We cultivate trust with customers

Source Present-CTP website (https://www.ctp.co.jp/company/) and interviews with the CEO of Present-CTP

clients. In other words, facilitation-driven consulting services were provided to help their clients learn the skills needed to achieve corporate changes within their companies. This consulting approach allowed the company to differentiate from its competitors. To make maximum use of the CTP RAD, the six core values, FROGBB, and Cambridge Magic were essential elements for CTP. The six core values and FROGBB also worked as an engine for cooperation within the group across borders and were the sources of global competitive advantages for Global-CTP. These unique cultural features helped the company expand their business in the global market. Based on this culture, a new discipline of consulting, Cambridge Magic was created (Table 10.4).

Cambridge Magic was the ground rule for any project team, which aimed to resolve the issues of the client companies. The members of a project team were the project managers and business consultants of Global-CTP and clients who experienced issues in their companies. When starting a project, all members agreed on the goal of the project, project procedure, and confirmed the role of each project member. Then, project members practiced the discipline of Cambridge Magic regardless of their specific role in the project to achieve a successful outcome. By construing a shared culture within the group, Cambridge Magic was practiced by project members in day-to-day tasks. This practice generated imitable advantages for Global-CTP, and the company transferred its business model and culture to all operations through HRM practices designed to enhance global competitiveness.

Table 10.4 The Cambridge magic

1. Don't sell bad business	11. Don't be afraid to stretch
2. Leadership is the most valued skill	12. Listen to your colleagues
3. We all succeed or fail together	13. Know how to call a "time out"
4. Raise a yellow flag before it's too late	14. Be cautious of "rat-holes"
5. Never go into a client meeting unprepared	15. See the forest for the trees
6. Get it or get out	16. Think outside the box
7. Don't make excuses	17. Under-commit and over-deliver
8. Be flexible	18. Work smart not hard
9. Learn by doing, not by thinking	19. Have fun doing the work
10. Give immediate feedback	

Source Present-CTP website (https://www.ctp.co.jp/company/) and interviews with the CEO of Present-CTP

10.5.2   HRM Practices and Knowledge Transfer from CTP HQs to the Japanese Subsidiary[8]

CTP utilized a job grading system and competency management as a standardized HRM system for all operations. HRM practices developed at the HQs were adopted by Novell-CTP Japan and Present-CTP. During the second phase, Novell and Novell-CTP Japan used different HRM practices due to differences between their business models, and CTP had a unique culture based on mutual and complementary ties.

10.5.2.1   Training and Development

Global-CTP had two types of training and development programs. The first was the New Employee Orientation Program conducted at the HQs in the U.S. The purpose of this program was to help employees grasp CTP's core business values and methodologies by providing focused training on their respective service lines or functional areas. By conducting a monthly "boot camp" at the HQs, Global-CTP assured the consistency of candidates' vision, corporate culture, and methodology.[9] New consultants could not start their career at CTP until they were familiarized with the Cambridge Culture and CTP RAD. The second type of training consisted of programs for existing executive consultants, such as the CTP RAD, consulting, leadership, and project management trainings. The consulting training and courses

---

[8]The descriptions of HRM practices at Global-CTP are based on interviews with the CEO of Present-CTP and are referenced from past versions of the website accessed via Internet archive: http://web.archive.org/web/*/http://www.ctp.com.

[9]When Global-CTP acquired other companies, it also conducted an on-site "jumpstart" program that featured extensive service, methodology, and cultural training.

covered topics such as facilitating consensus and were aimed at enhancing employees' ability to work with clients and project teams to craft world-class solutions. Consultants were also required to learn various topics ranging from risk management to scheduling and cost control to ensure on-time delivery and on-budget projects. CTP ensured the quality of its consultants through these training and development programs, which helped support the business model within the group and build interrelations with consultants from other operations.

With the launch of the Japanese overseas subsidiary in 1997, the start-up members of CTP Japan attended a one-month training program at the HQs. The purpose of the program was to put CTP's culture and methodology into practice at the Japanese office. This required cultivating Japanese trainers capable of practicing the culture and business model for facilitation-driven consulting, which was a low-profile consulting methodology in Japan, in the same way as practiced at the HQs of Global-CTP in the U.S. After the completion of the programs, the start-up members worked at the HQs for one year to understand and embody the company's culture and business model. These training and development programs not only helped enhance project success rate and ensure consultants' quality globally but also transferred the HQs' business model and culture, which were well-integrated into the Japanese office. Although the start-up members of CTP Japan received training at the HQs, when CTP Japan was founded, consultants from other subsidiaries supported the operation of CTP Japan's business to replicate the HQs' culture and business model.

As I noted earlier, the HQs took the initiative to develop consulting solutions and transferred them to their subsidiaries even though different solutions were required for subsidiaries' business environments to secure the completion of each solution service. For example, most projects for which CTP Japan took the initiative advanced through cooperation with other subsidiaries, which had already developed careers and knowledge in the CRM consulting area. When CTP Japan began offering new solution services and consulting methodologies in Japan, it requested the dispatch of training coaches. A subsidiary that requested the dispatch of consultants and training coaches from other subsidiaries and affiliates absorbed the charges as part of its inter-company transactions. The CTP group had an enterprise service line management team that operated the knowledge management cycle and controlled consultants' information at each subsidiary. These support systems facilitated cooperation within the group and knowledge transfer among inter- and intra-organizations because all CTP consultants shared the company's culture through training and development programs.

10.5.2.2   Career Development

The careers of CTP's consultants were developed in accordance with job grading and competency levels, especially for new employees who were expected to experience every career track, including engagement, project management, business analysis, creative/cognitive, technologist, and architect. Each track comprised various levels. For example, in the engagement track, consultants were trained in sales that emphasized the ability to build client relationships; in the creative track, consultants were trained as advertising experts; and in the cognitive track, they were trained as experts who designed user interfaces in terms of cognitive psychology. After experiencing each career track, new consultants developed their career in line with their interests. The HQs rigidly managed the job grades and competency requirements worldwide until 2006. After 2006, CTP Japan revised the job grades and competency requirements in simpler forms because the number of employees had decreased. However, Present-CTP has not changed the essence of the career development or job grades and competency management that were used at Global-CTP.

CTP had a program to support each consultant's career development, called the "Resource Manager Program." Each consultant was assigned a resource manager (currently called a resource advisor) who helped the employee achieve their career goals. The resource manager took responsibility for the junior consultant's project assignments, career planning, mentoring, and performance appraisal. This unique program trained consultants into professionals with a deep understanding of Global-CTP's culture, and helped secure consultants' quality on a global basis. It also worked as a way to transfer tacit and explicit knowledge from senior to junior consultants through on-the-job training and collaborative work.

10.5.2.3   Performance Appraisal

Performance appraisal was designed to improve consultants' performance and encourage them to continue developing their career and collaborate with colleagues by engaging in team work. Global-CTP basically evaluated its consultants' performance and competency in terms of common competencies across all career tracks and competencies by each career track.

CTP had two types of review: a project review and an annual review. Project members received feedback every day from their project members. The project review served as on-the-job training and contributed to enhancing consultants' facilitation competency. After project completion, project managers evaluated the performance of each project member in terms of teamwork, job knowledge, judgment, problem-solving, and flexibility, which were core competencies across all career tracks. In addition, each consultant was reviewed in terms of the specific competencies that were required in each career track. Based on the results of the project review, executives at each local office evaluated each consultant's performance. As I mentioned above, resource managers were assigned to each mentored consultant, and they played a significant role in submitting an objective recommendation regarding consultant's promotion and salary raise to the executives at each local subsidiary. By using a system in which resource managers took complete responsibility for their junior consultants' development, CTP executed a unique business model group-wide.

To keep the turnover rate below the industry average, CTP hired bright, energetic people and assigned them challenging work in an environment where their efforts were rewarded. Equity was shared with employees through stock options and an employee stock purchase plan that distributed year-end bonuses based on the company's overall performance. As a result, employees recognized a direct correlation between their performance and the company's achievements and were committed to CTP's success. This contributed to satisfying clients and led to clients' success.

As described above, HRM practices were designed not only to secure the degree of completion of CTP's business model but also to integrate it with its culture and assure the compatibility between these components.

10.5.3 HRM Practices and Knowledge Creation in Present-CTP

Present-CTP was founded after the IT bubble burst and was affiliated with Nihon Unisys in 2006. At the time, some subsidiaries spun off from Novell-CTP, and the former CEO of Novell-CTP Japan and many talented consultants left the company. Only 32 employees remained at Present-CTP. These start-up members had an off-site meeting aimed at rebuilding CTP. They decided that the Cambridge Culture would be retained in Present-CTP because all members shared and recognized the company's culture as its primary source of competitiveness. However, while their mission and vision were set by the HQs during the IT bubble, the start-up members set original objectives that reflected the Japanese market. Their mission consisted of two policies: facilitating fast corporate transformation toward the client's goal and doing the right thing for clients. Their vision consisted of changing the general expectation in the IT industry by achieving a high project success rate, enlarging the base of supporters by making an excellent impression on clients, and enjoying team work by providing many significant opportunities to grow. In addition, the start-up members achieved consensus on building and providing opportunities to learn consulting knowledge and skills since this was one reason for the loss of talented consultants at Novell-CTP Japan. Through the off-site meeting, the start-up members of Present-CTP decided that the facilitation-driven consulting approach, which was built at and enhanced the competitiveness of Global-CTP Japan, would evolve into one consulting service area that would carry on analysis/proposal and solution services. The main clients of Present-CTP are business companies, but the firm also delivers facilitation-driven consultancy to non-profit organizations to support regional innovation. In addition, by using the consultancy, Present-CTP provides workshops for promoting work-style reforms led by the Japanese government to remedy the long hours of work in 2016. This indicates that the scope of Present-CTP goes beyond the application of accumulated knowledge on the facilitation-driven approach of Global-CTP. The uniqueness of Present-CTP lies in pursuing fast corporate transformation by facilitation-driven consulting "with clients." Present-CTP places utmost priority on helping clients become self-sustaining, and the company is willing to turn down a job offer from a client who lacks the motivation to transform and be self-sustaining. To offer facilitation-driven consulting, Present-CTP enhanced the HRM practices inherited from Global-CTP and built a new knowledge creation system. In 2018, it won the best company award from Great Place to Work® Institute in the mid-size company category (number of employees from 100 to 999) as a result of the knowledge creation activities supported by its HRM practices.[10]

10.5.3.1 Training and Development

Present-CTP uses training programs inherited from Global-CTP but which are customized to the Japanese business context. The company developed these programs through its own projects, which generated know-how in facilitation-driven consulting. Until 2006, Present-CTP used training programs that were developed at the U.S. HQs by translating them into Japanese. However, some training programs did not fit the Japanese situation; hence, Present-CTP developed original programs, which now number over 250, ranging from fundamentals to advanced levels. These are all geared toward sharing know-how with its consultants and clients. Present-CTP uses unique training programs for in-house training. All CTP consultants, including the CEO, are required to participate in training courses that are provided as standard programs. These courses appear on employee boards with red, yellow, and white pins to enhance participants' awareness of the features of the courses and trainers. Consultants with red pins have already taught the program, while a yellow pin means that consultants can teach the program but have not yet done so. A white pin means that consultants have completed the program. Present-CTP offers semi-monthly training programs. However, consultants can acquire knowledge and skills on their project by directly contacting the trainer thanks to the company's project-based work style. Present-CTP intends to improve its methodology through this training system because consultants also need to realize the theory by practicing and teaching. This method is taught to clients through a workshop. Moreover, Present-CTP provides 200,000 yen per person as a self-improvement subsidy for all consultants to enhance their knowledge. Consultants can use the subsidy for external training.

The company developed knowledge creation systems through training and development practice, such as workshops, networking with its clients, and off-site meetings. The Cambridge Project Facilitation Association (CPFA) is a semi-annual workshop held by CTP's consultants and clients. In the workshop, consultants provide training programs requested by their clients and share their know-how on

---

[10]Great Place to Work® Institute Japan website (https://hatarakigai.info/ranking/japan/). Present-CTP has been nominated for the best company award thrice times in the past.

facilitation-driven consulting, while clients introduce their business cases and experiences to CTP consultants. This is not only the place for consultants to acquire new knowledge but also for clients to learn the know-how and tacit knowledge that CTP possesses. The clients and Present-CTP create training programs together through mutual learning, as one team, and the team offer workshops for training and development to other CTP clients. Likewise, CTP holds networking events to share its know-how on facilitation methods with its clients, while clients share their business agendas with CTP. This is an opportunity for CTP to acquire new knowledge and develop long-term relationships with its clients. Another knowledge creation system requires CTP to hold off-site meetings within the company every year. For start-up members of Present-CTP, the off-site meeting was the place to begin building the new CTP in 2006. As of 2019, CTP features over 100 consultants, but still holds this meeting as an annual brain-storming session on how to improve the company with the help of all consultants. The organizer is selected through an open allocation system and can decide the topic of the off-site meeting for that year. For example, the Cambridge Grand Prix (the so-called C-1 Grand Prix) was held in 2016. The topic of the meeting was making No. 1 products that CTP can be proud of. CTP consultants published books on facilitation methods for meetings and developing project leaders, and released mobile applications related to their business. This meeting is where new knowledge and business is generated for Present-CTP.

10.5.3.2 Performance Appraisal

Present-CTP customized the performance assessment criteria based on the concepts inherited from the Global-CTP era because the old system was unfit for Present-CTP's business size. A resource advisor is assigned to each junior consultant, similar to the practice in Global-CTP, and its roles have been retained. The company carries out a two-phase performance appraisal process: project review and annual review, as in the Global-CTP era. Consultants receive feedback on their daily project business every day from their project members. This is on-the-job training for project facilitation management. The result of the daily feedback is reflected in the project review. Present-CTP devotes plenty of time to close communication with project members and leaders to cultivate consultants who can embody its culture and business model. More concretely, consultants are reviewed by a performance evaluation (comprehensive evaluation of quality, productivity, behavior, and training experience) and competency evaluation (logical thinking, delivery, project management, and task domain and IT knowledge). A project review is conducted four times a year. Based on the results of the project review, the management executives of the company discuss each consultant's performance until they reach an agreement. The annual review is used to assess the results of the project reviews and the "Plus One Activity" (more on this later). In this manner, Present-CTP spends 84 days providing feedback to each employee in order to enhance each consultant's abilities.

10.5.3.3 Incentives

Global-CTP had a culture in which people's efforts, ideas, and achievements were recognized. CTP offered a variety of recognition programs such as the Recognize and Value Everyone (RAVE) cards, Employee Appreciation Day (EAD), and awards that recognized the contributions of individuals and teams. Present-CTP honors exceptional individuals as most valuable person (MVP) and teams (MVT). Moreover, the company encourages consultants to carry out "Plus One Activity" to contribute to the company. For example, Present-CTP began a workout activity in 2017 as a plus one activity. In this activity, top executives make decisions and take quick actions with budgets for ideas proposed by consultants for improving CTP. A bonus is paid for individual performance and profit share. In the assessment of the bonus, the company reviews each consultant's "Plus One Activity," which includes contributions for the improvement of CTP's brand (e.g., system development within the company). For example, one consultant's individual performance result was 309%. He had developed hands-on training programs on technical skills for junior consultants and a knowledge creation server through which consultants could conduct their search not only on the knowledge server but also on the intranet, emails, and social networking service (SNS) chats. His contribution was to integrate the knowledge server and other media within the company.

The knowledge creation system at Present-CTP may be summarized by using the socialization, externalization, combination, and internalization (SECI) model as follows (Fig. 10.1).

Figure 10.1 shows the knowledge creation system at Present-CTP. Regarding the process of socialization, the company uses a resource manager or advisor program to train consultants as professionals with a deep understanding of its culture and business model. It also works as a way to transfer tacit knowledge from senior to junior consultants through on-the-job training and collaborative work. This program also contributes to externalizing tacit knowledge into explicit knowledge. The externalization process works by conceptualizing and describing the culture and business model of the company into the code of conduct for consultations; however, in order to embody them, CTP has a resource manager program and has developed 250 training programs to transfer tacit and explicit knowledge from senior to junior consultants. Regarding the combination process, Present-CTP adapts the idiosyncratic facilitation-driven consulting methodology to provide services for non-profit organizations and training programs for project facilitators to its clients. In addition, the company elevated its knowledge server by integrating the server and other media within the company to enable searches for internal knowledge in a broad context. With regard to the internalization process, the company holds CPFA and networking activities with its clients. Moreover, the company works aggressively to create new knowledge through plus one activities including workout activities and off-site meetings within the company. As such, it thoroughly manages its knowledge creation system.

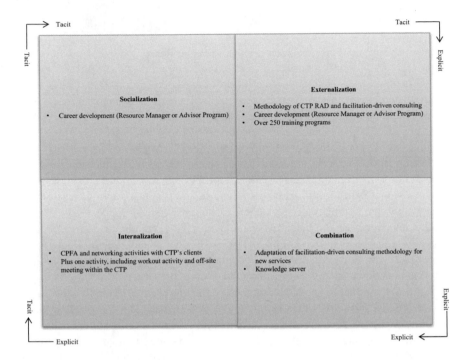

Fig. 10.1 Knowledge creation system at present-CTP. *Source* Author-made, based on Nonaka and Takeuchi's (1995) SECI model

Present-CTP plans to globalize its business. The company primarily offers facilitation-driven consulting services to Japanese MNCs in Japan but its global clients have requested services for their overseas operations. In addition, Present-CTP would like to attract talented new university graduates as candidates, but is struggling to do so because many are interested in working abroad. As a first step toward globalizing its business, the company is building collaborative networks with companies in North America, Southeast Asia, and China. It is using these networks to deliver services to Japanese MNCs in foreign countries with the aim of setting up its local operations.

Moreover, Present-CTP offers project-based consulting services at the client's company. As I noted earlier, the company published books on their facilitation-driven methodology as its original services developed by evolving Global-CTP's methodology. CTP offers learning and training for facilitation-driven consulting with its clients (e.g., the CPFA), but it also began providing training and education in facilitation consulting services to meet its clients' growing demand. This service is employed to build courses at corporate universities now.

10.6 Concluding Remarks

This study explored the role played by CSMs and HRM practices in inter- and intra-unit knowledge flows and creation systems in MNCs. Previous studies on MNC knowledge transfer and the role of HRM practices showed that MNCs should use CSMs to eliminate barriers to knowledge transfer and sharing among inter- and intra-organizations and to develop interpersonal networks to enhance communication and smooth knowledge transfer within the group. In the context of HRM, human capital theory shows that the value of corporate human capital can be increased through HRM practices. Research on social capital theory emphasizes that firms should encourage employees to learn with their colleagues and build social ties through shared corporate philosophy, project groups, and participation in international training and corporate programs. These two approaches indicate the importance of HR configurations and training and development, along with the related HRM practices such as performance appraisal, incentives, and career development, for encouraging employees to develop knowledge and social ties with colleagues. The relationships between knowledge transfer, socialization mechanisms, and HRM practices have been primarily addressed via a quantitative approach. However, to interpret these relationships, a qualitative approach is appropriate and has many advantages. Therefore, in this study, I adopted a qualitative approach and a longitudinal case study focusing on one PSF.

The results of the case study suggest the following implications. CTP was established with an unprecedented business model during a paradigm shift in the consulting industry driven by the advancement of IT. At Global-CTP, sharing the corporate philosophy within the group served as an infrastructure for transferring knowledge such as their unique business model, Fixed Time/Fixed Price based on the CTP RAD, and HRM practices, from the HQs to the Japanese subsidiary. Global-CTP globalized its business by replicating the business model for all operations. Moreover, the CSM played a role in facilitating collaborations among operations. These results accord with previous studies (Ghoshal and Bartlett 1988; Tsai and Ghoshal 1998; Lengnick-Hall and Lengnick-Hall 2003; Björkman et al. 2004). The distinctiveness of Global-CTP was that its business model and consulting methodology could be assimilated with its corporate culture. This was the main reason Novell-CTP Japan spun off from the Novell group. HRM practices were designed for practicing this idiosyncratic business model. In particular, training programs played a key role in transferring Global-CTP's competitiveness to the Japanese subsidiary. In these programs, understanding, sharing, and cultivating the HQs' culture and business model were emphasized to deliver high-quality services in each country. This allowed for collaboration between other subsidiaries' consultants and the Japanese subsidiary to support its business and to transfer the knowledge and technology that the Japanese subsidiary did not possess.

Moreover, Global-CTP had a resource manager program that enhanced tacit and explicit knowledge transfer from senior to junior consultants within the group. Career development was designed to develop "Cambridge Consultants."

Performance appraisal was designed for the generation of high project success rates, which involved giving feedback to consultants on their projects in order to evaluate and develop consulting abilities for business model delivery. This practice also encouraged consultants to collaborate with colleagues by engaging in team work. These practices were based on mutual and complementary relationships in line with previous studies (Tsai and Ghoshal 1998; Björkman et al. 2004; Kang et al. 2007; Kaše et al. 2009; Yamao et al. 2009).

Present-CTP evolved the facilitation-driven consulting approach into a methodology. This approach can be interpreted as knowledge creation at Present-CTP. Although the proposed approach differentiated the company from its competitors in Japan during the Global-CTP era, Present-CTP now offers a new service on facilitation consulting for training and education along with standard consulting services. Based on this methodology, the company enlarged its client base to non-profit organizations. HRM practices are designed for creating new knowledge and enhancing the CTP business model and methodology. As I explained earlier, Present-CTP has a well-integrated knowledge creation system that may be explained by the SECI model. New knowledge is created through workshops and by networking with Present-CTP's clients through training programs that are assimilated with its culture and business model. These activities aim to acquire new outside knowledge. In addition, new knowledge is created through HRM practices inherited from Global-CTP and those developed at Present-CTP to encourage employees to create knowledge through the off-site meeting and Plus One Activity at Present-CTP as part of training and development. Likewise, performance appraisal is designed to enhance consultants' project knowledge and knowledge creation by giving feedback on their projects on an average of 84 days per employee. Incentive schemes have also been designed to encourage consultants to create new knowledge.

One of the contributions of this study is that it shows the processes through which one PSF has implemented corporate socialization within the company and how it has gone about creating new knowledge in practice in terms of the relationships between knowledge transfer and creation, and HRM. The findings from this case study are in line with previous studies on the relationships between knowledge transfer and HRM.

This study has several limitations. First, it has biased research objectives in the longitudinal case study. The study's objectives are limited to Japan because Global-CTP and Novell-CTP are no longer in business. I was therefore unable to access anyone who worked at the HQs of Global-CTP and Novell-CTP. Moreover, I cannot show inter-unit knowledge transfer and sharing at Present-CTP because the company has not globalized its business yet. Second, this study's external validity could be enhanced by conducting comparative case studies involving both foreign and Japanese PSFs, as well as PSFs and manufacturing MNCs, to clarify the uniqueness and commonalities of knowledge transfer and creation, socialization mechanisms, and HRM practices. These are future research agendas that should be pursued.

References

Adler, P. S., & Kwon, S.-W. (2002). Social capital: Prospects for a new concept. *The Academy of Management Review, 27*(1), 17–40.

Alvesson, M. (2004). *Knowledge work and knowledge-intensive firms*. New York: Oxford University Press.

Ambile, T. M., Baker, G. P., & Beer, M. (1995) Cambridge technology partners (A), HBS case 9-496-005 (Rev. April 11, 1996). Harvard Business School.

Argote, L., & Ingram, P. (2000). Knowledge transfer: A basis for competitive advantage in firms. *Organizational Behavior and Human Decision Processes, 82*(1), 150–169.

Bartlett, C. A., & Ghoshal, S. (1989). *Managing across borders: The transnational solution.* Boston, Mass: Harvard Business School Press.

Birkinshaw, J., & Hood, N. (1998). Multinational subsidiary evolution: Capability and character change in foreign-owned subsidiary companies. *Academy of Management Review, 23*(4), 773–795.

Björkman, I., Barner-Rasmussen, W., & Li, L. (2004). Managing knowledge transfer in MNCs: The impact of headquarters control mechanisms. *Journal of International Business Studies, 35* (5), 443–455.

Boxall, P. (1996). The strategic HRM debate and the resource-based view of the firm. *Human Resource Management Journal, 6*(3), 59–75.

Burt, R. S. (1992). *Structural holes: The social structure of competition.* Cambridge, Mass: Harvard University Press.

Burt, R. S. (1997). The contingent value of social capital. *Administrative Science Quarterly, 42*(2), 339–365.

Cabrera, E. F., & Cabrera, A. (2005). Fostering knowledge sharing through people management practices. *The International Journal of Human Resource Management, 16*(5), 720–735.

Cohen, W., & Levinthal, D. (1990). Absorptive capacity: A new perspective on learning and innovation. *Administrative Science Quarterly, 35*(1), 128–152.

Coleman, J. S. (1988). Social capital in the creation of human capital. *The American Journal of Sociology, 94,* 95–120.

Collins, C. J., & Smith, K. G. (2006). Knowledge exchange and combination; The role of human resource practices in the performance of high technology firms. *Academy of Management Journal, 49*(3), 544–560.

Ghoshal, S., & Bartlett, C. A. (1988). Creation, adoption, and diffusion of innovations by subsidiaries of multinational corporations. *Journal of International Business Studies, 19*(3), 365–388.

Ghoshal, S., & Bartlett, C. A. (1990). The multinational corporation as an interorganizational network. *Academy of Management, 15*(4), 603–625.

Gompers, P., & Catherine, C. (1997a). Cambridge technology partners: 1991 start up. HBS case9-298-044 (Rev. November 23, 1998). Harvard Business School.

Gompers, P., & Catherine, C. (1997b). *Cambridge technology partners: Corporate venturing.* HBS case9-297-033 (Rev. January 8, 1999). Harvard Business School.

Grant, R. M. (1996). Toward a knowledge-based theory of the firm. *Strategic Management Journal, 17*(Issue S2), 109–122.

Gupta, A. K., & Govindarajan, V. (2000). Knowledge flows within multinational corporations. *Strategic Management Journal, 21*(4), 473–496.

Gupta, A. K., & Singhal, A. (1993). Managing human resources for innovation and creativity. *Research Technology Management, 36*(3), 41–48.

Hansen, M. T., Nohria, N., & Tierney, T. (1999). What's your strategy for making knowledge? *Harvard Business Review, 72*(2), 106–116.

Hatch, N. W., & Dyer, J. H. (2004). Human capital and learning as a source of sustainable competitive advantage. *Strategic Management Journal, 25*(12), 1155–1178.

Hedlund, G. (1986). The hypermodern MNC: A heterarchy? *Human Resource Management, 25* (1), 9–35.

Kang, S.-C., Morris, S. S., & Snell, S. A. (2007). Relational archetypes, organizational learning, and value creation: Extending the human resource architecture. *Academy of Management Review, 32*(1), 236–256.

Kaše, R., Paauwe, J., & Zupan, N. (2009). HR practices, interpersonal relations, and intrafirm knowledge transfer in knowledge-intensive firms: A social network perspective. *Human Resource Management, 48*(4), 615–639.

Kostova, T., & Ross, K. (2003). Social capital in multinational corporations and a micro-macro model of its formation. *Academy of Management Review, 28*(2), 297–317.

Kubr, M. (2002). *Management consulting: A guide to the profession* (4th ed.). Geneva: International Labour Office.

Lado, A. A., & Wilson, M. C. (1994). Human resource systems and sustained competitive advantage: A competency-based perspective. *Academy of Management Review, 19*(4), 699–727.

Lengnick-Hall, M. L., & Lengnick-Hall, C. A. (2003). HR's role in building relationship networks. *Academy of Management Executive, 17*(4), 53–63.

Lepak, D. P., & Snell, S. A. (1999). The human resource architecture: Toward a theory of human capital allocation and development. *The Academy of Management Review, 24*(1), 31–48.

Lepak, D. P., & Snell, S. A. (2002). Examining the human resource architecture: The relationship among human capital, employment, and human resource configurations. *Journal of Management, 28*(4), 517–543.

Lødendahl, B. R. (2005). *Strategic management of professional service firms* (3rd ed.). Copenhagen, DK: Copenhagen Business School Press.

Maister, D. H. (2003). *Managing the professional service firm.* New York: Free Press.

Mäkelä, K., & Brewster, C. (2009). Interunit interaction contexts, interpersonal social capital, and the differing levels of knowledge sharing. *Human Resource Management, 48*(4), 591–613.

Martin, J. (1991). *Rapid application development.* New York: Mcmillan Publishing.

Minbaeva, D., Pedersen, T., Bjorkman, I., Fey, C., & Park, H. (2003). MNC knowledge transfer, subsidiary absorptive capacity and HRM. *Journal of International Business Studies, 34*(6), 586–599.

Minvaeba, D., Pedersen, T., Björkman, I., Fey, C. F., & Park, H. J. (2014). MNC knowledge transfer, subsidiary absorptive capacity and HRM. *Journal of International Business Studies, 45*(1), 38–51.

Minvaeba, D. B. (2005). HRM practices and MNC knowledge transfer. *Personnel Review, 34*(1), 125–144.

Nahapiet, J., & Ghoshal, S. (1998). Social capital, intellectual capital, and the organizational advantage. *Academy of Management Review, 23*(2), 242–266.

Nishii, S. (2013). *Chishiki shuuyaku gata kigyo no global senryaku to business model (Global strategy and business model in KIFs: The generation, development, and evolution in the management consulting firms).* Doyukan Publishing Company (in Japanese).

Nohria, N., & Ghoshal, S. (1994). Differentiated fit and shared values: Alternatives for managing headquarters-subsidiary relations. *Strategic Management Journal, 15*(6), 491–502.

Nonaka, I. (1994). A dynamic theory of organizational knowledge creation. *Organization Science, 5*(1), 14–37.

Nonaka, I., & Takeuchi, H. (1995). *The knowledge-creating company: How Japanese companies create the dynamics of innovation.* New York: Oxford University Press.

Peltokorpi, V., & Vaara, E. (2014). Knowledge transfer in multinational corporations: Productive and counterproductive effects of language-sensitive recruitment. *Journal of International Business Studies, 45*(5), 600–622.

Pettigrew, A. M. (1990). Longitudinal field research on change: Theory and practice. *Organization Science, 1*(3), 267–292.

Punch, K. F. (2005). *Introduction to social research: Quantitative and qualitative approaches* (2nd ed.). London: Sage.

Reed, K. K., Lubatkin, M., & Srinivasa, N. (2006). Proposing and testing an intellectual capital-based view of the firm. *Journal of Management Studies, 43*(4), 867–893.

Senge, P. (1990). *The fifth discipline: The art and practice of the learning organization.* New York: Doubleday/Currency.

Szulanski, G. (1996). Exploring internal stickiness: Impediments to the transfer of best practice within the firm. *Strategic Management Journal, 17*(Winter Special Issue), 27–43.

Szulanski, G. (2000). The process of knowledge transfer: A diachronic analysis of stickiness. *Organizational Behavior and Human Decision Processes, 82*(1), 9–27.

Tsai, W. (2001). Knowledge transfer in intraorganizational networks: Effects of network position and absorptive capacity on business unit innovation and performance. *The Academy of Management Journal, 44*(5), 996–1004.

Tsai, W., & Ghoshal, S. (1998). Social capital and value creation: The role of intrafirm networks. *Academy of Management Journal, 41*(Issue 4), 464–476.

Tsoukas, H. (1996). The firm as a distributed knowledge system: A constructionist approach. *Strategic Management Journal, 17*(S2), 11–25.

Watkins, K. E., & Marsick, V. J. (1993). *Sculpting the learning organization: Lessons in the art and science of systemic change.* San Francisco: Jossey-Bass.

Yamao, S., De Cieri, H., & Hutchings, K. (2009). Transferring subsidiary knowledge to global headquarters: Subsidiary senior executives' perceptions of the role of HR configurations in the development of knowledge stocks. *Human Resource Management, 48*(4), 531–554.

Youndt, M. A., Subrammaniam, M., & Snell, S. A. (2004). Intellectual capital profiles: An examination of investments and returns. *Journal of Management Studies, 41*(2), 335–361.

Zahra, S., & George, G. (2002). Absorptive capacity: A review, reconceptualization, and extension. *Academy of Management, 27*(2), 185–203.

Zander, U., & Kogut, B. (1995). Knowledge and the speed of the transfer and imitation of organizational capabilities: An empirical test. *Organization Science, 6*(1), 76–92.

Tamiko Kasahara (Ph.D., Kobe University of Commerce) is lecturer in International Business at School of Management and Information, University of Shizuoka, Japan. She received her Ph.D. and M.S. From Kobe University of Commerce which was integrated into University of Hyogo in 2004. Her primary research interest is in the area of global talent management at multinational corporations, recently focusing on professional service firms. Her book "Global Human Resource Management in Japanese Multinational Corporations (in Japanese, Hakuto Shobo, Tokyo, 2014)" won the Best Book Award for Young Researchers from Japan Academy of Multinational Enterprises (JAME), 8th Annual Conference in 2015. She is a board member of JAME.

Chapter 11
Redefining the Internationalization of R&D Activities: How Far Have the Firms' R&D Members of US and Japanese Companies Been Diversified?

Takabumi Hayashi

Abstract This paper examines that the role played by foreign researchers and engineers engaged in R&D activities in the US and the overseas R&D activities of US multinational corporations are no longer negligible. This paper focuses on the fact that foreign scientists and engineers residing in the US make considerable contributions as inventors of US patents, and it examines the extent to which the internationalization of R&D by US companies would result if the outcomes of their activities in the US were included in the internationalization of R&D. Specifically, the level of internationalization of R&D is verified by studying the nationality of the inventor's institution (i.e., IBM), which has consistently been the top US patent collector from 1993 to 2017. Additionally, looking at Canon Inc., which has been exceptionally ranked in the top five from 1985 to 2017 in both the US and Japan, this paper examines the inventors and the nationalities of the organizations to which the inventors work, thereby confirming the internationalization of the company's R&D in the same way. Finally, we examine how much the internationalization of R&D activities differs between IBM and Canon.

Keywords Internationalization of R&D · US Patents · Japanese Patents · IBM Corp. Canon Inc. · Nationality of the Inventor's Affiliation · Foreign Researchers in the US

T. Hayashi (✉)
Rikkyo University, Tokyo, Japan
e-mail: takabumi@rikkyo.ac.jp

Tokyo Fuji University, Tokyo, Japan

© Springer Nature Singapore Pte Ltd. 2019
J. Cantwell and T. Hayashi (eds.), *Paradigm Shift in Technologies and Innovation Systems*, https://doi.org/10.1007/978-981-32-9350-2_11

11.1   Introduction

With new digital technology innovations, information and knowledge is now mobilized and shared across time and space more than ever before. This has not only created a technological foundation for promoting the international and geographical divergence of scientific and technological knowledge and knowledge production capability, it also has contributed to the concentration of research and development (R&D) capabilities in specific enterprises. These changes in the technological environment and the globalization of the market have led to the internationalization of production activities by companies in each country and the internationalization and transformation of R&D activities. Under these circumstances, conventional internationalization of R&D has been vigorously discussed, as companies, particularly multinational enterprises, have become more inclined to build R&D centers overseas.

Discussion of internationalization of R&D in the 1970–1980s involved empirical research on overseas R&D activities, mainly by US multinational corporations (Mansfield and Romeo et al. 1979, 1984; Mansfield and Teece et al. 1979; Creamer 1976; Behrman and Fischer 1980). However, much debate over international R&D activities during this period focused on the improvement of R&D capability abroad from technology transfer from multinational headquarters to overseas affiliations (Hirota 1985, 1986; Komoda 1987; Hayashi 1987, 1989). However, it was also learned that IBM Corp., for instance, had already established an international R&D system and that reverse technology transfers of new knowledge from overseas R&D operations in developed countries to their home countries were taking place (Leroy 1978; Mansfield and Romeo et al. 1984; Shanrokhi 1984; Hayashi 1989).

Since the 1990s, there have been many studies on R&D activities at overseas R&D centers by US and European-based multinational corporations (Patel and Pavit 1991; Pearce and Singh 1992; Cantwell 1995). More detailed research on overseas R&D activities by Japanese companies has also been published as a book (Iwata 1994, 2007; Takahashi 2000; Nakahara 2001). Moreover, these issues have been increasingly referred to as "R&D linkage" or "R&D networks," having been formed as a result of the interactive flow of technological knowledge between the headquarters and oversea R&D centers and among overseas R&D sites (Asakawa 1996, 2001; Medcof 2001; Roberts 2001; Serapio and Hayashi 2004; Hayashi and Serapio 2006).

As the creation and transfer of scientific and technological knowledge progressed through the movement of people, the geographical divergence of R&D capabilities has advanced not only among developed countries but also on a global scale, including emerging countries. In particular, the issue of "brain circulation from brain drain" from the emerging countries has drawn attention (Saxenian 2005; Hayashi 2007).

As of 2010, the total R&D expenditures of US multinationals was $252.0B, of which R&D expenditures by overseas majority-owned subsidiaries was $39.5B. Consequently, R&D expenditures at the overseas subsidiaries accounted for about 16% of the total (NSB and S&E Indicators 2014). The same term for 1985 was 6% (NSB 1996). As reflected in the increase of about 10 points in overseas R&D expenditures by US multinational corporations in the last 25 yrs, from the input perspective, it can be said that the internationalization of R&D activities has strengthened. This trend is particularly evident in countries that are more dependent on foreign markets than on domestic markets, such as Switzerland with 130% overseas R&D ratio to the total R&D expenses, Sweden with more than 40%, and Germany with more than 20% as of 2007 (European Commission 2012).

The research on the relationship between the utilization of overseas R&D capabilities of major Japanese companies in Europe and the US and the R&D capabilities in Japan has also reflected the trend of major Japanese companies establishing overseas R&D departments as global R&D centers (Asakawa 2001; Song et al. 2011).

In the 2010s, discussions emerged concerning the internationalization of R&D activities and the improvement of R&D capabilities of multinational companies' (MNC) overseas subsidiaries (Iguchi 2011; Cantwell and Mudambi 2005). However, despite the ongoing international development of such R&D activities, the international development of R&D activities by Japanese overseas subsidiaries has not progressed much. R&D activities by these companies have been limited to joint research within affiliated companies in Japan (Cantwell and Zhang 2006).

Similarly, factors, such as scale and scope economies, coordination costs associated with overseas R&D activities, and the degree of protection of intellectual property rights have been defined as factors that cannot be ignored when R&D activities are biased in the home country (Belderbos et al. 2013). Analysis of US pharmaceutical companies has also indicated that firms transferring knowledge from select locations may be able to achieve the necessary foci and internal resources to successfully integrate new knowledge gained from these resources for innovation (Kotabe et al. 2007, p. 275).

Conventional research on the internationalization of R&D activities has predominantly focused on inputs, such as R&D expenditures, the number of R&D personnel involved in R&D activities, outputs as results, the quality level of R&D activities based on field surveys, and management and organization of knowledge creation. However, with these viewpoints, it is important to note that the advancement of internet search engines has enabled the transfer and creation of knowledge beyond time and space. Simultaneously, the importance of software in the technological architecture of industrial infrastructure has increased. Thus, the paradigm shift in technological structures and systems are a factor that decentralizes the geographic transfer and creation of knowledge. However, it is also a factor that concentrates R&D capabilities of specific companies and countries using it strategically (Hayashi 2018).

This paper examines, based on the above points, that the role played by foreign researchers and engineers engaged in R&D activities in the US and the overseas

R&D activities of US multinational corporations is no longer negligible. If scientists and engineers from overseas continue to stay in the US after studying in the home country of an MNC (i.e., US) and continue to engage in R&D activities as highly skilled professionals in the field of natural sciences, the need for US MNCs to set up and strengthen R&D facilities overseas beyond their borders in search of superior overseas R&D personnel is reduced.

This paper focuses on the fact that foreign scientists and engineers residing in the US make considerable contributions as inventors of US patents, and it examines the extent to which the internationalization of R&D by US companies would result if the outcomes of their activities in the US were included in the internationalization of R&D.

Specifically, the level of internationalization of R&D is verified by studying the nationality of the inventor's institution (i.e., IBM), which has consistently been the top US patent collector for 24 yrs (1993–2017). Additionally, looking at Canon, Inc., which has been ranked in the top five for 32 yrs (1985–2017) in both the US and Japan, this paper examines the inventors and the nationalities of the organizations to which the inventors work, thereby confirming the internationalization of the company's R&D in the same way. Finally, we examine how much the internationalization of R&D activities differs between IBM and Canon.

11.2   Methodology

As a method of quantitatively examining the degree of internationalization of R&D activities by enterprises, there have been many viewpoints provided about expenditures, numbers of R&D staff, etc., of specific enterprises and industries overseas and from outputs, such as S&T papers and patents. This paper examines the latter from the perspective of output. For patent data, this paper focuses on data from the US Patent and Trademark Office (USPTO) and the USPATFUL database. Additionally, as a new attempt to examine the internationalization of R&D by US companies, various datasets regarding foreign scientists and engineers residing in the US is relied upon for S&E indicators of the National Science Board.

The primary purpose of examining patent data is to identify the nationality of the institutions to which inventors belong from the patent specification. Nationality is therefore indicated not by the passport nationality of these inventors but by the nationality of their institutions. However, via this method, if a researcher or engineer belonging to an overseas R&D department of a US MNC is the only subject of R&D internationalization, and if a foreign researcher or engineer belonging to an R&D facility in the US is the inventor, it is treated as an R&D activity in the US. Thus, these R&D activities are excluded from the scope of globalization.

In this paper, we examine the degree of internationalization of R&D for IBM when considering the number of foreign scientists and engineers engaged in R&D activities at US facilities. Similarly, we compare the degree of internationalization of R&D of IBM with Canon by examining patents obtained in the US and Japan.

11.3 Redefinition of Internationalization of R&D Activities by US Companies

11.3.1 Nationalities of Organizational Affiliations and Passport Nationalities of Foreign Scientists and Engineers in the US

How many foreign scientists and engineers are engaged in R&D institutions in the US? Around 2,000, more than one-third of Silicon Valley's highly skilled engineers, were mainly Asian foreign nationals (Saxenian 2005). According to the research by Sana (2010), of the number of employed scientists and engineers in US, foreign born accounted for 13.3% in 1994, and 23.4% in 2006 (Sana 2010, 805–806).

When estimating the number of foreign scientists and engineers engaged in advanced R&D work, the number of students studying natural sciences at US universities, especially graduate schools, from overseas can be used as a reference, because they tend to stay in the US after obtaining degrees, especially doctorates, in these sciences. For example, 70% of the 31,600 persons who received a doctorate in science and engineering (S&E) in 2005 continued to reside in the US in 2015, 10 yrs later. In particular, 90% of the 10,700 Chinese doctoral degree recipients, 85% of the 3,500 Indians, and 56% of the 3,000 South Koreans stayed in the country 10 yrs longer, engaged in related work (NSB and S&E Indicators 2018). When they still want to keep residing in the US, after finishing master or doctorate course, they apply H1-B visa. The number of H1-B visa approved in the US has been around 200–300 thousands after 2000[1].

The percentage of those born overseas among those engaged in S&E related work in the US in 2015 was 21.1% for bachelor's degrees, 38.2% for master's degrees, and 45.3% for doctoral degrees. The higher the degree, the more likely it is they will stay in the US and engage in S&E related jobs (see Table 11.1). Moreover, the ratio is gradually increasing, as shown in Table 11.1. It can be assumed that non-negligible number of authors of scientific papers and inventors of patented technologies published in the names of research institutions of US universities and companies were accounted for by researchers and engineers born outside the US.

In the case of computer and mathematics fields in business/industrial sectors, the percentage of persons with master's and doctoral degrees engaged in research and engineering work born overseas was about 50.4 and 58.2%, respectively, as of 2013 (Table 11.2).

Therefore, if approximately 30,000 foreign students in the US receive a degree in natural sciences every year, and about 70% continue to stay for 10 + yrs to engage in R&D work as highly skilled professionals in the field of natural sciences, the ratio will probably increase further (NSB 2018).

---

[1]The number of H1-B visa approved includes that from overseas applicants, of which India accounts for about 71% in 2015 (USCIS 2017).

Table 11.1 Foreign-born workers in science and engineering occupations in the US by education level (percent)

	1993	2003		2015	
	SESTAT	SESTAT	ACS	NSSCG	ACS
Total college graduates	15.8	22.6	25.2	30.0	28.8
(Bachelor's degrees)	11.4	16.4	18.7	21.2	21.1
(Master's degrees)	20.7	29.4	32.0	40.6	38.2
(Ph.D.s)	26.8	36.4	38.7	42.3	45.3

Note Data are cited from the Scientist and Engineers Statistical Data (SESTST), National Sur-vey of College Graduates (NSCG), and American Community Survey (ACS)
Source NSB, Science & Engineering Indicators (2018)

Table 11.2 Percentage share of non native-born* employed scientists and engineers in the US by education level in business/industrial sectors and computer/mathematical scientists of these sectors: 2013

	Business/industry	(Computer/mathematical)
All college graduates	25.8	30.7
Bachelor's	17.9	21.7
Master's	38.2	50.4
Doctorate	46.5	58.2

Note *Non-native born means naturalized US S&Es and non-US S&Es residing in the US.
Source NSF, *Scientists and Engineers Statistics Data System Surveys*, Survey (2013)

11.3.2 Nationalities of Institutions to Which Inventors Belong and Their Passport Nationalities

Since 2007, there have been between 180,000 and 350,000 US approvals (about 197,000 in 2017) (US Citizenship and Immigration Services 2018) for H1-B visas issued to highly skilled foreign workers, including those who continue to work in R&D-related institutions, including US companies, as postdoctoral fellows. The degree of internationalization of R&D differs considerably depending on the method of how to quantify nationalities of the inventors of patented technologies who engage in R&D-related work as highly skilled professionals under H1-B visas.

Even if the passport nationality of an inventor at a US company is other than US, it is indicated by the nationality of the organization to which the inventor belongs. This means two things. First, the need for cross-border R&D is lower for US companies than for companies in countries having few foreign scientists or engineers. The second point is that, if the number of US patents claimed by foreign researchers and engineers belonging to US companies in the US are included in the ratio of internationalization of R&D, the ratios of US companies increase considerably.

In the next section, we examine the degree of internationalization of R&D, based on the ratio of the US inventions of the top 35 US patenting companies, and the degree of internationalization of R&D, based on the abovementioned number of foreign researchers and engineers in the US, focusing on the case of IBM.

11.4 Internationalization of R&D Activities by the Top 35 US Patenting Companies and IBM

11.4.1 Degree of Internationalization of R&D of Top 35 US Companies with the Largest Number of US Patents

We examine trends in the ratio of overseas inventions by 35 US companies with the largest number of US patents. Figure 11.1 shows the average ratio of inventions outside the US for the top 35 US patenting companies and the same ratio of inventions for major information technology (IT) companies.

As shown in Fig. 11.1, the average ratio of inventions outside the US for the top 35 US companies rose from 15.2% in 2011 to 17.0% in 2015. Similarly, the ratio of IT-related companies, which account for a large portion of top companies, has risen at the 2015 level from the 2011 level, with the exception of GE. As described in the caption, the USPTO definition of the ratio of inventions outside the US is as follows; for each patent, the number of listed inventors who are identified as residing outside of the United States at the time of grant are divided by the total number of inventors for that patent to determine the foreign inventor share for that (one) patent.

The ratio of the overseas inventions of the top 35 US companies, shown in Fig. 11.1, rose from 15.2% of the 2011 average to 17.0% in 2015, whereas the ratio of domestic inventions declined from 84.8% in 2011 to 83.0%. If 38.2–46.5% of researchers and engineers with master's degrees or higher were foreign scientists and engineers in the US, these ratios corresponded to 31.7–38.6% of patents.

When accepting the definition of R&D internationalization is "born outside the US.", around 48.7 to 55.6%, adding 17.0 to 31.7 and 38.6%, in a broader sense, was the ratio of internationalized R&D in 2015. Additionally, the overseas invention ratio of the top nine IT-related US companies, excluding GM (8th place) shown in Fig. 11.1, increased from 12.1% in 2011 to 17.5% in 2015. In the case of foreign-born workers engaged in R&D at IT-related companies in the fields of computers and mathematics, the percentage of master's and doctoral degree holders was approximately 50.4 and 58.2%, respectively, as of 2013. In a broad sense, the ratio of internationalization of R&D in the same calculation method was around 54.0–59.7%, adding 17.5% to 36.5–42.2%.

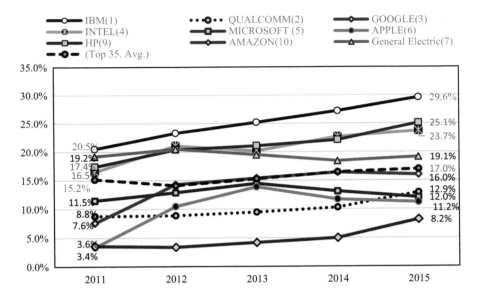

Fig. 11.1 Trends in the ratio of inventions outside the US by the top 35 US patenting companies and major information technology (IT) companies *Note 1* The USPTO definition of overseas invention outside the US is "for each patent, the number of listed inventors who are identified as residing outside of the United States at the time of grant are divided by the total number of inventors for that patent to determine the foreign inventor share for that (one) patent".
Note 2 Each numerical value attached to the company indicates the ranking order in terms of the number of US patents in 2015. Eighth is GM Global Technology.
Source USPTO

When we look at the trends in the overseas invention ratio of IBM, which ranks the top in the number of patents, the company's ratio of inventions outside the US rose from 20.5% in 2011 to 29.6% in 2015. In the next section, we examine in more detail how the degree of internationalization of R&D of IBM changes when the nationality of institutions to which joint inventors belong and foreign scientists and engineers in the US are taken into account.

11.4.2 Internationalization of R&D of IBM

In this section, we examine how the internationalization of R&D activities changes when looking at US patents when researchers and engineers belonging to the company are described as inventors. We consider the percentage of foreign researchers and engineers in the US and the nationalities of the institutions to which the first-named inventors and other co-inventors belong.

Table 11.3 shows the number of patents that the company applied for and obtained from the USPTO between 1980 and 2015, classified according to the

Table 11.3 Internationalization of R&D of IBM

		1980	1990	2000	2005	2010	2015
1=(2+3)	Number of US patents	399 (100.0)	649 (100.0)	2958 (100.0)	6852 (100.0)	7704 (100.0)	10411 (100.0)
2	Sole invention in the US: (2)/(1)	357 89.5%	545 84.0%	2492 84.3%	5534 80.8%	5550 72.0%	6671 64.1%
3= (4+5+6)	Overseas invention/ Total US patents: (3)/(1)	10.5%	16.0%	15.7%	19.2%	28.0%	35.9%
4	Overseas Sole invention: (4)/(1)	10.3%	12.5%	9.5%	12.0%	17.2%	19.0%
5	Joint invention between US and overseas: (5)/(1)	0.5%	2.6%	5.2%	7.0%	10.9%	16.3%
6	Joint invention between overseas: (6)/(1)	0%	0.9%	0.9%	1.6%	1.7%	3.9%
7	Number of nationality of overseas sole invention	5	8	12	18	24	26
8	Total number of nationality of overseas sole and joint invention, including US.	7	11	25	32	38	57

Note 1: The definition of Internationalization of R&D is the ratio of the number of patents which overseas inventors are listed, to the total number of patents.

Note 2: Total share of (4) (overseas sole invention), (5) (joint invention between US and overseas), and (6) (joint invention between overseas and overseas) is 39.2%, which is bigger than (3) (joint invention between US and overseas) in 2015. This is because the number of nationalities involving joint inventions is more than three affiliations' nationalities. For example, in the case of the joint invention by one US, one German, and one Israeli, the number is counted as one between US and Germany, one between US and Israel, and one between Germany and Israel. Accordingly, (5) (joint invention between US and overseas) is doubly counted. These overlapped numbers reach nearly 340.

Source Calculated from data by USPATFULL searching

nationality of the institution to which all the inventors belong, and it shows the ratio by year of publication.

In 1980, 89.5% of the company's US patents were based solely on the nationality of the US inventor, and the remaining 10.5% were from overseas (non-US) inventors. Of the 10.5% patents associated with this foreign inventor, 10.3% were patents issued by overseas sole inventors, and the remaining 0.5% (see Note[5]) were joint patents between US and overseas IBM. Of the company's US patents, the ratio of US IBM's sole inventions has declined steadily since then, reaching 64.1% in 2015. Conversely, as mentioned earlier, the ratio of patents by inventors belonging to overseas institutions, including joint inventors, to the number of US patents increased from 10.5% in 1980 to 35.9% in 2015.

Conversely, the ratio of patents by inventors belonging to overseas institutions, including joint inventors, to the number of US patents increased from 10.5% in 1980 to 35.9% in 2015. Particular attention should be paid to the fact that the number of nationalities of institutions to which inventors belong, including the US, listed in the lower column (7) of the same table, has gradually increased from 7 countries in 1980 to 57 countries in 2015. Looking at these trends in the company's international R&D activities from the perspective of output, it nearly agrees with the increasing trend of diversification and divergence of R&D capabilities examined in Chap. 4. Thus, IBM has strengthened its R&D capabilities by taking advantage of its superior brainpower, by diversifying its R&D, and by strengthening its joint international R&D system with other domestic and overseas institutions, centered on its intra-firm international R&D systems.

When the definition of internationalization of R&D is the ratio of foreign inventors of the total number of inventors, the internationalization of R&D was 29.6% (see Fig. 11.1). When the number of patents which are solely or jointly invented by overseas of the total number of patents, the ratio was 35.9%. The point to be kept in mind is that the number of nationalities of overseas IBM excluding the US to which foreign inventors belong increased from 6 in 1980 to 56 in 2015. As is revealed in Fig. 11.2, IBM has strengthened its R&D capabilities, leveraging globally dispersed and diversified R&D capabilities, centering on India and Israel among others.

Furthermore, when the number of US-patented inventions by foreign scientists and engineers belonging to US domestic institutions is included, we seek another degree of internationalization of R&D. As we have seen, approximately 50 and 58%

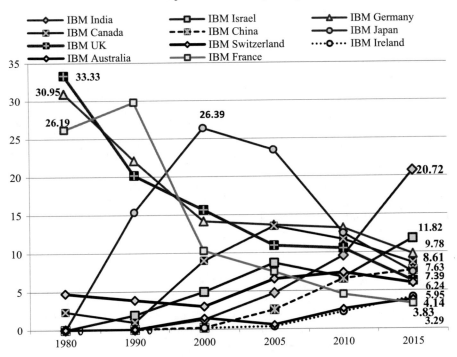

Fig. 11.2 The ratio of the number of US patent obtained by each major overseas IBM to the total number of US patents obtained by all overseas IBM *Note* Also refer to Notes of Table 11.2. *Source* Calculated from data by USPATFULL searching

of master's and doctoral degree recipients, respectively, in the US, specializing in computing and mathematics were born overseas, as of 2013 (NSF 2018). Therefore, assuming this ratio corresponds to the number of researchers and engineers involved in IBM's R&D department in the US, the company's 2015 US patents owned by foreign scientists and engineers belonging to their US R&D department were respectively 50–58% or 32.1–37.2% of the 64.1%. If the results of R&D activities by foreign researchers and engineers engaged in R&D in the US are counted as the internationalization of R&D activities, the internationalization of IBM's R&D activities by US-patented inventions was approximately 68–73%, comprising 35.9% of R&D activities outside the US and 32.1–37.2% of R&D activities of foreign researchers and engineers in the US. If this figure reflects the actual situation, the ratio of internationalization of R&D, based on the ratio of overseas inventions of IBM, as published by the USPTO, is significantly underestimated.

11.5   Internationalization of R&D Activities by Top 35 Japanese Patenting Companies and Canon Inc

11.5.1   Degree of Internationalization of R&D by Top 35 Japanese Companies with US Patents

Figure 11.3 shows trends in the ratio of inventions in the US to the average number of US patents held by the top 35 Japanese companies, those of major electronics companies, and those of the electrical equipment Japanese manufacturer, Hitachi, Ltd. However, the ratio of inventions in the US to the number of US patents of these Japanese companies accounts for a part of the ratio in foreign countries, excluding Japan. Therefore, the former value is smaller than the latter. However, considering that the US R&D department of these companies tends to occupy a more important position among Japanese overseas R&D departments and that they make patent applications in the US, the difference between the US invention ratio and the overseas invention ratio among Japanese companies' US patents may not be significant. This point will be examined in the case of Canon, described later.

Importantly, the average US invention ratio of 35 Japanese companies has not risen from 3.7% since 2011. It declined slightly to 3.3% in 2015. Among the top 35 Japanese companies, the company with the highest US invention ratio was Honda, ranked 12th in the US patent number by Japanese companies in 2015, at 16.4%. However, the ratio of Honda has declined from 20.9%, reached in 2011. Of the nine major Japanese electronics and electronics companies shown in Fig. 11.2, Sony Corp. showed the highest US invention ratio. The same ratio in 2015 was the same at the 9.1% level. Furthermore, it cannot be said that the ratio has been on an upward trend since 2011. Fujitsu is the only Japanese company in the electronics field that continuously increased its ratio of inventions in the US from 4.0% in 2011 to 8.3% in 2015. However, the ratio is still below 10%.

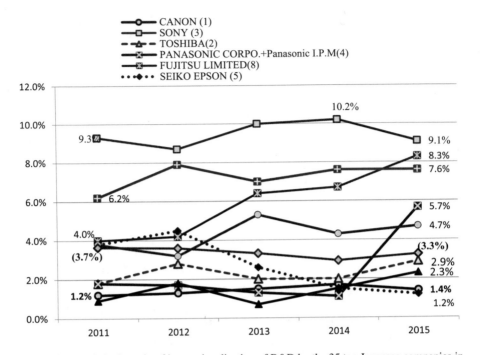

Fig. 11.3 Trends in the ratio of internationalization of R&D by the 35 top Japanese companies in the number of US patents *Note 1*: The definition of the ratio of US inventions includes the nationality of affiliations to which first-named inventors belong is US.
Note 2: Each number attached to the company is a ranking order in terms of the number of US patents in 2015. The ratio of Toyota, 7th place, was 0.9% in 2015. Brother Industries was 9th place at 0%, and Semiconductor Energy Institutes was 10th place, also at 0%.
Source: Calculated from the USPTO database

Therefore, the ratio of inventions in the US by the top eight Japanese electrical and electronics companies was clearly lower than that of the eight US IT companies shown in Fig. 11.1, where the ratio of inventions overseas was 10% or more.

In 2015, Canon was the top Japanese company in terms of the number of US patents granted. However, the company's US invention ratio by the definition of USPTO, was only 1.4%. Furthermore, it has not risen from 1.3%, reached in 2011. In the next section, we analyze in more detail the degree of internationalization of R&D by considering the number of patents solely or jointly invented by foreign inventors in the case of Canon Inc., which ranks at the top for the number of patents granted as a Japanese company both in Japan and the US.

The ratio of the number of patents solely or jointly invented by overseas inventors to the number of patents granted to Japanese companies in Japan was even lower than the ratio of the number of overseas inventions, excluding Japan, to the number of US patents granted them. The number of patent applications filed overseas, not just by Japanese companies, tended to be limited to innovative technologies with more strategic significance, mainly stipulated in the costs of applications, registration, renewal, etc.

With the lower cost of domestic patents in Japan, the number of patents for defense and domestic applications includes those with an uncertain future, regarding whether or not they will be used. Thus, the number of applications for patents overseas tends to decrease relative to the number of domestic patents. Therefore, the ratio of the number of patents invented outside the US to the number of US domestic patents brought by US companies was lower than that of patents filed outside the US to the number of patents applied for in the US to major overseas countries.

Therefore, when comparing the ratios of Figs. 11.1 and 11.2 on the same basis, it is necessary to calculate and compare the number of US patents applied to the US by the Japan based companies to be analyzed, including US inventions and all foreign invention patents from outside Japan. It is also necessary to calculate and verify the ratio of the number of foreign invention patents to the number of Japanese patents applied for and granted to the companies in Japan. This paper examines these necessities for Canon in 2015.

11.5.2   Internationalization of R&D Activities by Canon Inc

Comparing the number of patents granted in the US in Figs. 11.1 and 11.3, we examine patents invented outside the US, excluding inventions in Japan, for patents granted to Canon in the US in 2015.

11.5.2.1   The Number of Overseas Inventions and the Ratio of Canon's US Patents

As shown in Fig. 11.3, only 1.4% of Canon's US patents were invented in the US. Thus, the number of overseas inventions (other than Japan) and their ratio were calculated focusing on the nationality of the organization to which the overseas inventor belonged, based on the USPTO definition. Table 11.4 shows the number and the ratio of US patents granted to the company in 2015 by nationality of the organization to which the inventor belonged.

Of the 3,470 patents Canon obtained in the US in 2015, 3,298 were invented solely in Japan. Therefore, 95.0% of the company's US patents were sole inventions of Japan. The company's overseas R&D operations contributed to 126 patents. The largest number of overseas sole inventions was made in the US with 59 patents, followed by France with 32, Australia with 26, China with 3, Germany with 2, India with 1, the Netherlands with 1, Taiwan with 1, and the UK with 1. Therefore, the ratio of overseas sole inventions to the number of patents published in the same year was 3.6%. Additionally, there were 41 patents for joint inventions between Japanese institutions and overseas institutions in 10 countries. There were 21 patents with the US, 8 with Taiwan, 2 with the UK, 2 with Poland, 2 with India, 2 with France, 1 with Germany, 1 with China, 1 with Austria, and 1 with Australia.

Table 11.4 The number and ratio of US patent inventions by nationality of institutional affiliations of inventors in Japan and overseas by Canon Inc. (2015 US patents)

	N	%
Sole Invention in Japan	3,298	95.0
Overseas Invention (1 + 2 + 3)	172	5.0
1: Overseas Sole*	(126)	(3.6)
2: Joint between Japan and Overseas**	(41)	(1.2)
3: Joint between overseas***	(5)	(0.1)
Total	3,470	100.00

Note*: The number of US patents by overseas sole inventions: AUS, 26; FRA, 32; US, 59; CHN, 3; DEU, 2; IND, 1; NLD, 1; TWN, 1; UK, 1
Note**: Joint between Japan and overseas: JPN–US, 21; JPN–TWN, 8; JPN–FRA, 2; JPN–IND, 2; JPN–POL, 2; JPN–UK, 2; JPN–AUS, 1; JPN–AUT, 1; JPN–CHN, 1; JPN–DEU, 1
Note***: Joint between overseas: US–AUS, 2; US–CAN, 1; US–FRA, 1; US–IND, 1
Source calculated from USPATFUL search

Among the company's overseas R&D institutions, the US was the only country with joint inventions with overseas partners, and there were five joint inventions with each of four countries: Australia, France, Canada, and India. Therefore, 41 jointly invented patents between Japanese and foreign inventors accounted for 1.2% of the total. Therefore, the ratio of patents for inventions belonging to overseas institutions to the number of US patents (3,470 patents), including joint inventions with overseas institutions, was 5.0%.

As with IBM's US domestic patents, we next examine the number and ratio of overseas inventions among patents filed in Japan by Canon and published in 2015.

11.5.2.2 Overseas Inventions as a Percentage Share of Canon's Japanese Patents

Canon obtained a total of 13,039 patents in Japan in 2015. The total number of Japanese patents by overseas inventions, including joint inventions with Japan, was 57, as shown in Tables 11.5 and 11.6. Of these, 33 were inventions in the US, of which 26 were sole, and 7 were joint with Japan. Out of 15 in Australia, 14 were sole, and 1 was joint with Japan. There were 4 in France, 2 in Poland and China, and 1 in Italy.

As shown in the same table, the ratio of overseas inventions to the number of patents granted to Canon in Japan, including the number of joint inventions with Japan, was only 0.4%.

The ratio of internationalization of R&D of Canon Inc. in terms of the number of patents published in Japan was only 0.4%. Conversely, the ratio of inventions in the home country was 99.6%.

As shown in Table 11.6, of Canon's Japanese patents in 2015, 57 were by overseas inventions, whereas 172 of their Canon's US patents were by overseas inventions, as shown in Table 11.4. These included sole and joint inventions. Thus,

Table 11.5 Number of Japanese patents by nationality of affiliated institutions of inventors of Japanese Canon Inc. (2015 Japanese patent: ratio to the number of duplications in parentheses)

CANON (13,039)	US	AUT	FRA	POL	CHN	ITA	JPN	Total
US	26 (0.0020)	0	0	0	0	0	7 (0.0005)	33 (0.0025)
Australia (AUT)		14 (0.0011)	0	0	0	0	1 (0.0001)	15 (0.0011)
France (FRA)			4 (0.0003)	0	0	0	0	4 (0.0003)
Poland (POL)				2 (0.0002)	0	0	0	2 (0.0002)
China (CHN)					2 (0.0002)	0	0	2 (0.0002)
Italy (ITA)						0	1 (0.0001)	1 (0.0001)
Japan (JPN)							12,991 (0.9949)	13,000 (0.9956)
Total								13,048 (1.0000)

Note (1) Each number indicates the number of Japanese patents per sole and joint inventions. For ex ample, the total 33 of the US indicates 26 sole inventions by the US and 7 joint inventions between the US and Japan. The first-named inventors of six out of seven joint inventions are Japanese. The first-named inventor of one joint invention by Australian and Japanese was also Japanese. Therefore, the number of overseas sole inventions by the first-named inventor was 48 out of 57 overseas, including joint Japanese
(2) There were 20 Japanese patents invented by foreign researchers or engineers residing in Japan, of which 18 were jointly invented with Japanese, and 2 were invented solely. The nationalities of these two sole inventions were Korean and Chinese
(3) 13,048 total number of patents excludes 9 duplicated number of patents
Source Calculated from Japanese published patent search

Table 11.6 Number and ratio of patents by nationalities of institutions to which inventors belong: Japanese Patents of CANON Inc. in 2015

	N	%
Inventions in Japan	1,2991	99.6
Overseas Inventions (1 + 2 + 3)	57	0.4
1: Overseas Sole Inventions*	(48)	(0.4)
2: Joint Inventions (Japan + Overseas)**	(9)	–
3: Joint Invention (Between Overseas)	(0)	–
Total	1,3048	100.0(100.0)

Note 1: Overseas Sole*: US, 26; AUT, 14; FRA, 4; POL, 2; CHN, 2
2: Joint Inventions (JPN + Overseas)**: JPN-US, 7; JPN-AUS, 1; JPN-ITA, 1
Source Calculated from Japanese published patent search

foreign inventions, particularly US inventions, contributed more to US patents than to Japan's.

The ratio of internationalization of R&D of Canon Inc. in terms of the number of patents published in Japan was only 0.4%. Conversely, the ratio of inventions in the home country was 99.6%.

Of the 57 patents in Japan involving overseas inventions, 48 were made solely by researchers and engineers belonging to the company's overseas R&D department. Only nine were jointly invented by R&D departments in Japan and overseas, and no joint inventions were found between overseas R&D departments.

As described in the footnotes to the table, there were only 20 patents invented solely or jointly by inventors who are considered foreign nationals, regarding the passport nationality of the inventors belonging to the company's Japanese institution. Therefore, 77 total patents, comprising 20 invented by foreign inventors residing in Japan and 57 at overseas inventions, accounted for only 0.6% (=77/ 13048), in terms of the ratio to the number of Japanese patents published.

11.6 Discussion and Conclusion

The degree of internationalization of R&D activities by IBM Corp. and CANON Inc, most typical US and Japanese patenting companies has been examined by considering the overseas sole invention rate, and joint invention rate with overseas. The paper has also examined research output of foreign scientists and engineers residing in the US and Japan, where there is a greater degree of difference than in conventional output paradigms. We reconfirm this point using Tables 11.7 and 11.8.

Table 11.7 Ratio of internationalization of R&D in terms of foreign patent inventors of IBM and Canon (in 2015) (each ratio in parentheses shows that of Japanese patents: %)

	IBM Corp.	CANON Inc.
Overseas invention ratio by the USPTO definition[1]	29.6	1.4
Overseas invention ratio based on the number of patents by foreign inventors[2]	35.9 (68.2)	5.0 (0.4)
Overseas invention ratio including inventions by foreign researchers and engineers in the US[3] and Japan[4]	68.0–73.1 (84.1–86.6)	−(0.6)

Note Nationalities are classified by those of institutions to which inventors belong.
(1) Refer to the attached note1 of Fig. 11.1
(2) Other ratios beside the USPTO definition are based on the definition in terms of the number of patents by foreign inventors (refer to the attached note of Table 11.1)
(3) Overseas invention ratio, including US inventions by foreign researchers and engineers residing in the US by IBM is the estimated value based on NSF, Scientist and Engineers Statistical Data (2013), which is shown in Table 11.2
(4) In the case of CANON, each ratio is calculated from 13,057 published CANON patents at the Japanese patent office
Source Calculated from the USPATFUL database and Japanese published patents

As shown in Table 11.7, the ratio of inventions by overseas researchers and engineers to the number of US patents granted to IBM and Canon in 2015 was estimated to be 29.6 and 1.4% respectively based on the USPTO definition. The ratios based on the definition 2 are 35.9 and 5.0%. The ratios of the number of overseas invented patents to the number of Japanese patents are 68.2% and 0.4% respectively, and between 68–73% and 84–87% when considering inventions by foreign researchers and engineers residing in the US.

Compared with IBM's US patents, Canon's overseas inventions accounted for 0.4% of the total Japanese patents. Their patent inventions of foreign nationalities residing in Japan accounted for 0.6%. All of them were below 1%.

Additionally, as was shown in Table 11.3, the number of IBM's US patent inventors' institutional nationality comprised 26 countries in the case of overseas sole, and it comprised 56 countries in the case of joint inventions with overseas. On the other hand, in the case of Canon, as was indicated of Notes of Tables 11.4 and 11.6, the former was 9 countries, and the latter was 12, both at IBM's 34.6% level and 21.4% level, respectively. Comparing the numbers of inventions by foreign researchers and engineers, as is revealed in Table 11.8, Canon's numbers were only 4.6%, 25.8%, and 0.5–0.6% to IBM's, respectively.

On the basis of these results, the following three points can be made to account for the differences in contributions found in patented inventions by foreign scientists and engineers between Japanese and US companies.

First, if the output of foreign scientists and engineers in the US is included in the concept of R&D internationalization, the degree of internationalization of R&D by US companies will increase significantly. Second, the ability of foreign R&D personnel who enjoyed advanced research and educational opportunities in the US to be utilized in the US greatly reduces negative factors, such as coordination costs with foreign researchers and engineers in a cross-cultural environment associated with the cross-border R&D (Asakawa 2001; Belderbos et al. 2013). Third, in the case of Japanese companies, output by cross-border internationalization of R&D in

Table 11.8 Comparison of the number of patents invented by overseas and foreign nationals of IBM and Canon in US & Japanese patents

	IBM(A)	Canon(B)	(B)/(A)
Overseas invention (of US patent)	3737	172	0.046
Overseas invention (of Japanese patent)	221	57	0.258
Invented by foreign researchers and engineers in the US[1] or Japan[2]	3,335–3,869[3]	20[4]	0.005–0.006

Note (1): Overseas invention ratio, including US inventions by foreign researchers and engineers residing in the US by IBM is the estimated value based on NSB (2018)

(2): In the case of CANON, each ratio is calculated from 13,057 published CANON Japanese patents issued by the Japanese patent office

(3): and (4) Refer to Note (1) & (2) of Table 11.5.

The numbers of US patents in 2015 by IBM and CANON are 10,411 and 3,468 respectively, and those of Japanese patents by these companies are 362 and 13,039 respectively

Source Calculated from USPATFUL database and Japanese published patent data

terms of patent inventions was extremely low, as is shown in the case of Canon, which has been in the top position in terms of the number of patents. The coordination costs associated with the utilization of cross-border R&D personnel becomes significantly higher owing to the significant limitation of opportunities for foreign scientists and engineers in Japan. Therefore, the greater the opportunities for US companies to leverage foreign scientists and engineers at home, the less the need for them to build cross-border R&D systems. In the case of Japanese companies, on the other hand, the necessity is greater than US companies to a considerable extent.

When R&D activities are examined from the perspective of the output of patent inventions, we must conclude that the R&D system of Japanese companies is a closed innovation system in the sense that it lacks an extremely international character, as was examined in the case of Canon Inc., a Japanese top patenting company.

As the business environment becomes turbulent with innovations in IT and digital technologies and as the value chain assumes an international character through collaborations, alliances, and outsourcing among enterprises and businesses that can flexibly utilize the superior business resources of other institutions, the negative elements embedded in those closed R&D systems seem to have an even greater weight in the age of world-wide dispersed R&D capabilities.

References

Asakawa, K. (2001). Organizational tension in international R&D management: The case of Japanese firms. *Research Policy, 30*(5), 735–757.

Behrman, J. N., & Fischer, W. A. (1980). *Overseas R&D activities of transnational companies.* Cambridge: Oelgesclagert, Gunn & Hain.

Belderbos, R., Lete, B., & Suzuki, S. (2013). How global is R&D? Firm-level determinants of home-country bias in R&D. *Journal of International Business Studies, 44,* 765–786.

Cantwell, J. (1995). The globalization of technology: What remains of the product cycle model? *Cambridge Journal of Economics, 19*(1), 155–174.

Cantwell, J. A., & Mudambi, R. (2005). MNE competence–creating subsidiary mandates. *Strategic Management Journal, 26*(12), 1109–1128.

Cantwell, J. A., & Zhang, Y. (2006). Why is Internationalization in Japanese firms so low? A path —dependent explanation. *Asian Business and Management, 5,* 249–269.

Creamer, D. B. (1976). Overseas Research and development by the US multinationals, 1966–1975; Estimates of expenditures and a statistical profile, New York, The Conference Board.

Hayashi, T. (1989). *Takokuseki kigyo to chiteki shoyuken (Multinational enterprise and intellectual property rights).* Tokyo: Moriyamashoten.

Hayashi, T. (2004). Globalization and networking of R&D activities by 19 electronics MNCs. In M. Serapio & T. Hayashi (Eds.), *Internationalization of research and development and the emergence of global R&D networks* (pp. 85–112). Oxford: Elesvier.

Hayashi, T. (2007). Higashi ajia ni okeru kokkyo wo koeta komyuniti to chishiki kyoso no mekanizumu (Transnational community in East Asia and the knowledge co-creating mechanism). In T. Sakuma, T. Hayashi, & Y. Kaku (Eds.), *Idousuru ajia (Moving Asia)* (pp. 18–47, Chap. 1). Tokyo: Akashishoten.

Hayashi, T. (2018). Kenkyu Kaihatsu Nouryoku No Kokusaiteki Saihensei: IBM No Keisu Wo Chusintosite (International Reorganization of R&D System: Focusing on the Case of IBM Corp.). *Keieironso, 8*(1), 85–108.

Hayashi, T., & Serapio, M. (2006). Cross-border linkages in research and development: Evidence from 22 US Asian and European MNCs. *Asian Business & Management, 5,* 271–298.

Hirota, T. (1985). Beikoku kigyou no kenkyu kaihatsu Kkatsudo to senryaku (R&D Activities and strategies of US companies). *Kansai University Shougakuronshu, 30*(4–5), 1–42.

Hirota, T. (1986). Nihon to beikoku kigyou no gijutsu kaihatsu (Technology development of Japanese and US companies). *Kansai Univ. Shougauronshu, 30*(6), 1–67.

Iguchi, C. (2011). Globalisation of R&D by TNC subsidiaries: The case of South-East Asian countries. *Asian Business & Management, 11*(1), 79–100.

Iwata, T. (1994). *Kenkyu kaihatsu no gurobaruka (Globalization of research and development).* Tokyo: Bunshindo.

Iwata, T. (2007). *Gurobaru inobeishon manejimento (Management of global innovation).* Tokyo: Chuoukeizaisha.

Komoda, F. (1987). *Kokusai gijutsu iten no riron (Theory of international technology transfer).* Tokyo: Yuhikaku.

Kotabe, M., et al. (2007). Determinants of cross-national knowledge transfer and its effect on firm innovation. *Journal of International Business Studies, 38,* 259–282.

Leroy, G. P. (1978). Transfers of technology within the multinational enterprise. In M. Gehrtman & J. Leontiades (Eds.), *European research in international business.* Amsterdam: North Holland Publishing.

Mansfield, E., Romeo, A., et al. (1979). Foreign trade and US research and development. *Review of Economics and Statistics, 61*(1), 49–57.

Mansfield, E., Romeo, A., et al. (1984). Reverse transfers of technology from overseas subsidiaries to American firms. *IEEE Transactions on Engineering Management, EM (31)*3, 122–127.

Mansfield, E., Teece, D., et al. (1979). Overseas research and development by US based firms. *Economica, 46,* 187–196.

Medcof, J. (2001). Resource-based strategy and managerial power in networks of internationally dispersed technology units. *Strategic Management Journal, 22*(11), 999–1012.

National Science Board, *Science & Engineering Indicators.* (2018).

Patel, P., & Pavitt, K (1998). Uneven technological accumulation among advanced countries. In G. Dosi, D. J. Teece, & J. Chytry (Eds.), *Technology, organization, and competitiveness* (pp. 289–317). Oxford, NY: Oxford University Press.

Pearce, R., & Singh, S. (1992). *Globalizing research and development.* London: MacMillan.

Roberts, E. (2001). Benchmarking global strategic management of technology. *Research Technology Management, 44*(2), 2536.

Sana, M. (2010). Immigration and natives in US science and engineering occupations, 1994–2006. *Demography, 47*(3), 801–820.

Saxenian, A. (2005). From brain drain to brain circulation: Transnational communities and regional upgrading in India and China. *Studies in Comparative International Development, 42* (2), 35–61.

Shanrokhi, M. (1984). *Reverse licensing.* Santa Barbara: Praeger.

Song, J., Aasakawa, K., & Chu, Y. (2011). What determines knowledge sourcing from host locations of overseas R&D operations? A study of global R&D activities of Japanese multinationals. *Research Policy, 40*(3), 380–390.

Takahashi. (2000). *Gurobaru R&D Nettowahk (Global Network of R&D).* Tokyo: Bunshndo.

Takabumi Hayashi (Ph.D. in Economics, Rikkyo University) is Professor Emeritus of Rikkyo University, Tokyo. He successively filled the position of senior lecturer at Fukuoka University, associate professor and professor of International Business at Rikkyo University, and Professor at Kokushikan University, Tokyo. His recent research arears are innovation systems and R&D

management, focusing on knowledge creation and diversity management. His works have been widely published in books and journals. His book "Multinational Enterprises and Intellectual Property Rights" (in Japanese; Moriyama Shoten, Tokyo, 1989.)" is widely cited, and "Characteristics of Markets in Emerging Countries and New BOP Strategies" (in Japanese; Bunshindo, Tokyo, 2016) received the award from Japan Scholarly Association of Asian Management (JSAAM) in 2018. He has been sitting on the editorial board of several academic journals.

Correction to: Knowledge Transfer and Creation Systems: Perspectives on Corporate Socialization Mechanisms and Human Resource Management

Tamiko Kasahara

Correction to:
Chapter 10 in: J. Cantwell and T. Hayashi (eds.),
Paradigm Shift in Technologies and Innovation Systems,
https://doi.org/10.1007/978-981-32-9350-2_10

In the original version of this chapter, the following belated correction has been incorporated in Table 10.1 of Chapter 10:

2016 • Present-CTP was nominated for the best company award from Great Place to Work® Institute
2017 • Present-CTP was nominated for the best company award from Great Place to Work® Institute
2018 • Present-CTP was nominated for the best company award from Great Place to Work® Institute

The erratum chapter has been updated with the change.

The updated version of this chapter can be found at
https://doi.org/10.1007/978-981-32-9350-2_10

Printed in the United States
By Bookmasters